Smart Cities - Their Framework and Applications

Edited by Anuar Mohamed Kassim and Lutfi Al-Sharif

Published in London, United Kingdom

IntechOpen

Supporting open minds since 2005

Smart Cities - Their Framework and Applications
http://dx.doi.org/10.5772/intechopen.87707
Edited by Anuar Mohamed Kassim and Lutfi Al-Sharif

Contributors
Juan Dols, Enrique Alcalá, Luis Martinez, Naoufal El Youssfi, Rachid El Bachtiri, Jonathan Beebe, Ahmad Hammoudeh, Lutfi Al-Sharif, Asaf H. Hajiyev, Yuri K. Belyaev, Kuo-Chi Chang, Kai-Chun Chu, Hsiao-Chuan Wang, Fu-Hsiang Chang, Yuh-Chung Lin, Tsui-Lien Hsu, Alok Kumar Mishra, Shibani Mishra, Dimitrios Thomas, Evangelos Kotsakis, Sun-Yuan Hsieh, Chih-Wei Hsu, Nancy Edith Ochoa Guevara, Sabyasachi Tripathi, Mohammad Irshaad Oozeer, Simon Haykin, Bitange Ndemo, Ben Mkalama, Kumar Avinash Chandra, Juan Sanchez Arteta, David Almesiga Riaño, Diego Sarmiento Vargas, Bryan Tunarosa Naranjo, Sandra Patricia Ochoa Guevara, Hector Edmundo Davila

First published in London, United Kingdom, 2021 by IntechOpen
IntechOpen is the global imprint of INTECHOPEN LIMITED, registered in England and Wales, registration number: 11086078, 5 Princes Gate Court, London, SW7 2QJ, United Kingdom
Printed in Croatia

British Library Cataloguing-in-Publication Data
A catalogue record for this book is available from the British Library

Additional hard and PDF copies can be obtained from orders@intechopen.com

Smart Cities - Their Framework and Applications
Edited by Anuar Mohamed Kassim and Lutfi Al-Sharif
p. cm.
Print ISBN 978-1-83962-294-6
Online ISBN 978-1-83962-295-3
eBook (PDF) ISBN 978-1-83962-296-0

For EU product safety concerns:
IN TECH d.o.o., Prolaz Marije Krucifikse Kozulić 3, 51000 Rijeka, Croatia, info@intechopen.com or visit our website at intechopen.com.

Meet the editors

Ir. Ts. Dr. Anuar bin Mohamed Kassim received his bachelor's degree in Electrical and Electronic Engineering in 2006 from Ehime University, Japan. At the beginning of his career, he spent two years as an R&D Engineer in the Optical Disc Drive Department in Panasonic Communications Co. Ltd, Japan. In 2010, he received a master's degree in System Innovation Engineering from Tokushima University, Japan. Later that year, he was appointed lecturer at Universiti Teknikal Malaysia Melaka and was promoted to senior lecturer in 2013 due to vast achievements in his research. In 2017, he received his doctorate in System Innovation Engineering from Tokushima University, Japan, and continued his postdoctoral work at Ghent University, Belgium. There, he worked in the field of Precision Agriculture through the FWO Odysseus project awarded by Belgium. In this position, he worked in the Site-Specific Management (SiteMan) project within the Precision Scoring Group to develop the precision measurement of soil, crop-sensing devices, geospatial mapping, and analysis. His research interests include wireless sensor networks, assistive technology, system innovation, robotics, automation, the Internet of Things (IoT), precision agriculture, and smart systems. He has also completed 25 research projects and is currently working on another 10 such projects. Finally, he has received international and national awards and filed industrial design, copyright, and patents for various products.

Lutfi Al-Sharif is currently the dean of Engineering Technology and professor of Electrical Engineering at Al-Hussein Technical University in Amman/Jordan, and jointly professor of Building Transportation Systems at the Department of Mechatronics Engineering, the University of Jordan. He received his Ph.D. in elevator traffic analysis in 1992 from the University of Manchester, U.K. He worked for 10 years for the London Underground, United Kingdom, focusing on elevators and escalators. He has over 50 papers published in peer-reviewed journals and conferences in vertical transportation systems and is the co-inventor of four patents, coauthor of the 2nd edition of the Elevator Traffic Handbook, and author of the "Indoor Transportation" chapter in the Elsevier Encyclopaedia of Transportation.

Contents

Preface

Smart Cities - Their Framework and Applications provides the overall concepts as well as the application of the technology for developing smart cities. This book will provide the platform for researchers, academics, scientists, policymakers, and governments to design, develop, analyze, evaluate, and implement smart systems in the smart cities' development process. Smart cities consist of various smart systems that will be integrated to develop the interlinkage between these smart systems. Smart systems are possible due to changes in the telecommunication world, which the ubiquitous mobile internet is advancing on a worldwide basis. The usage of the 5G telecommunication system changes the behavior of every sector in this world, such as education, healthcare, energy, transportation, etc. In addition, the development of cutting-edge sensing technology within telecommunication systems that communicate with each other by applying the Internet of Things (IoT) concept to smart cities will provide a great boost in advancing smart systems.

The four sections in this book demonstrate the capability of smart systems in smart cities to solve scientific and engineering problems of varying degrees of complexity. The first section introduces various frameworks for developing smart cities, including a new business model for developing smart cities. The contribution of smart cities to the country's overall urbanization also will be discussed.

Section 2 focuses on smart building and architecture where the model approach for institutional buildings is explained. In addition, the universal methodology for elevator design, including the necessary calculations and simulations beneficial in the development of smart cities, is described scientifically.

Section 3 focuses on smart energy in the development of smart cities. Energy management and optimal power scheduling are also discussed, with an emphasis on uncertain conditions. This section also discusses the compression of data strategies used in advanced metering infrastructure networks to manage the great amounts of smart meter data deployed in smart cities. The cognitive dynamic system for AC state estimation and cyber-attack detection in the smart grid also will be modeled and discussed comprehensively.

Section 4 details the integration of smart mobility and transportation systems in the development of smart cities. This chapter also discusses the architecture of a telemonitoring system as a component of the smart city supported by Internet of Things technology to improve the mobility of those in wheelchairs. Section 4 also reviews smart growth and transit-oriented developments for new urbanization, including finance and execution models necessary to develop a smart city in India. The transportation planning pursuit of more sustainable, economical, congestion-free, and pollution-free cities requires new designs in urbanization modeling. In addition, this section includes estimation and efficiency indices for operating vertical transportation systems and ensuring the safety of children using public transportation.

I would like to express my appreciation to IntechOpen for its support and guidance throughout the editorial process of this book. I would also like to express my gratitude to the author service manager, Dr. Lutfi Al-Sharif as the coeditor, and all the authors of the book chapters whose work has allowed me to successfully publish this book. I offer my regards and blessings to all of those who supported me in any respect during the completion of this book.

I hope that readers from various backgrounds such as researchers, academics, scientists, industry players, policymakers, and governments will benefit from realizing the development of smart cities all over the world.

Anuar Mohamed Kassim
Faculty of Electrical Engineering,
Universiti Teknikal Malaysia Melaka (UTeM),
Malacca, Malaysia

Lutfi Al-Sharif
Al-Hussein Technical University,
Jordan

Section 1

Smart Cities Framework and New Business Model

Chapter 1

Application of Advanced Framework Technology in Smart Cities to Improve Resource Utilization

Kai-Chun Chu, Kuo-Chi Chang, Hsiao-Chuan Wang, Fu-Hsiang Chang, Yuh-Chung Lin and Tsui-Lien Hsu

Abstract

Nowadays, the application technology and demand are growth; there have been millions of solutions for user communication in smart cities. However, the quality of the autonomy of handheld devices and the information exchange of applications are functions of requesting services or participating in communications. Therefore, it is very difficult and tedious to implement resource management and control in such an environment. This study here proposes distributed cyber-physical systems (CPS) for agent-based middleware framework (AMF) using to achieve technology, thereby improving the reliability of environmental communication in smart cities. The technical solution has the characteristics of avoiding the problem of data source interruption because of the proxy technology of the linear calculation model. The aforementioned agents are independent and autonomous of each other in terms of providing seamless resource sharing and response scheduling, and have nothing to do with communication time and request queries. In this study, the architecture mainly uses the best linear calculation model to classify overlapping agents, and then allocates non-overlapping resources, and finally analyzes the overall architecture operation performance by responding to processed queries, storage utilization and resource usage, pause time and response.

Keywords: smart city, resource allocation, cyber physical system, linear optimization, agent technology

1. Introduction

1.1 Smart city development

Currently the most famous smart city development organization in the world is Intelligent Community Forum (ICF), Since 1999, the most representative smart cities in the world have been selected for awards and publicity every year. ICF is the American World Teleport Association it belongs to a non-business policy research organization with members from more than 40 advanced regions and countries, including the United Kingdom, France, the United States, Russia, Canada, Singapore, Belgium, etc. Headquartered in the United States, it aims to study job creation and economic development in the broadband economy. It has research and promotion targets regardless of the size of cities, developing or developed countries. ICF believes that the broadband economy creates a whole new industry, which enables small companies to become global exporters. Including exports of skills and knowledge that have never been transported between borders, It can ensure that remote areas and internal schools have access to the latest information tools and reference materials. By improving the economy and society of the community, broadband can reduce the aggressive, ICF also assesses whether the city has three major success factors [1, 2]:

1. Collaboration: focus on the mutual cooperation among industry, government, academia, and research.

2. Leadership: partnership with industry, government, academy, research and construction, and has a focus on improving the social economy and society.

Vienna University of Technology developed an assessment tool for evaluating European medium-sized cities in 2007; Rank the medium-sized cities according to the Smart city model built by this study. The model results show that smart cities are cities that perform well in key areas of urban development. It is based on the "Smart" combination of self-government, independence and knowledgeable citizens and assignment and behavior [3, 4].

The ranking is determined based on 6 characteristics and their respective components **Table 1**.

1.2 Smart manufacturing

The industrial revolution is represented by the technological and scientific breakthroughs in which automatic machinery replaces labor, and production and operation in the factory are replaced by manual labor production lines. An important evolution in the structure of the industrial economy that can allow personnel production can change the overall economic dimension. **Figure 1** shows the evolution from Industry 1.0 to Industry 4.0 [5].

The development of intelligent manufacturing technology revolution is to integrate CPS into fiercely competitive technologies, including physical phenomena, to

Item	Description
Smart mobility	It generally includes technical content such as international/national accessibility, sustainability of transportation systems, local transportation systems, and information and communication technology infrastructure.
Smart economy	The content includes important parts such as international integration, entrepreneurship, innovation, growth, city image and labor market.
Smart people	Including lifelong learning, education, open minded and racial diversity.
Smart environment	Including air quality, sustainable resource management and eco-environmental awareness.
Smart governance	Including efficient and transparent administrative management, political and democratic consciousness, public management and social services, etc.
Smart living	Including personal safety, environmental sanitation conditions, tourism planning, hotel room quality, educational facilities, social solidarity and cohesion, historical culture and leisure sports facilities.

Table 1.
Six characteristics of smart cities.

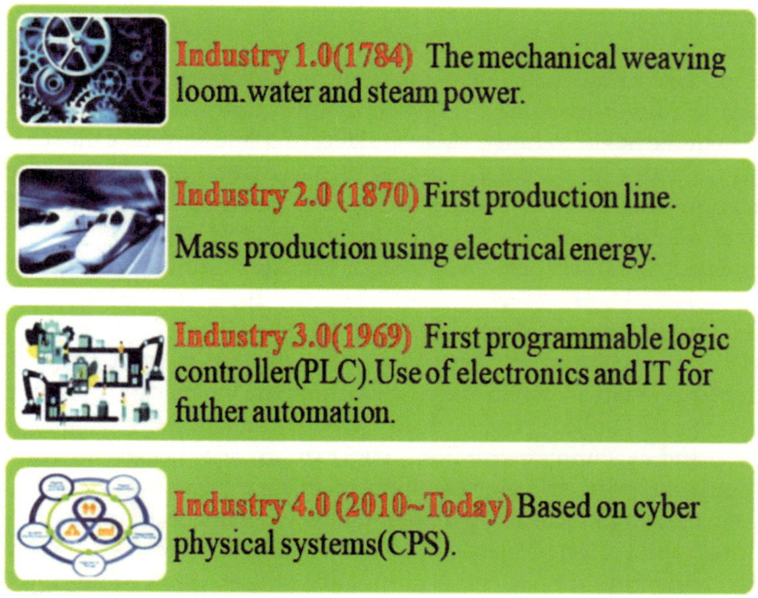

Figure 1.
Industry 1.0 to industry 4.0 evolution.

digitize physical (virtual) technology, Internet of Things (IoT), etc., and to develop necessary adaptability, resource integration efficiency and ergonomics The smart factory learned can also contact industry professionals in the manufacturing process and commercial value manufacturing process, innovative products, and customer-specific supply service functions.

There are five main levels of intelligent manufacturing automation including field level, programmable logic controller (PLC), supervisory control and data acquisition (SCADA/HMI) and mechanical equipment preventive sensing function manufacturing execution system (MES), enterprise resources planning (ERP) has become an optimized and integrated smart factory in the cloud [6, 7].

To build a smart factory based on the production process, the production machines must be intelligently optimized to improve efficiency, and the smart optimized system must be strengthened to assist in the deployment and application as shown in **Table 2**.

1	Sensor has many types and a wide range of applications
2	Data information collection framework (such as Web SCADA)
3	Communication protocol (MQTT or Ethercat)
4	Mechanical robot (flexible serial connection with production machine interface)
5	Bidirectional IoT/M2M (communication between production goods and processing equipment)
6	MES machinery and equipment preventive sensing function (inspecting the health of production machinery and equipment and process units; process scheduling procedures)
7	Scheduling items and process management of the production line.
8	Big data (big data analysis efficiency production and industrial to smart manufacturing).
9	MES/ERP serial connection and integration.

Table 2.
Intelligent automated systems.

1.3 CPSs and IoT

Two important production line manufacturing process environments for Industry 4.0 are machine-to-machine (M2M) communication and IoT. Strengthening CPS is the main key skill to integrate manufacturing and service value chains.

CPS is a collective model for evaluating the calculation process. These mass communication platforms that use collectives are absorbed as independent individuals. From miniature sensors in the environment to large data storage physical systems to detect the input process. Through various communication and information technologies, CPS provides users with security application access and support, service and data sharing privileges [8–10].

CPS will serially connect different physical evaluation computing facilities and processing components in a distributed environment, which can confirm the direction of user-centric applications and forward data and analysis to share. In addition to processing and execution capabilities, CPS can provide the best communication performance to support services, thereby supporting the sharing of information and data between users and facilities in different locations. In accordance with the use of mutual controllability, the properties and stability can be stretched and shortened to improve the reliability, service and performance of user applications, the CPS that evaluates and calculates tightness has been deployed in a large environment. In the CPS system, the physical network platform collected by the best method can be used to implement the resource co-allocation and communication security sharing in the production system and manufacturing, smart city transportation system and e-commerce. The current situation of smart cities is to provide people with the most practical and preferential sharing and communication services anytime, anywhere. The smart living environment integrates ICT and people as a whole [11–13].

For the smart city database environment with large amounts of data, it is necessary to immediately query various resources and execute processing to share aspects. These aspects require the use of many well-known technical capabilities to operate and manage. The CPS collection and synthesis in the smart city living environment enhances optimized user management to obtain applications and resources through distributed operation processing. Through the distribution of various resources from the distributed intelligent living environment, to meet the needs of users of different classes. Distributed living environments include multiple technologies, such as mobile edge computing, fog, and Internet of Things (IoT) clouds. These are examples of physical technologies. The CPS in the smart city has been deployed to inherit various technologies and resources. In the shared manual and resource allocation, it is a distributed smart environment to improve the reliability of applications and services. A process of multi-agent sharing and resource distribution is proposed. The agency skills in resource distribution and sharing have been widely used in various processes to reduce time and complexity [14, 15].

Provide distributed and combined execution for the structure and system of the agent to enhance the reliability of the system. For the agent's architecture and system to provide distributed and combined execution processing to enhance the system's reliable program. As described above, the complex processing and distributed purchase model implemented by CPS are used to process various resources sharing, distribution, and adjustment between users. In order to meet the increasing consulting needs of users and the requirements of facility setting density, it is necessary to perform optimal distribution and adjustment of data in a distributed smart living environment. CPS is based on the distribution of distribution storage resources and evaluation calculations from the distributed intelligent living environment, relying on sending and combining processing to provide services for user consultation and immediate response. The overall service provider level of resource reflection and collaborative perception, request execution, access control and adjustment levels are all stored in CPS. Distributed computing is jointly used by CPS, virtualized, and shared the best advantages of inspection and physical sensing to meet user requirements. In this kind of smart environment, the degree of resource distribution and adjustment is for the use of a dedicated operation process to build a purchase model to execute a management service platform [16, 17].

Reduce errors in the data sharing of smart telecommunication networks; it is necessary to provide a message transmission and playback classification framework. The use of collection and synthesis networks and telecommunication networks reduces the failure of data distribution and sending applications. Certain scholars developed a diagnostic system (FDDS) and model-free fault detection, used in a large number of cyber- physical systems. This communication isolation can be enhanced by performing autonomous learning on the temporal and spatial reasons of CPS. Elshenawy et al. research and develop the collection and synthesis of intelligent transportation systems used in smart cities and assist in adjusting the support structure. This architecture uses speech knowledge to display, establish a model for the collection and synthesis service operation and collection and synthesis service planning, and use the operation traffic adjustment degree and service demand in this in-vehicle application. Liu et al. in CPS, an event-driven tree model is adopted to improve faults [18–20].

In the distributed intelligent environment, the fault problem is handled according to the sorting process. The purpose of this study is to extend the hesitation model and the Internet network series model. The prevention evaluation calculation has been used for mitigation resolution and fault handling detection. This method can improve the accuracy of detecting the fault range. Past introduced a distributed computing model (DCMSP) for shared processing. Identifying the distribution task and estimating the terminal are the most important tasks of this operational model. According to the terminal's efficiency, it assists in dispatching and dispatching tasks [18–20].

Past research use CPS low-power wide area network (LPWAN) to manage radio resources. The analysis of the transmission model, according to the needs of a large number of CPS, executes the distribution of resources. Distributed CPS can effectively support two-way operation data conversion between LPWAN and extended cloud. The assimilation platform and related technologies perform the forwarding, exchange and processing of user codes. The scheduling model is used in the actual status control roles defined by the priority level. This model is suitable for pre-measurement of the model and online monitoring visits to establish a framework to solve scheduling and execution issues. The evaluation model closes the gap between physics and the network space to optimize the integration level of the estimation process. With bilateral directional quantification and network analysis, the performance of the network system can be counted. Past introduces a system (UPES-CPS) that applies a unified process to execute CPS to solve the work flow in actual data processing. The imported system can communicate, exchange and select through collection and synthesis services, and perform various texture operations with reliability. The virtual machine scheduling (QVMS), use promotion cloud system to assist CPS performance [21–24].

2. Methodology and study procedure

2.1 Smart cities physical system framework of agent-based cyber

The agency CPS is specifically designed to allow smart city resources to share and provide seamless services. Through CPS, users' applications and seamlessly integrated services are shared. Through the agent efficiency technology, it is endowed with the functional requirements of the synchronization of the smart city and the resource management of the user decentralization. It is proposed that for the agency CPS architecture, distributed task distribution and data resources can be used to manage efficiency. In the next section, you need to understand first, first explain this architecture. The agent-oriented architecture has been developed as an intermediate component between the platform layer and the application program, which can provide better interoperability. The intermediate component is composed of various agents, for example, the degree of adjustment of the visit control machine, the program of the resource management processor, etc. An agent is an integrated or mini program of applicable rules. This agent obtains the hardware and software of the facility that must be connected to the decision. In this architecture, the agents described above must be considered to solve the function of sharing and distributing the service resources of the data source together. Therefore, the intermediate component research and development concentric CPS specializes in serving in the smart city living environment. The CPS architecture of the agent is shown in **Figure 2** [9, 25, 26].

For intermediate components, the protection agent repository collects and generates agent processes, can provide reliable services to respond and can operate the application level to handle users. The agent is a context sensor developed with multiple functions to execute the model building. The multi-functional R&D design allows consulting agents to execute consultations and respond to actual cases in different time. But it can only pass when the agent needs to be released from the distribution process. Distribute to responding agents and consulting operations to exchange with the RM platform level. Therefore, the exchange of agents is of different levels, but it also handles multiple interruptions in the services of the user's smart city living environment. Agency performance technology assists in the implementation of multiple consultations on re-downloading functions and distribution methods. Unlike the conventional consulting execution system, this architecture focuses on the degree of adjustment and resource distribution. Since the agency operates in a combined delivery and distribution method, this process is already optimized for joint [27–32].

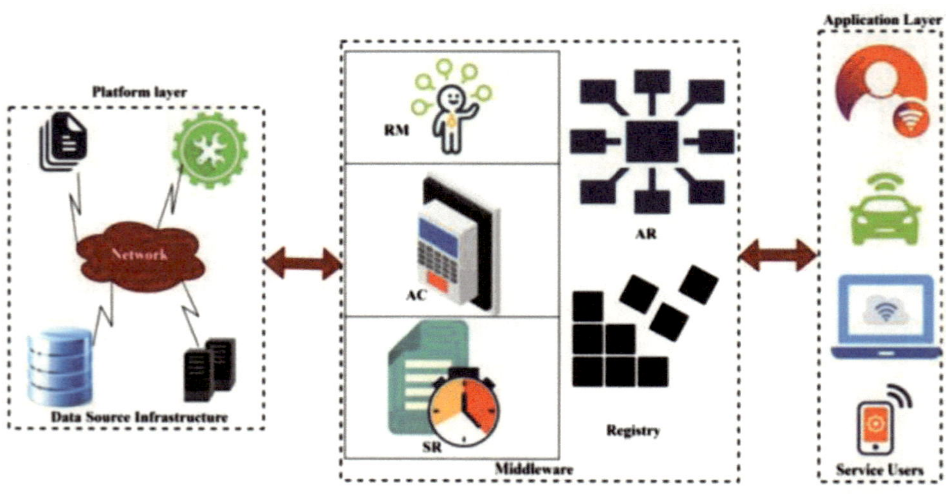

Figure 2.
CPS of agent-based.

2.2 Problem formation

The command $\rho_{i,j}$ represents the probability that user $\boldsymbol{i}$ has been assigned agent $\boldsymbol{j}$, then

$$\left.\begin{array}{c} \rho_{ij} = \begin{cases} 1, \textit{if } i^{th} \textit{ user is assigned with } j^{th} \textit{ resource} \\ 0, \qquad\qquad\qquad\qquad \textit{otherwise.} \end{cases} \\ \textit{such that} \\ f(x) = \min\left\{t \cup q_d\right\}, \forall \rho_{a(i,j)} = 1 \end{array}\right\} \tag{1}$$

In the formula (1), $f(x)$ represents a linear combination, which represents t is delay and qd drops. If ρij = 1, let qd and dq become the maximum in the time interval Δt. The problem of using a proxy is an important goal, the definition is like below.

$$\frac{\sum_{a\in A}\rho^a_{i,j}}{|A|} < \rho_{i,j} \le \sum_{a\in A}\rho^a_{i,j}, \forall i \in user \textit{ and } j \in resource \textit{ and a connects i with j} \tag{2}$$

The meaning representative is that if the user's consultation is executed by the agent a $\in A$, then the user is used to distribute the resources. On the contrary, in a single process, an agent is distributed to the user to prevent the same agent from overloading the load, defined as like below.

$$\sum_i \rho^a_{ij} \le 1 \forall j \in R \& a \in A \tag{3}$$

Let R represent the distribution and distribution of the ith user's valuable resource through the RM agent.

2.3 Resource allocation

The distribution of resources in the distributed intelligent environment is performed by the source of the database that does not require complicated execution and the largest connection. The commands Pt and Δc represent the processing execution time and the connection of R resources and the demand for first-in-first-out distribution. The execution time of processing of Δc and R are suitable for equation evaluation calculation. Formula (4) is show in below.

$$\left.\begin{array}{c} P_t = \rho_{i,j}\sum_n (t_a - t_s) \\ \Delta c = \frac{\sigma_n}{|n|} \end{array}\right\} \tag{4}$$

The main σn is the activity consultation that can be noticed at ta, and at (ta-ts) is the time to search for services and acceptance. The service time is to search for the response through the distribution and distribution resources. The enhancement of resource distribution and distribution rate optimizes resource utilization. On the contrary, the search response will be affected by both qd and t. If qd is high, it will continue to destroy resources and increase downtime. The reduced quantity estimate is calculated as like below.

$$q_d = \left(1 - \frac{\Delta c}{S_r}\right) + r_a \left(\frac{\Delta c}{S_r} - \frac{t_s}{t}\right) \tag{5}$$

The ***ra*** is the search consultation arrival rate. The combined optimization uses a separate method to separate the linear display patterns classified into t and ***qd***. Linear model display of ***qd*** show like below.

$$S(o) = -f(x)_{q_d} + P_t . \frac{\Delta c}{S_r} \forall \begin{matrix} i \in quires\ and \\ j \in R \end{matrix} \tag{6}$$

The sequence of the formula (6) Relatively the system output ***S(o)*** is promoted and developed as like below.

$$S(o) = -\begin{bmatrix} q_{d11} & q_{d12} & \cdots & q_{d1j} \\ q_{d21} & q_{d22} & \cdots & q_{d2j} \\ \vdots & \vdots & & \\ q_{di1} & q_{di2} & \cdots & q_{dij} \end{bmatrix} + \begin{bmatrix} P_{t1} \\ P_{t2} \\ \vdots \\ P_{ti} \end{bmatrix} \frac{1}{S_r} \begin{bmatrix} \Delta c_1 \\ \Delta c_2 \\ \vdots \\ \Delta c_i \end{bmatrix}$$
$$\forall \frac{P_t}{(t_a - t_s)} \leq \rho_{ij}.S_r \tag{7}$$

In the formula (7), the linear display of ***S(o)*** is simply $\frac{1}{S_r} P_{ti} \Delta c_i - q_{dij}$. As stated earlier, in the formula (2) and (3) will allow users to control the load download of the limited agent when they are in different processes. This means that $\{q_{d11}, q_{d22}, \dots q_{dij}\}$ is the only collection of non-load downloads, show in below.

$$S(o) = \begin{cases} -q_{dij} + \dfrac{P_{ti} \Delta c_i}{S_r}, \forall \dfrac{P_t}{(t_a - t_s)} < \rho_{ij}.S_r \forall i = j \\ \left. \begin{matrix} -q_{di+1j} + \dfrac{P_{ti} \Delta c_i}{S_r} \\ -q_{dij+1} + \dfrac{P_{ti} \Delta c_i}{S_r} \end{matrix} \right\}, \forall \dfrac{P_t}{(t_a - t_s)} < \rho_{ij}.S_r \forall i \neq j \forall S_r = S_r + 1 \end{cases} \tag{8}$$

From formula (7), it is shown that Δc_i_i and P_{ti} are available in the S_r and $S_r + 1$ slots of ***R***. Through the largest slot in S_r of, the ***R*** resource management component agent will connect to other information, intermediate component of CPS. Communication must be executed in ***R*** before it can be executed $\frac{P_t}{(t_a - t_s)} > \rho_{ij}.S_r$

i. You can use ***R*** to reduce the waiting time for exchange search.

ii. The new ***R*** must meet the requirements in the balance formula (2) and (3) are to prevent overload downloading.

In the formula, the information distribution process of the above two cases. (7) is shown in **Figure 3(a)** and **(b)**.

In response to the request of the access mark to identify the overload downloading agent process to obtain the overflow, it is necessary to perform the verification and verification of the information distribution process. The structure shown in **Figures 2** and **3** analyzes this information distribution and distribution. In **Figure 4(a)** and **(b)**, the progress process overload download has been divided into

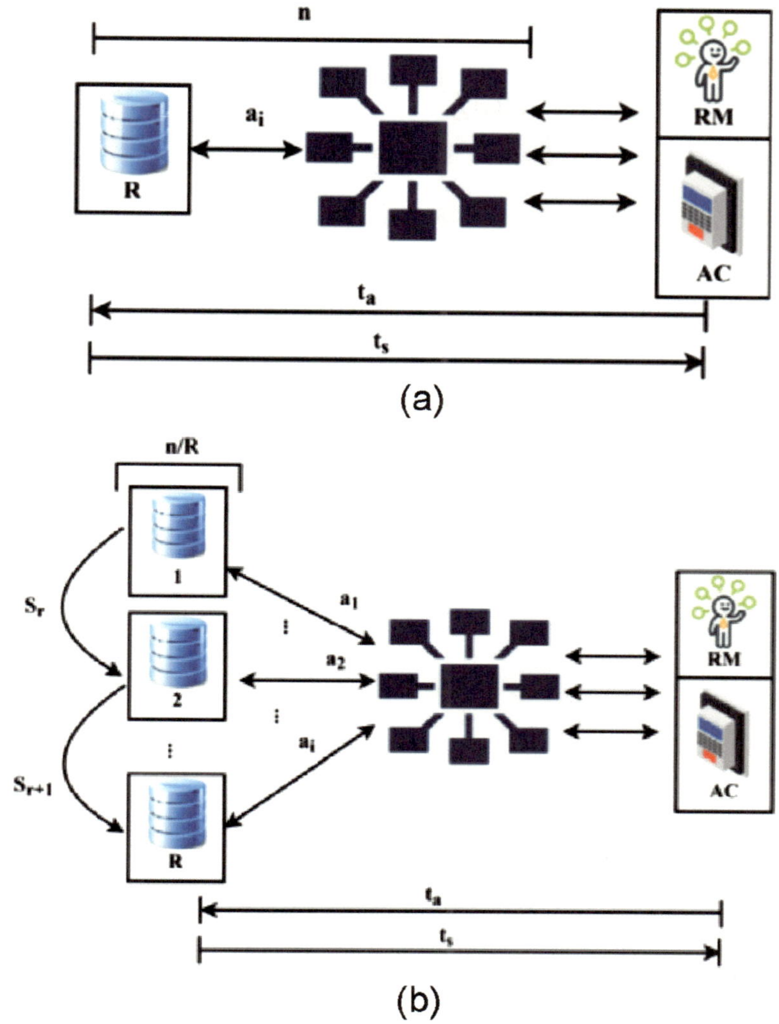

Figure 3.
(a). i = j *condition* (w.r.t qdij). *(b).* i≠ j *condition* (w.r.t qdij).

separate categories to prevent q_d. If resource distribution is not seamless, q_d cannot be restricted by control. Relative to t, the system output is linearized, for example in the formula (9)

$$S(o) = \frac{S_r}{(t_a - t_s)} + \frac{\rho_{ij}.a}{t} \tag{9}$$

Such as formula (10) is the expansion in the below.

$$S(o) = S_r \begin{bmatrix} \frac{1}{t_{a_1} - t_{s_1}} \\ \frac{1}{t_{a_2} - t_{s_2}} \\ \vdots \\ \frac{1}{t_{a_i} - t_{s_i}} \end{bmatrix} + \rho_{ij} \begin{pmatrix} a_1 \\ \vdots \\ a_2 \\ \vdots \\ a_i \end{pmatrix} \begin{bmatrix} \frac{1}{t_{11}} & \frac{1}{t_{12}} & \frac{1}{t_{ij}} \\ \frac{1}{t_{21}} & \frac{1}{t_{22}} & \frac{1}{t_{2j}} \\ \vdots & \vdots & \vdots \\ \frac{1}{t_{i1}} & \frac{1}{t_{i2}} & \frac{1}{t_{ij}} \end{bmatrix} \tag{10}$$

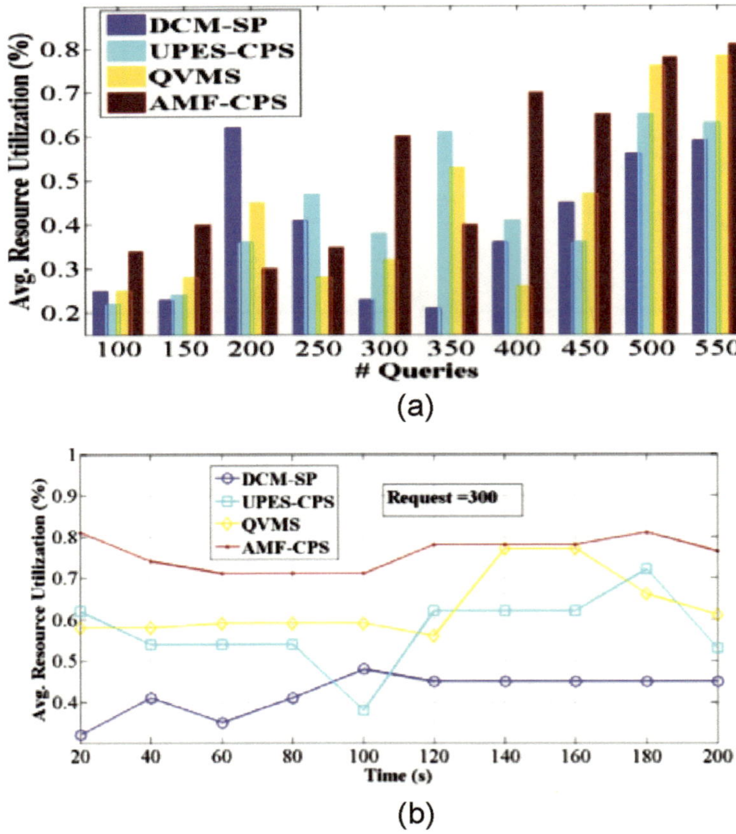

Figure 4.
(a). Average resource utilization vs. queries. *(b).* Average resource utilization vs. time.

As mentioned earlier, new resources are distributed and distributed to the overloaded downloading agents.

$$S(o) = \begin{cases} \dfrac{S_r}{(t_{a_i} - t_{s_i})} + \rho_{ij}.\dfrac{a_i}{t_{ij}}, if\ i = j \\ \dfrac{S_r}{(t_{a_i} - t_{s_{i+1}})} + \rho_{ij}.\dfrac{a_i}{P_t + t_{ij}}, if\ i \neq j \end{cases} \tag{11}$$

The output from the formula (11), the optimization requirements in the formula (1) If the classification reflects the conditions of t and q_d for $i = j$ and $i \neq j$ condition, then formula (2) and (3) are met. The illustration of $S(o)$ for t and $(t_{a_i} - t_{s_i})$ is shown in **Figures 1** and **2**, as shown in **Figure 5(a)** and **(b)**.

From this point of view, the equations solved can be used to check and verify the conditions for optimal distribution in formula (8) and (11),

$$\left.\begin{array}{r} \dfrac{P_{t_i}\Delta c_i}{S_r} - q_{d_{ij}} = \dfrac{S_r}{(t_{a_i} - t_{s_i})} + \rho_{ij}.\dfrac{a_i}{t_{ij}}, if \forall i = j \\ \dfrac{P_{t_i}\Delta c_i}{S_r} - q_{d_{ij+1}} = \dfrac{S_r + 1}{(t_{a_i} - t_{s_{i+1}})} + \rho_{ij}.\dfrac{a_i}{\sum P_{t_R} + t_{ij}}, \forall if\ i \neq j \end{array}\right\} \tag{12}$$

In the formula (12), the first condition meets the non-loaded download search, and the second condition shows the loaded download search. The search and

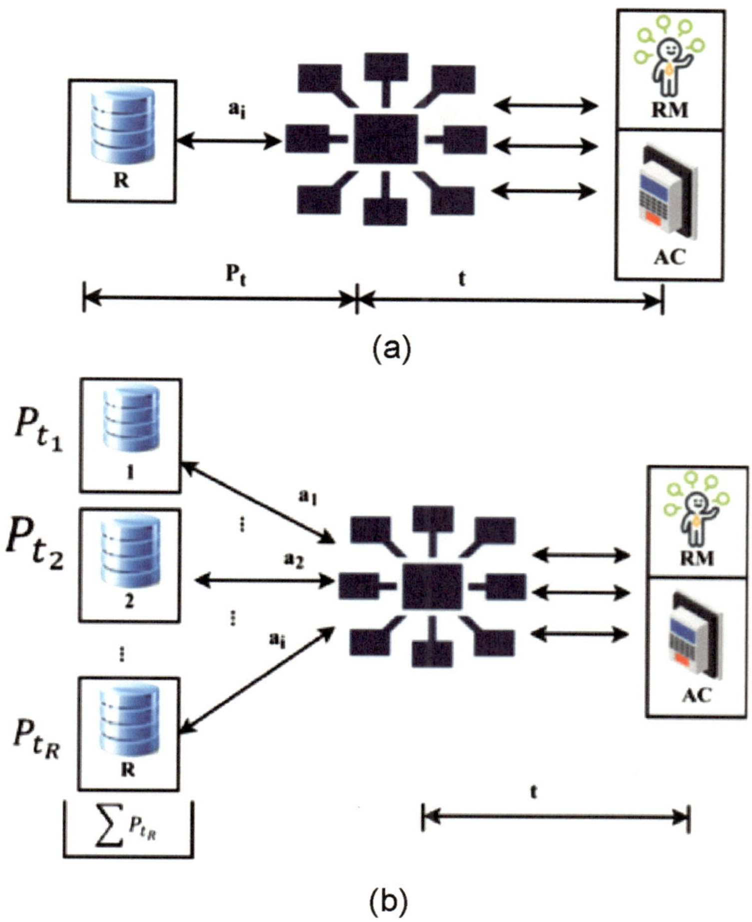

Figure 5.
(a). i = j condition (w.r.tPt). *(b).* i ≠ j *condition* (w.r.tPt).

distribution under heavy load will have new $a \in A$ and R, which will gradually increase S_r. The same via t_s (through combined sending execution, $P_{t_R} = t_s$, therefore,

$$\left.\begin{array}{c}\sum P_{t_R} + t_{ij} = \sum t_s + t_{ij} \\ \sum t_s + (t_a - t_s)(n - k)t_s - t_a\end{array}\right\} \tag{13}$$

It is the time to perform k searches for overload download. By identifying q_d to maximize the feasible and available R in t_a, the downtime can be reduced. The downtime observed in the proposed architecture will be compared with the current job execution in **Figure 6(a)** and **(b)**.

2.4 Scheduling of query response

RM and AC are responsible for resource information management through distribution. The other item, AC and adjustment level is the response to the user scheduling search consultation. The model for CPS has been established as intermediate component architecture, so the response will be provided as needed. The agent will complete the task and continue to release it to enable users to search and

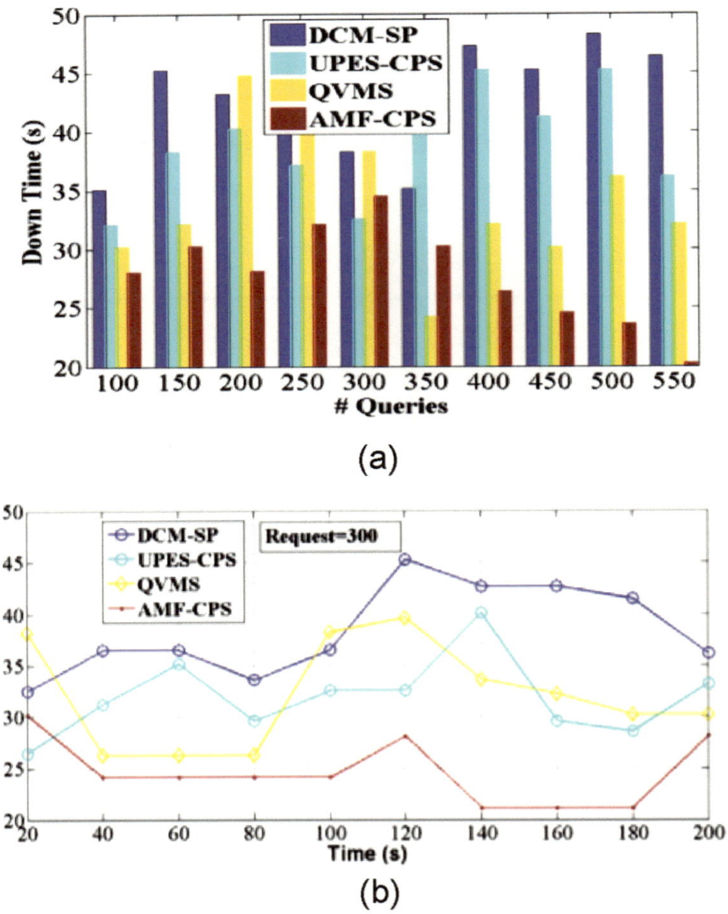

Figure 6.
(a). Queries vs. down time. *(b).* Time vs. down time.

distribute intelligent resources. Under the response, storage management is a challenging process, because the response can be used to obtain different large and small data messages. The optimized method is to use storage to help limit the response beyond the limit and prolong the hesitation. Therefore, the joint combination of AC and SR has been designed and developed to respond with an optimized method that uses available storage to distribute and distribute.

The adjustment level of this setting is different from the conventional first-in first-out process, because the searched t_a and P_t will vary with the user. From this time on, the agent will enhance the utilization efficiency of t storage. For all n requirements for distribution, the response waiting time is evaluated as $t + P_t$ (minimum value) and $t + [(n - k)t_s - t_a]$ (maximum value). The criteria in formula (1) need to be considered; you can evaluate and calculate the maximum value response waiting time. Seamless is the resource information supply, through the S_r or $S_r + 1$ slot to distribute and distribute $t + P_t$ **or** $t + [(n + k)t_s - t_a]$**.** The final evaluation calculation Search execution. The successful completion of the framework is evaluated and calculated based on the response time and storage utilization efficiency. Storage is a linear, first-in-go, first-out search system, where the first entry (response message) is based on the verification time (t_v) for confirmation. Therefore, storage must first conform to the $t_v = t + P_t \forall i = j$ **and** $t_v = t + [(n - k)t_s - t_a) \forall i \neq j$ of the agent distribution. Therefore, the linear model for t in formula (9)–(11) must be considered to optimize storage utilization efficiency. Let S_s display put into $'m'$ fan-shaped area to use the storage size and quantity of storage utilization efficiency is affected, causing response time delay. To deal with this problem, the response adjustment degree and storage utilization efficiency are modeled for the least amount of free time. In this architecture, regular idle time scheduling is not used. To meet the purpose of formula (1), the repeated stacking time is analyzed on the basis of the interval, and the equation of the model is established under the coordination of t_a and t_s, and the model is established for the idle time t_{s_i}. A comparison will be made in **Figures 1** and **2** for different search and time scenarios. See **Figure** 7**(a)** and **(b)** respectively.

Use formula (14) to evaluate and calculate the response schedule time (t_{dis}) like below.

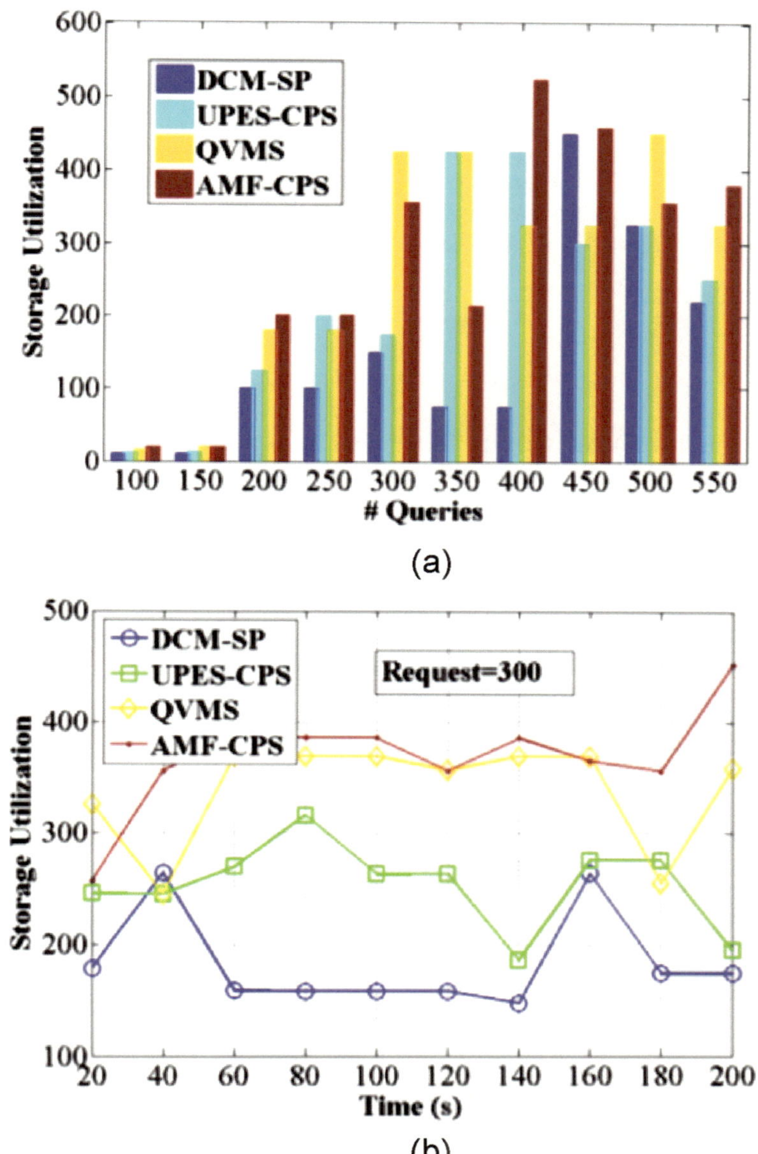

Figure 7.
(a). Storage utilization vs. queries. *(b)*. Storage utilization vs. time.

$$\boldsymbol{t_{dis}} = \boldsymbol{P_t} - \boldsymbol{t_a}, \forall \boldsymbol{i} = \boldsymbol{j}\ \boldsymbol{and}\ \boldsymbol{i} \neq \boldsymbol{j} \tag{14}$$

When using $\boldsymbol{S_r}$ and $\boldsymbol{S_r + 1}$ to reflect the search query to $\boldsymbol{R}$, the scheduling time will be different. The sub-categories consulted for $\boldsymbol{t_a}$ and $\boldsymbol{k} \times \boldsymbol{t_a}$ have been identified as formula (15) that provides a linear distribution (τ_{t_a}).

$$\tau_{t_a} = \frac{1}{S_r}\left[\int_0^k r_a P_t dt + \int_k^n r_a(P_t + t)dt\right] \tag{15}$$

Among them, one level $\int_0^k \boldsymbol{r_a P_t dt}$ provides services through the available S_r time gap, so $\boldsymbol{t} + \boldsymbol{P_t} \leq \boldsymbol{t_v}$. On the contrary, the second level $\int_k^n \boldsymbol{r_a(P_t + t)dt}$ cannot meet the requirements of $\boldsymbol{t_v}$, so the classification must be able to have a better degree of adjustment. Within the time of $(\boldsymbol{k} \times \boldsymbol{t_a})$, the second-level derivative of $\boldsymbol{\tau_{t_a}}$ should be activated, and $\boldsymbol{P_t(n - k)} + \sum \boldsymbol{t}$ will be regarded as prolonged hesitation. In this case, it can be confirmed that at least $(\boldsymbol{n} - \boldsymbol{k})$ storage space $\boldsymbol{S_s}$ can be used to receive information responses, and the capacity in the storage space can be considered. If it is to perform virtualization/replication $(\boldsymbol{n} - \boldsymbol{k})$, the distributed nature of CPS will provide more information than that. For the estimated time, the information in $\boldsymbol{P_t}$ used to analyze the allocated demand in $\boldsymbol{t_a}$ can be applied. Therefore, the linear form of formula (8) with $\boldsymbol{i} \neq \boldsymbol{j}$ has been modified to

$$\left.\begin{array}{l}\dfrac{\alpha}{S_r} - \Delta = \dfrac{S_r + 1}{\left(t_{a_i} - t_{s_{i+1}}\right)} + \dfrac{\rho_{ij} a_i}{\sum P_{t_R} + t_{ij}} \\ where, \alpha = P_{t_i} \Delta c_i\ and\ \Delta = q_{d_{ij+1}}\end{array}\right\} \tag{16}$$

From formula (15) and (16) the linear class is only given to the second class, such that

$$\left.\begin{array}{l}\dfrac{\boldsymbol{1}}{\boldsymbol{S_r}} \boldsymbol{\tau_{t_a}(\alpha, \Delta)} = \dfrac{\boldsymbol{1}}{\boldsymbol{S_r}} \displaystyle\int_k^n \boldsymbol{r_a(P_t + t)dt} \\ \boldsymbol{\tau_{t_a}(\alpha, \Delta)} = \displaystyle\int_k^n \boldsymbol{r_a(P_t + t)dt}\end{array}\right\} \tag{17}$$

As mentioned above, as in formula (13), $\boldsymbol{Pt}$ and $\boldsymbol{t}$ need to be used instead like below.

$$\boldsymbol{\tau_{t_a}(\alpha, \Delta)} = \int_k^n \boldsymbol{r_a[(n - k)t_s - t_a + t_a - t_s]\ dt} \tag{18}$$

$$= \int_k^n \boldsymbol{r_a[t_s(n - k - 1)]dt} \tag{19}$$

The time obtained from formula (19) is the maximum response to prolong hesitation, so that $(\boldsymbol{P_t} + \boldsymbol{t}) \leq (\boldsymbol{\alpha}, \boldsymbol{\Delta}) < [(\boldsymbol{n} - \boldsymbol{k})\boldsymbol{t_s} - \boldsymbol{t_a}] + \boldsymbol{t}$. All the executed second-level consultation searches must meet this requirement. Therefore, $\boldsymbol{t_{dis}} = \boldsymbol{t_s} - \boldsymbol{t_a} - \boldsymbol{t_a} = \boldsymbol{t_s} - \boldsymbol{2t_a}$ is the execution time of the second level of the opening action. Therefore, the degree of adjustment is outside the time $[\boldsymbol{n} - \boldsymbol{k})\boldsymbol{t_s} - \boldsymbol{t_a}](\boldsymbol{t_s} - \boldsymbol{2t_a})$ and is processed in time, which can reduce the waiting time for users to reserve when responding to inquiries. In the end, the overall proposal will be compared with the current situation discussed in the previous section for the overall proposal for the perceived waiting time in the agent structure, and the consultation search and time have been changed (see **Figure 8(a)** and **(b)**) . Comparison and analysis have represented the proposed architecture by limiting the response time control to $(t_s - 2t_a)$, and set aside to extend the delay $[(n-k)t_s - t_a] + t$ has been limited.

Through the deployment of smart cities at different levels, users (in terms of application categories) are used to analyze the performance of the proxy CPS architecture using the OPNET simulator. The application scenarios need to include voice, multimedia, database and Http harsh users. The size and capacity of the application varies from 100Kb to 5 Mb, and it executes requests from users in the form of consulting queries. CPS has been deployed as an intermediate component that utilizes cloud and various other communication performance technologies. In **Table 3**, a detailed explanation and analysis of the settings and their values in the experiment are presented.

As mentioned in the foregoing, we have compared the downtime, resource utilization efficiency, storage utilization efficiency and response time of the

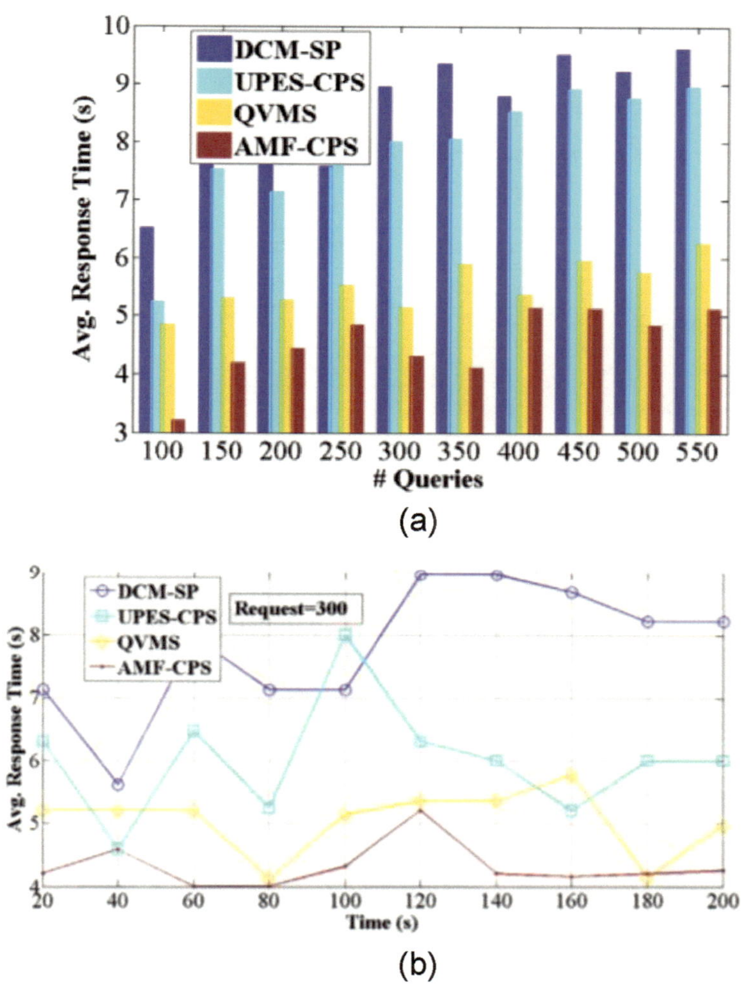

Figure 8.
(a). Response time vs. queries. (b). Response time vs. queries.

Experimental parameters	Values
Users	180
Application Type	Constant Bit Rate
CPS Middleware	8
AR Capacity	25
Application Rate	100Kb-5 Mb
Query Requests	30/sec
Storage Size	100 Mb
Slots/Storage	20

Table 3.
Experimental parameters and values.

Metrics	DCP-SM	UPES-CPS	QVMS	AMF-CPS
Down Time (s)	46.32	36.25	32.14	20.25
Resource Utilization	0.59	0.63	0.782	0.81
Avg. Storage Utilization	220.32	250.12	325.4	380.25
Avg. Response Delay (s)	9.62	8.96	6.25	5.12

Table 4.
Comparison and analysis of various query searches.

Metrics	DCP-SM	UPES-CPS	QVMS	AMF-CPS
Down Time (s)	36.14	33.22	30.21	28.14
Resource Utilization	0.45	0.53	0.61	0.763
Avg. Storage Utilization	174.2	195.24	358.41	452.32
Avg. Response Delay (s)	8.22	6.01	4.96	4.26

Table 5.
Different time comparison analysis.

proposed architecture to measure the degree of execution. In the comparison performed, it has been considered that the current way DCPSM, UPES-CPS and QVMS are regarded as indicators. In **Tables 4** and **5**, the results of comparative analysis for various different queries and time have been listed.

3. Conclusion and suggestion

This content manuscript mainly introduces the structure of the CPS intermediate component agency in the smart living city service. The operation process of resource information distribution and service provision is controlled and managed with the assistance of CPS and agency performance technology. A separate agent has already distributed and used for consulting query execution and resource information distribution and distribution without causing overload downloads. In addition to the task of solving and responding, the agent of the intermediate component can also handle the classification and time scheduling of the consultation query to reduce the downtime caused by the failure of the available resources. In the distributed CPS, the linear evaluation and calculation model is used to solve the problem of prolonged delay and loss. The improvement is transformed into a minimized joint optimization, which can respond to processing delays in the fastest time and improve resource utilization.

Author details

Kai-Chun Chu[1,2], Kuo-Chi Chang[1,3,4*], Hsiao-Chuan Wang[5], Fu-Hsiang Chang[6], Yuh-Chung Lin[1] and Tsui-Lien Hsu[7]

1 Fujian University of Technology Associate Professor, China

2 Department of Business Administration Group of Strategic Management from National Central University, Taoyuan, Taiwan

3 College of Mechanical and Electrical Engineering, National Taipei University of Technology, Taiwan

4 Department of Business Administration, North Borneo University College, Sabah, Malaysia

5 Institute of Environmental Engineering, National Taiwan University, Taiwan

6 Department of Tourism, Shih-Hsin University, Taipei, Taiwan

7 Institute of Construction Engineering and Management, National Central University, Taiwan

*Address all correspondence to: albertchangxuite@gmail.com

References

[1] Alemayehu, A. M., Dagne, H., & Dagnew, B. (2020). Knowledge and associated factors towards diabetes mellitus among adult non-diabetic community members of Gondar city, Ethiopia 2019. PloS one, 15(3), e0230880.

[2] Lu,C.C., Chang,K. C. and Chen,C.Y. Study of high-tech process furnace using inherently safer design strategies (IV). The advanced thin film manufacturing process design and adjustment, Journal of Loss Prevention in the Process Industries, 40, 2016, pp.378–395.

[3] Sut-Lohmann, M., Jonczak, J., Parzych, A., Šimanský, V., Polláková, N., & Raab, T. (2020). Accumulation of airborne potentially toxic elements in *Pinus sylvestris* L. bark collected in three Central European medium-sized cities. Ecotoxicology and Environmental Safety, 200, 110758.

[4] Chun-Yu Chen, Kuo-Chi Chang, Gwo-Bin Wang, Study of high-tech process furnace using inherently safer design strategies (I) temperature distribution model and process effect, Journal of Loss Prevention in the Process Industries, Volume 26, Issue 6, 2013, Pages 1198–1211, ISSN 0950–4230, https://doi.org/10.1016/j.jlp.2013.05.006.

[5] Chang, K. C., Pan, J. S., Chu, K. C., Horng, D. J., & Jing, H. (2018, December). Study on information and integrated of MES big data and semiconductor process furnace automation. In International Conference on Genetic and Evolutionary Computing (pp. 669–678). Springer, Singapore.

[6] Chang, K. C., Chu, K. C., Lin, Y. C., Sung, T. W., Zhou, Y. W., & Pan, J. S. (2019, November). Study on Health Protection Behavior Based on the Big Data of High-Tech Factory Production Line. In International Conference on Genetic and Evolutionary Computing (pp. 571–578). Springer, Singapore.

[7] Lu, C.C., Chang, K.C. and Chen. C.Y. Study of High-Tech Process Furnace Using Inhe-rently Safer Design Strategies (III) Advanced Thin Film Process and Reduction of Power Consumption Control. Journal of Loss Prevention in the Process Industries, 43, 2016, 280–291.

[8] K.C. Chang, K.C. Chu, H.C. Wang, Y. C. Lin and J.S. Pan, "Energy Saving Technology of 5G Base Station Based on Internet of Things Collaborative Control," in IEEE Access, vol. 8, pp. 32935–32946, 2020.

[9] Chang, K. C., Chu, K. C., Wang, H. C., Lin, Y. C., & Pan, J. S. (2020). Agent-based middleware framework using distributed CPS for improving resource utilization in smart city. Future Generation Computer Systems.

[10] Chun-Yu Chen, Kuo-Chi Chang, Chih-Cheng Lu, Gwo-Bin Wang, Study of high-tech process furnace using inherently safer design strategies (II). Deposited film thickness model, Journal of Loss Prevention in the Process Industries, Volume 26, Issue 1, 2013, Pages 225–235, ISSN 0950–4230, https://doi.org/10.1016/j.jlp.2012.11.004.

[11] Nikolakis, N., Maratos, V., & Makris, S. (2019). A cyber physical system (CPS) approach for safe human-robot collaboration in a shared workplace. Robotics and Computer-Integrated Manufacturing, 56, 233–243.

[12] Kuo-Chi Chang, Yuh-Chung Lin, Kai-Chun Chu. Mobile Edge Computing Technology and Local Shunt Design. The Frontiers of Society, Science and Technology (2019) Vol. 1 Issue 10: 135–140. https://doi.org/10.25236/FSST.2019.011017.

[13] Liang, W., Long, J., Weng, T. H., Chen, X., Li, K. C., & Zomaya, A. Y. (2019). TBRS: A trust based recommendation scheme for vehicular CPS network. Future Generation Computer Systems, 92, 383–398.

[14] Chang, K. C., Chu, K. C., Lin, Y. C., Sung, T. W., Zhou, Y. W., & Pan, J. S. (2019, November). Study on Hazardous Scenario Analysis of High-Tech Facilities and Emergency Response Mechanism of Science and Technology Parks Based on IoT. In International Conference on Genetic and Evolutionary Computing (pp. 191–199). Springer, Singapore.

[15] Chun-Yu Chen, Kuo-Chi Chang, Chi-Hung Huang, Chih-Cheng Lu, Study of chemical supply system of high-tech process using inherently safer design strategies in Taiwan, Journal of Loss Prevention in the Process Industries, Volume 29, 2014, Pages 72–84, ISSN 0950–4230, https://doi.org/10.1016/j.jlp.2014.01.004.

[16] Zhou YW. et al. (2020) Study on IoT and Big Data Analysis of Furnace Process Exhaust Gas Leakage. In: Pan JS., Li J., Tsai PW., Jain L. (eds) Advances in Intelligent Information Hiding and Multimedia Signal Processing. Smart Innovation, Systems and Technologies, vol 156. Springer, Singapore.

[17] K.C. Chang, K.C. Chu, T. Chen, Y. W. Lee, Y.C. Lin and T. Nguyen, "Study of the High-tech Process Mechanical Integrity and Electrical Safety," 2019 14th International Microsystems, Packaging, Assembly and Circuits Technology Conference (IMPACT), Taipei, Taiwan, 2019, pp. 162–165.

[18] Negri, E., Fumagalli, L., Cimino, C., & Macchi, M. (2019). FMU-supported simulation for CPS digital twin. Procedia manufacturing, 28, 201–206.

[19] K.C. Chang, K.C. Chu, D. Horng, J. C. Lin and V. Yi-Chun Chen, "Study of wafer cleaning process safety using Inherently Safer Design Strategies," 2018 13th International Microsystems, Packaging, Assembly and Circuits Technology Conference (IMPACT), Taipei, Taiwan, 2018, pp. 218–221.

[20] Chang, K. C., Chu, K. C., Lin, Y. C., Nguyen, T. T., & Pan, J. S. (2019, October). Study of Inherently Safer Design Strategy Application for IC Process Power Supply System. In 2019 14th International Microsystems, Packaging, Assembly and Circuits Technology Conference (IMPACT) (pp. 158–161). IEEE.

[21] Mekki, K., Bajic, E., Chaxel, F., & Meyer, F. (2019). A comparative study of LPWAN technologies for large-scale IoT deployment. ICT express, 5(1), 1–7.

[22] Bozkurt, S., & Gligor, D. (2019). Customers' behavioral responses to unfavorable pricing errors: the role of perceived deception, dissatisfaction and price consciousness. Journal of Consumer Marketing.

[23] d'Amour, N. J., Chang, K. C., Li, P. Q., Zhou, Y. W., Wang, H. C., Lin, Y. C., ... & Hsu, T. L. (2020, October). Study of Region Convolutional Neural Network Deep Learning for Fire Accident Detection. In International Conference on Advanced Intelligent Systems and Informatics (pp. 148–155). Springer, Cham.

[24] Amesimenu, D. K., Chang, K. C., Sung, T. W., Zhou, Y., Gakiza, J., OMER, A. A. I., ... & Haque, S. M. O. (2020). Study of Smart Monitoring and Protection of Remote Transformers and Transmission Lines using GSM Technology.

[25] Negri, E., Fumagalli, L., Cimino, C., & Macchi, M. (2019). FMU-supported simulation for CPS digital twin. Procedia manufacturing, 28, 201–206.

[26] Amesimenu D.K. et al. (2020) Home Appliances Control Using

Android and Arduino via Bluetooth and GSM Control. In: Hassanien AE., Azar A., Gaber T., Oliva D., Tolba F. (eds) Proceedings of the International Conference on Artificial Intelligence and Computer Vision (AICV2020). AICV 2020. Advances in Intelligent Systems and Computing, vol 1153. Springer, Cham.

[27] Chang KC. et al. (2020) Study on Hazardous Scenario Analysis of High-Tech Facilities and Emergency Response Mechanism of Science and Technology Parks Based on IoT. In: Pan JS., Lin JW., Liang Y., Chu SC. (eds) Genetic and Evolutionary Computing. ICGEC 2019. Advances in Intelligent Systems and Computing, vol 1107. Springer, Singapore.

[28] Hong Tian, Kuo-Chi Chang, Jin Song Chen, Application of hyperbolic partial differential equations in global optimal scheduling of UAV, Alexandria Engineering Journal, 2020, https://doi.org/10.1016/j.aej.2020.02.013.

[29] Chang, K. C., Zhou, Y. W., Wang, H. C., Lin, Y. C., Chu, K. C., Hsu, T. L., & Pan, J. S. (2020, October). Study of PSO Optimized BP Neural Network and Smith Predictor for MOCVD Temperature Control in 7 nm 5G Chip Process. In International Conference on Advanced Intelligent Systems and Informatics (pp. 568–576). Springer, Cham.

[30] Chang, K. C., Zhou, Y., Ullah, H., Chu, K. C., Sajid, T., & Lin, Y. C. (2020, June). Study of Low Cost and High Efficiency Intelligent Dual-Axis Solar Panel System. In 2020 IEEE International Conference on Artificial Intelligence and Computer Applications (ICAICA) (pp. 336–341). IEEE.

[31] Chu, K. C., Chang, K. C., Wang, H. C., Lin, Y. C., & Hsu, T. L. (2020). Field-Programmable Gate Array-Based Hardware Design of Optical Fiber Transducer Integrated Platform. Journal of Nanoelectronics and Optoelectronics, 15(5), 663–671.

[32] Chang, K. C., Chu, K. C., Chen, T. L., Lee, Y. L. W., Lin, Y. C., & Nguyen, T. T. (2019, October). Study of the High-tech Process Mechanical Integrity and Electrical Safety. In 2019 14th International Microsystems, Packaging, Assembly and Circuits Technology Conference (IMPACT) (pp. 162–165). IEEE.

Chapter 2

Orchestrating Smart Cities, New Disruptive Business Models and Informal Enterprises

Ben Mkalama and Bitange Ndemo

Abstract

As the fourth industrial revolutions technologies intensify, cities are becoming smarter, new business models are emerging and informal enterprises are formalizing by default. Research demonstrates that the future of our world is decided by the quality of its future cities. As cities invest in information and communication technologies (ICTs) and embrace the Fourth Industrial Revolution (4IR) technologies to make life easier and solve many of the problems we face today, employment opportunities expand and citizens enjoy better lifestyle. This chapter will examine how the concept of smart cities is disrupting existing business models and creating new ones that have positively impacting Africa's informal enterprise sector. The chapter leverages abundance theory to explain the emerging phenomenon in the nexus between smart cities, new business models and informal enterprises in Sub-Saharan Africa. The study finds that indeed the concept of smart cities is indeed facilitating new business models that are formalizing the informal sector.

Keywords: smart cities, digital innovation, informal economies, gig economies, Africa

1. Introduction

A global phenomenon that is currently not in dispute is rapid urbanization with estimates suggesting that by 2030, over 60 per cent of the global population will be living in cities, increasingly concentrated in Africa, Asia and Latin America [1]. This inevitably places social and economic strain on the existing urban infrastructure. These strains are placed on physical factors such as deteriorating conditions in the environment, transport efficiency, utilities such as water and energy, as well as economic factors such as unemployment. As a result of this, there is an emerging informal sector, which though unproductive and lacking employee protections, continues to be the place where bulging African youth population finds its livelihood [2].

Digital technologies and internet connectivity are playing a major role in making better cities that they are now considered as a panacea for solving Africa's chronic unemployment [3] and other associated challenges. As a result of this thinking, and to address these and emerging challenges, the smart city concept offers unusual opportunities for diverse countries [4]. Intensified digitization is increasingly becoming integral part of everyday life, more data is being collected and as a result leading to the accumulation of large amounts of data which is in most cases used in several beneficial application domains. Effective analytics of these data and utilization of the same is a critical factor for success in emerging business and service domains, as well as the smart city domain [5].

The emerging digital phenomenon is disrupting and transforming the informal sector that what seemed impossible a few years back could be possible. Studies [6] from Nigeria, confirm that majority of the informal enterprises could be easily formalized. This chapter is guided by the question: Taking into consideration that connectivity is growing across the continent, could digitalization end the curse of informal enterprises in Africa? The chapter will seek to address the questions: Will the new business models address disrupt and destroy livelihoods?

We attempt to explain the emerging phenomenon of technology, new business models and disruption of informal enterprises through the theory of abundance. That what is happening with technology in SSA is an opportunity for different combinations of existing problems that can be solved with new entrepreneurial openings as countries gear up to develop smarter cities. The chapter therefore makes a number of contributions. First a research that is looking at how informal enterprises can be formalized (something that has troubled policymakers for ages). Second, is theorizing abundance. The basic premise is that the world is big out there with opportunities for everybody such that if you are willing to achieve your goals, simply learn and polish the discipline of your craft [7].

2. Literature review

2.1 Demystifying smart cities

There is no standardized commonly accepted definition of or set of terminologies for a smart city [4, 8]. Numerous scholars have defined the smart city concept differently, but these conceptual definitions however converge around three broad dimensions namely technology, people and institutions [9]. The three however have a nuanced effect on smartness of the cities. To have an impact, the policy directions and goals of a smart city should be ambitious and transformational [5–7, 10–16]. The focus of the city is to achieve an enriched quality of life for its citizenry whilst deliver tangible benefits by prudently employing the city's natural resources and technology. The chapter adopts the dimensions of a smart city as articulated by Deloitte [4] and is shown in **Figure 1**.

According to Deloitte [4], there are five layers that make up a smart city namely infrastructure, interconnected city systems, ecosystem, people and goals, aspirations and quality of life. In addition to this, according to UNDP [8], digital infrastructure is considered in the form of different supporting digital layers which create different opportunities, as follows:

a. Urban Utilization: The layer where physical and digital infrastructures meet. Examples include smart buildings, smart mobility, smart grids (for utilities such as water, electricity and gas) and smart waste management systems.

b. Sensor: This layer includes smart devices that measure and monitor different parameters of the city and its environment. This could include measurements in pollution in air or water.

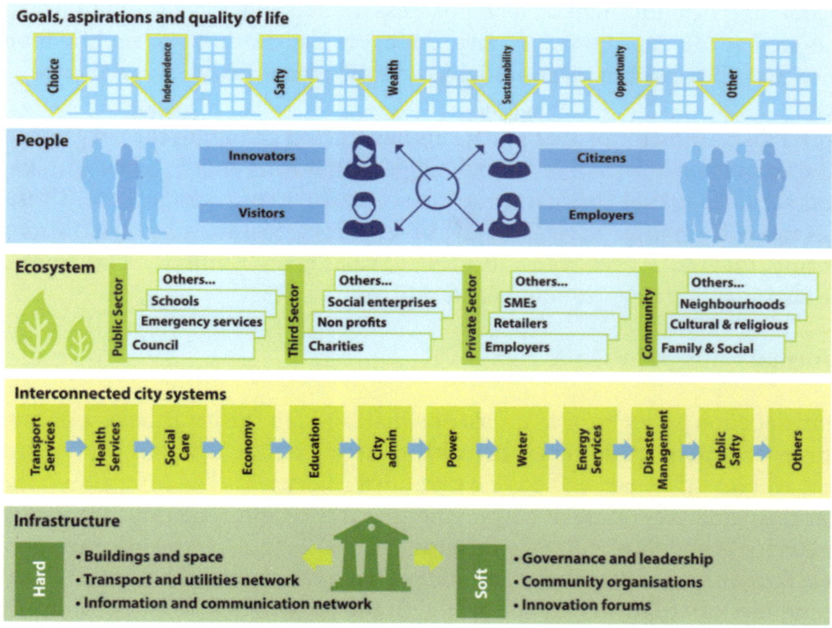

Figure 1.
Dimensions of a Smart City. Source: Deloitte [4].

c. Connectivity: This layer involves the transport of data and information from the sensor level to storage and to data aggregators for further analysis. This is supported by appropriate bandwidth and fiber networks.

d. Data analytics: This layer involves the analysis of data collected by different smart infrastructure systems, to help predict some events. This includes examples such as traffic congestion. It also includes digital health, whereby a programme determines on the basis of pre-determined symptoms on what the basic illness is likely to be and prescribes medicine.

e. Automation: The digital enabling interface layer that enables automation and scalability for a large number of devices across multiple domains and verticals.

Establishing a smart city is a continuously interactive process that entails a robust, reliable and affordable broadband network coupled with an efficient ecosystem for the internet of things (IoT) and the capacity to utilize the big data that will be generated. Governance and leadership support are very crucial also as they allow harnessing and tapping into the local innovation system [10]; Leadership can also support open data and open science models that would have less reliance on proprietary technology models and prop research collaborations and create further opportunities for innovation [17].

2.2 Disruptive business models

Disruption of business takes place when the traditional business models face a challenger who changes the game by offering greater value to the customer in a manner that existing firms are not able to match the offer or be able to compete. In other words, "detonation of the status quo" [11]. Technology has enabled unprecedented development of new business models that have brought greater value as well as increased productivity. For example, the entry of mobile money in Africa had changed business models, brought greater value and enhance efficiency.

Digital disruption in form of platforms has started changing the very nature of what it means to be informal or formal. Digital platforms enable firms with basic business services so that they can concentrate on their core competence. These basic services range from offer advice on how to set prices, customer service training, accounting, sales data, and even collection of sales taxes. Similarly, platforms can handle customer service, payments and returns.

2.3 Informal Enterprises in Africa

Informal enterprises refer to micro or small firms in the informal economy (unregistered with government, are mostly unregulated, employees have no formal contract and no safety net and pay no taxes to authorities). The informal economy is by far the principle source of employment in Africa and accounts for more than 70 per cent of employment in Sub-Saharan Africa [2, 12]. A large informal economy makes it harder to measure the economic performance hence the reason why many governments want to formalize the sector. In the hope of achieving greater value and efficiency, informal enterprises have widely adopted technology. The informal economy is complex and to paraphrase Dungy and Ndofor (2019) [18], is the, "... the utopian and the dystopian, the connected and disjointed, structure and chaos, legitimate yet illegal, legal yet illegitimate all residing together in one big tent...". Furthermore, informal work brings freedom, flexibility, precarity and vulnerability into the lives of African gig workers [3].

2.4 The state of digital readiness in Africa

2.4.1 ICT indicators

Globally, there has been an eightfold growth in the number of individuals using internet over a period of less than twenty years from 495 million to over 4 billion people [19, 20]. This portends well for additional value add products and positioning of technological growth in the continent. Sub-Saharan Africa (SSA) has not been left behind and has seen rapid growth in internet penetration and related technology investment. The International Telecommunications Union (ITU) estimates that sub-Saharan Africans' individual internet usage increased from 2.7% in 2005 to 28.2% in 2019. The trends for the key ICT indicators are shown below in **Figure 2**.

As observed from **Figure 2**, the penetration rate of all ICT indicators has seen positive growth. The increase in internet usage has been accompanied by increased investments in data storage, processing power and innovation ecosystems. In spite this, there are still further opportunities in terms of the population that is not digitized. This has resulted in many large multinational information technology firms viewing Africa as their next frontier of growth [4].

Since 2009, Sub-Saharan Africa (SSA) has witnessed massive investment in digital development and creating enabling conditions but there is still work to be done. With several (TEAMS, SEACOM, EASSy and LION) high-capacity undersea fiber optic connectivity into the continent and boosting capacity to more than 36 Terabits per second, SSA has made tremendous strides. Prior to 2009, the entire continent used a mere 1 Giga Byte per second from satellite. The challenge now lies with last mile coverage.

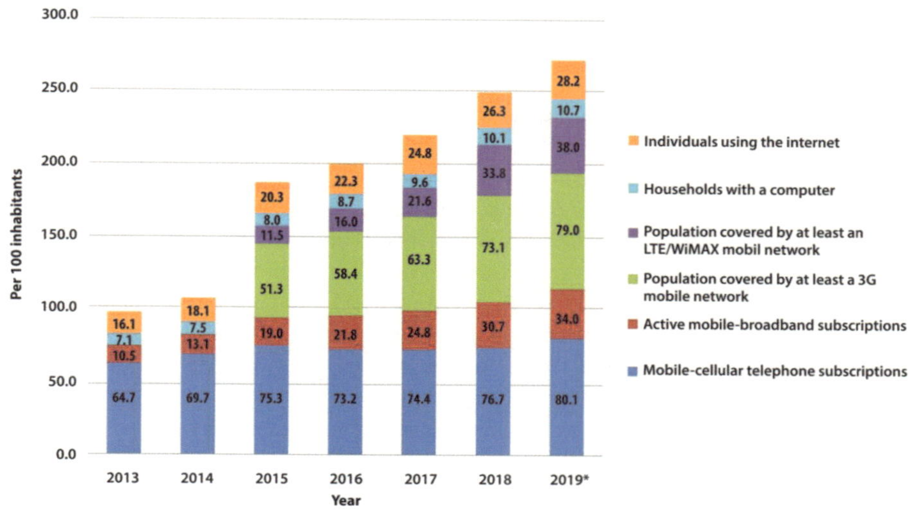

Figure 2.
Sub Saharan Africa key ICT indicator penetration rate (2013–2019). Source: ITU 2019 [20].

2.4.2 Improving access through 4G digital technologies

The last mile coverage especially in advanced systems like fourth generation (4G) network is a major strength of access to digital technologies. 4G network is an advanced network to replace 2G and 3G systems that were used for communication across the world. Its introduction was celebrated as important in SSA because many people first accessed the internet on their mobile phones. As such, 4G with its higher download speeds, sometimes as fast as high-speed fixed broadband, greatly improved user experience. Its introduction has helped to improve productivity. Investments in 4G have largely been in urban areas with much of the rural areas using the older generations. To unlock Africa's digital potential to stimulate enterprise, each country must work toward improving access and affordability.

2.4.3 Broadband affordability for improved access to digital technologies

Affordable broadband improves access to digital technologies and facilitates economic growth. Recent research however, questions if the internet has any effect on economic development. A study by McKinsey Global Institute (2011, p.7) [13] noted that "much of the impact of the internet and the way that it contributes to growth and raising standards of living have gone unmeasured". Follow up studies like Vanags and Grāvelis, 2015 [16], indicated that investment in broadband positively impacted on GDP gain as well as employment. The investment in 4G for example, was meant to improve broadband penetration but comparison between 4G coverage and broadband penetration revealed that without affordability, access alone had no impact. For example, 4G rollout in Rwanda is almost 100% compared to Kenya's 53% but Rwanda's broadband penetration (11.3%) is less than half of Kenya's (47.8%). The difference is explained by affordability, access to devices and human development index.

2.5 Technology-enabled business models

Technology has created a range of new opportunities in the gig economy, a demand-driven independent, short-term or a task-by-task economic activities that payment is received upon the completion of the assignment. The opportunities are available to anyone in the world and provided for anyone who demands them [3]. The operating models of the online gig platforms can be divided into 'on-demand' work like ride hailing drivers like Uber taxis for instance and 'crowd-work' which includes for instance a Kenyan online gig worker, providing translation services for a client based anywhere in the world through Upwork. These platforms may also include homestay hosts like Airbnb, e-commerce logistics like LoriSytems, e-commerce sellers like Jumia, and business-to-business marketplace platform like Twiga Foods. These enterprises would not be possible without investment in ICT infrastructure.

Digital progression enables informal businesses to successfully make the progressive transition from start-up to formal enterprise making each step at less cost and lesser risk [2]. This is exhibited in **Figure 3**.

As observed in **Figure 3**, an informal economy gradually gets used to mobile money and other additional steps. Unlike the Digital business progression model, the ascent in the traditional business progression model is very steep. The use of mobile money has enabled improved financial inclusion in the informal economies [1]. Indeed, according to GSMA [19], Sub Saharan Africa leads in the uptake of mobile money across the globe. This is shown in **Figure 4**.

The growth of mobile money in Sub Saharan Africa has been phenomenal thereby creating massive opportunities for integration. In due course, the firms get

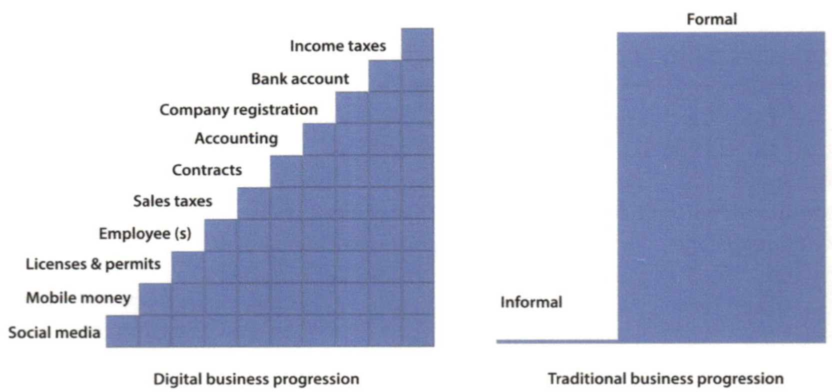

Figure 3.
Digital vs. traditional formalization process. Source: Ng'weno and Porteous (2018) [2].

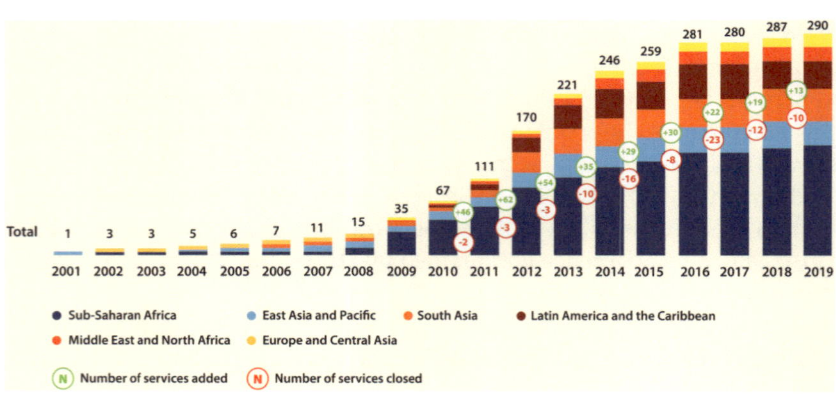

Figure 4.
Evolution of the global Mobile money landscape, 2001–2019. Source: GSMA (2019) [19].

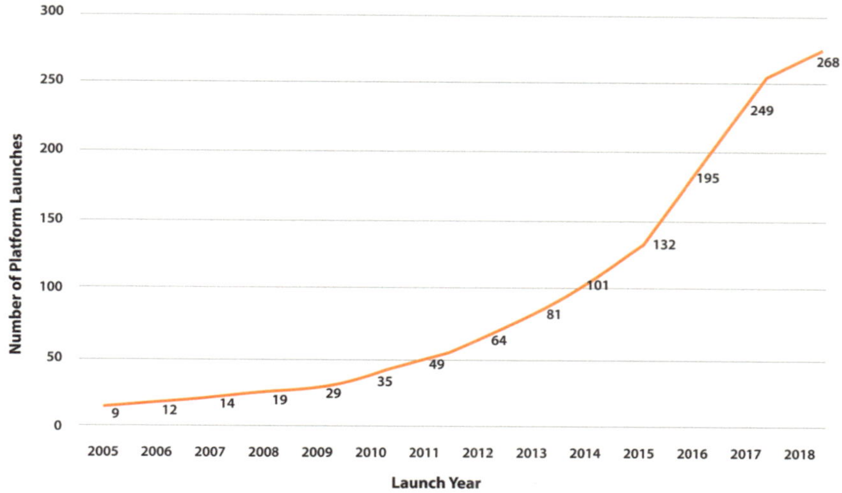

Figure 5.
Platform launches across eight African countries. Source: Smit et al. (2019) [23].

absorbed into the formal economy, paying taxes as required amongst other formal activities. This creates massive opportunities for the formalization of the national economies. In Kenya for example, a study by Genesis Analytics Limited [21] suggested that the total size of the online Kenyan gig economy as at 2019 was $109 million and employed more than 36,000 workers and was projected to grow by over 33% over the next 5 years. Additionally, an estimated 4.8 million African workers reported having derived an income from online gig work in seven surveyed countries namely Ghana, Kenya, Nigeria, Rwanda, South Africa, Tanzania and Uganda [22]. These digital labour markets have a potential for future growth [1]. In addition to this, Smit, Johnson, Hunter, Dunn and van Vuuren (2019) [23] established that the growth of platform launches across Africa has been steady and this is shown in **Figure 5**.

In spite all this, there have been challenges that are associated with working conditions in platforms and online gigs [2, 3, 21, 24]. These challenges range from low remuneration, social isolation as a result of having to work alone, working unsocial and irregular hours to meet strict deadlines, overwork, sleep deprivation and exhaustion as a result of the gig workers having to balance the gigs and their normal responsibilities [24].

Innovation hubs create various pan-sectoral initiatives that promote beneficial ecosystems where entrepreneurs and other stakeholders can collaborate and promote their ideas. To a large extent, these are mostly driven by technology. Hubs characteristically provide in-kind support that includes trainings, advice and facilities as well as financial support programmes. The number of identified tech-hubs in Africa, have seen a phenomenal growth from 314 in 2016 to 643 in 2019. This is represented by **Figure 6**.

According to the survey by Briter Bridges and Afrilabs, as shown in **Figures 6**, 41% of the tech-hub facilities are incubators, 24% are innovation hubs, 14% are accelerators and 39% offer coworking space [25].

2.6 Digitization of informal settlements and inclusivity

The use of information and communication technologies (ICT) and geographic information systems (GIS) to map informal settlements, and by openly providing spatial maps, has led to improved conditions for the poor people living in slums.

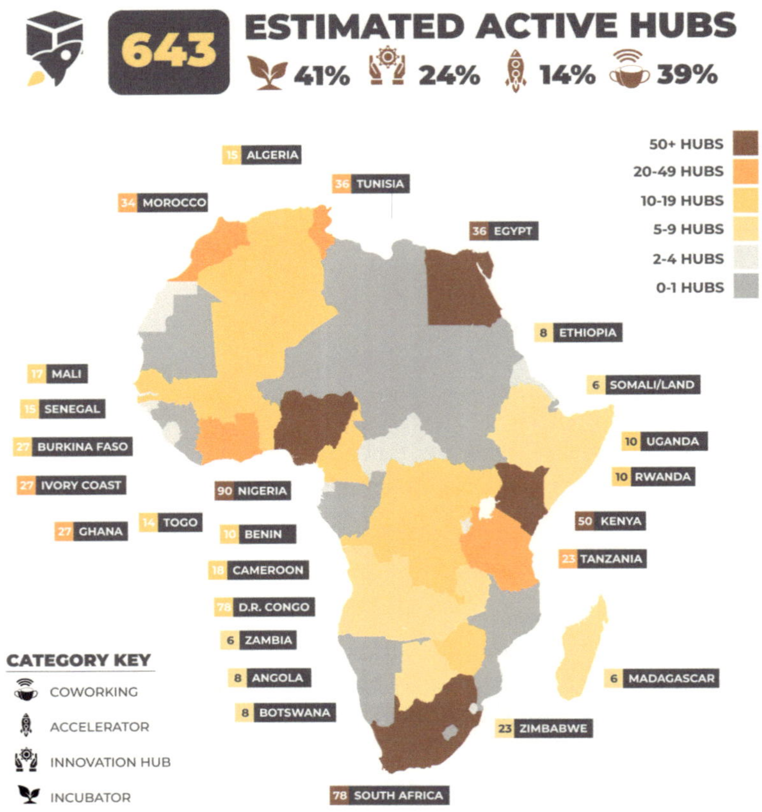

Figure 6.
Tech hubs in Africa, 2019. Source: Afrilabs and Briter Bridges (2019) [25].

Furthermore, ICTs and GIS has forced policymakers to apply much-needed changes of urban renewal, by beginning to pay attention to the plight of the poor in slums to not only identify their slum assets but become but of the growing e-commerce that has brought greater inclusivity improving both livelihood and security. Whereas those living in slums have complex situations, increased transparency through open mapping has provided a platform for sustainable renewal and created new enterprises [15].

3. Methods

In this chapter, we study the impact of smart cities on the creation of new business models and informal enterprises in Sub-Saharan Africa (SSA). SSA provides the best environment to conduct such a study since digitisation is under way, the informal economy is large and the emerging technologies are enabling disruptive business models. Over a period of three years we developed an innovative, inductive method of identifying policymakers from some of the most progressive countries in information and Communications Technologies (ICTs) in the continent for qualitative interviews. We eventually interviewed 18 policymakers as respondents at conferences and more notably at the Transform Africa Summit, 2019 that was held in Kigali, Rwanda. These were qualitative interviews with the key respondents. In addition to this, secondary data, where appropriate was used in the analysis. The findings have been discussed based on the thematic dimensions identified by Deloitte [4].

4. Findings and discussions

Based on observations and interviews, we found many responses to be largely positive in terms of attitude toward technology and what it can do to facilitate economic development in the continent. Almost all the respondents used the term leapfrog at least five times through the interview. There is a can-do attitude even in countries that have not developed sufficient human resource capacity. It confirmed the abundance theory that is keeping policymakers hopeful that it will guide Africa's competitiveness. Psychologists suggest that this theory explains a world in which any person "with the correct attitude, training, or spiritual alignment can acquire personal abundance which should lead to material abundance: wealth regardless of economic or social circumstances."

"With many young people in Africa and the spread of ICTs, nothing will stop Africa from leapfrogging. I have noted that Rwanda has attracted top learning institutions from across the world offering Africa the opportunity to develop global standard labour force..." a senior African Union Official said in the opening ceremonies of Transform Africa in Kigali).

4.1 Building the infrastructure

All of the participants were aware that building of the ICT infrastructure to support smart cities is critical but expensive and as such they are creating new embracing new models of infrastructure development, To build especially the hard infrastructure such as building and space; transport and utilities network; information and communication network they need to leverage public private partnerships (PPPs). Indeed, virtually all of the countries that had started major infrastructure development of new smart cities, leveraged on PPPs. Infrastructure has always been a major problem in Africa but the continent has extensively dealt with it. The current state of infrastructural requirements especially the undersea cables is in place. However, only a handful of countries that have developed or are in the process of developing smart cities (see Table below) as infrastructural bases for smart development.

The African transport infrastructure performance quality which has over time dropped in quality can be replaced with smart urban transport systems that combine and integrate the use of big data, AI and other multiple technologies. Apart from this, other infrastructural requirements can also be applied in areas of energy production and distribution through smart grids. Smart public and private health management systems through AI, blockchain and big data analysis can also be developed and devise innovative and efficient ways for disaster management. The use of additive manufacturing brings affordability and efficiency in production. Subsequently, smart cities leverage on e-commerce platforms to formalize the commercial entities. and education.

Many of these new cities are envisaged to create thousands of new employment opportunities in the emerging technology sector in the continent. Some of the cities at the advanced stages of development are seen in **Table 1**.

The soft requirements of infrastructure fall in the second category and include governance and leadership; community organizations and innovation forums. This can easily be the weakest link because technology is incessantly being developed and "...we love tech so much (that) it inhibits our ability to judge its pros and cons..." [26]. There is a need for public policy framework that enables the infrastructural base to provide an integrated support to the other dimensions of the smart city [27]. This will require leveraging on the regional pan-country like the Smart Africa Initiative [27] as well as other in country initiatives.

Name	Location	Financing model
True Wakanda	Ethiopia	USD 3 billion project in partnership with private developer.
Hope City	Ghana	$10 billion public private partnership
Konza Silicon Savanah City	Kenya	$10 billion in public-private partnership, with national government providing 10% of the total funding (mainly in infrastructure).
Ebène Cybercity	Mauritius	Loans guaranteed by the Indian government.
Eko Atlantic	Nigeria	$6 billion public private partnership
Kigali Innovation City	Rwanda	USD 2 billion project funded by the Rwandese Government and Africa50
Waterfalls	South Africa	$1.2billion Private sector funding expected to be completed in 2025.

Source: Compiled by Authors from Secondary Data.

Table 1.
Financing model of African smart cities.

"If Africa is to succeed in the digital economy, there will be need to develop a single digital market that will support scalability of local innovations across the continent. Small markets find it difficult to compete with such nations as China, United States of America and India. There is also need to create a technology-monitoring mechanism to understand changes in technology, future workforce demands and developing the necessary capacities to remain informed of the different varieties of opportunities for the future of work. For example, Africa must be ready for the emerging 4th industrial revolution and its technologies in order to remain competitive..." United Nations Economic Commission for Africa executive noted.

4.2 Considerations for the future smart cities in Africa

The fourth industrial revolution (4IR) describes the ongoing global conversion of labour-intensive processes to the use of information technology. It is not only ubiquitous but is also happening dramatically. Some of the pillars of 4IR include robotics, artificial intelligence (AI), internet of things (IoT), big data, customer service personalization, cloud computing, and other forms of digital innovation. The resulting shifts and disruptions imply that we live in a time of both great promise and great peril. As the 4IR technologies intensify, cities are becoming smarter, new business models are emerging and informal enterprises are continually under disruption. Experts have argued that 4IR has the potential to rejuvenate Africa's economy, enhance its productivity and improve its global competitiveness [27]. As a result of this, emerging technologies support the development of smart cities. As shown in **Figure 1**, the dimensions of a smart city require an interactive process between people, ecosystems, interconnected systems, and responsive infrastructure.

4.2.1 Goals, aspiration and quality of life

African cities are uniquely advantaged to have a competitive edge for the future. Some of these advantages include limited legacy drawbacks; youthful consumer population; urbanization; entrepreneurial culture; connectivity; overarching government leadership strategically positioning ICT as an enabler [4]. Furthermore, the rise of the gig economy has created employment opportunities that has impacted upon poverty reduction in the emerging economies [1, 28] thereby improving on the quality of lives. With a population that has increasingly been well educated and exposed to different cultures and lifestyles, the goals, aspiration and quality of life for the urban individuals has improved over time.

Lifestyle is a key driver for smart cities. The broad areas of concern in the individual's lifestyle include government efficiency, employment, transport, education, healthcare, energy, environment and public security and safety. As a result of the adapted lifestyle, individuals are ambitious to have freedom to choose without compromising on their feeling of independence. Furthermore, the same individuals are desirous of being in control and feeling safe and secure. To meet these objectives, the individuals need to have opportunities for creation of wealth and sustenance of their lifestyles.

4.2.2 People

The dimension of people relates to the nexus between the opportunity seekers who double up as the innovators, the residents who desire the quality of life, the employers who utilize the opportunities and visitors of the urban areas. A smart city makes use of the pragmatic ideas by creative people to provide smart scientific solutions that address the lifestyle concerns around the city. As a result of the solutions, the residents of the smart cities have access to a number of features that range from smart homes, smart buildings, smart offices and larger smart ensembles like airports, shopping malls hospitals or university campuses which are fitted with a multitude of mobile terminals and embedded devices as well as connected sensors that are monitored and programmed for certain decisions. In addition to these, there are a whole range of interconnected logistical support and services through various platforms that provide background support to the lifestyle solutions. These functionalities are enabled through technology by IoT and robotics.

"The deployment of sensors across the city of Nairobi has lessened crime especially carjacking that was rampant prior to installation of ICT infrastructure to monitor transport...." a Kenyan delegate told the conference.

"Our endgame with ICTs in Kigali is to see smart services especially Smart city tourism that will be enhanced through Smart ticketing, Smart-security services, intelligent crown management, improved transport services, virtual reality, linguistic services or even smart city bots to guide visitors around. Through IoT, additional features can be availed to the discerning tourist on the basis of their smart phones. These services would require the local city's residents to be adequately skilled to be able to perform the back ground support tasks..." a Rwandan official noted while explaining their future plans for Kigali.

Already employers are leveraging on a smart city infrastructure and have many routine tasks accomplished at decentralized locations. This sets a basis for online gig and platforms. A challenge that however that subsists is that whereas online platforms will not always have individual workers' interest as their priority, there is increased agitation for increased regulation of this sector. Secondly, by its very definition, online gig tasks can be carried out in a borderless fashion with minimum regard for local regulation on working conditions.

4.2.3 Building future ecosystems

Digital technologies have enabled inclusion, efficiency and innovation of opportunities. A smart ecosystem is a conceptual extension of smart space from the personal context to the larger community and the entire city. It straddles the public and private sector and the broader community. It encompasses policies, laws, regulations and processes that are weaved together to obtain a desired smart outcome. An enabling ecosystem is based on a skilled and equally aspirational citizenry as well as transformational leadership. To enable continuous innovation, smart cities proactively encourage innovation programmes that include labs, training, skills development and partnerships with different academic, vocational and research institutions. A Smart city ecosystem facilitates the integration of data and information. This is done by initiatives that support open data, analytical services as well as monetization framework. The ecosystem needs to encourage data sharing whilst protecting privacy and what and how data is generated. Herein lies the digital age paradox, where transnational firms have comprehensive information on individuals' lives and can trade it in the global marketplace, whereas individual citizens struggle to get rudimentary information on growth in income and wealth at a macro level [1].

4.2.4 Interconnected systems

The world is more connected than ever [19, 20, 28]. Similarly, as shown in **Figure 1**, a smart city has interconnected systems that cover different facets of the city's activities. Smart cities strive to make strategic choices that attain transformational leaps in the quality of life within its region of operation. There are many ways in which technology connectivity can be used in pursuit of urban management. For example, through the use of IoT, roads can be equipped with LED street lamps that sense pedestrian movement, and consequently dimming and brightening in accordance with the movement. These IoT enabled gadgets also sense and collect pollution data in the air, and send this information to a data base. Additionally, the IoT gadgets can detect humidity and weather conditions in an area and advised to the consumers of this information.

Furthermore, IoT can be used to analyze the traffic on roads and adjust parking metre fees accordingly. This feeds to an automated urban traffic management system. IoT can also identify weak infrastructure like potholes on roads thus helping authorities prioritize their budget for urgent repairs. Remotely monitored close circuit cameras, will take pictures of an accident scene and send it to some database, but machine learning and AI would be required to translate this data into actionable information that can trigger emergency services to save lives. Finally, the IoT enabled street lamps can also act as free WiFi routers to nearby citizens.

Such interconnected systems allow the smart cities to remove inefficiencies that come about as a result of manual monitoring and intervention. The role of human intervention would now be directed to higher skilled tasks. This calls for a re-think into our skills development in the continent by facilitating creation of mechanisms supporting school-to-work transition in each country by (i) developing content around career choices and investing in counselors to help students navigate the transition (ii) revamping TVET institutions to meet the new demand for jobs (iii) changing the negative perception of TVETs and (iv) collaborating with the private sector to provide internship to graduating students as part of the transition process.

5. Conclusion

Smart cities allow the leveraging of digital transformation in a shift to abandon traditional paradigms and create a novel globally entangled experience and lifestyle. As the cities become smart, formalization of the informal sector improves. New digital businesses are forcing the shift into formalization which benefits the economy by making it possible to measure the economy and better worker safety. Through the use of technology, and the democratization of information, smart cities enable transnational innovation processes that would be targeted at universal, grand challenges. Our contention is that the absence of legacy systems and the advent of affordable broadband has set SSA on a roller-coaster of change that may see the formalization of informal enterprises by new business models offering better value. The combination of informality, technology and disruptive business models brings a new territory of change and discovery [14]. However, not every country in SSA has had the chance of seeing change and discovery. Several other factors stand in the despite the fact that SSA has the necessary infrastructure to enable greater productivity through technology.

6. Policy recommendations

Not every African country that has embarked on making cities smart. As such each country needs some policy interventions starting from building a national vision that is dedicated to national level commitments to developing smart infrastructure. This also will spell out the commitment to funding through PPPs and collaborating with institutions of higher learning to do studies on the impact of the changes. Smart cities require an urgent development of localized relevant skilled capacity as they can easily result in gigs being performed from remote locations thereby not actually addressing the local issues of unemployment. For better outcome, it is imperative that governments embrace Research and Development as well as education. Hold regular hackathons on innovation and competition in order to build the future infrastructure to support the smart cities.

Conflicts of interest

The authors declare no conflict of interest with respect to the research, authorship, and/or publication of this article.

Appendices and Nomenclature

AI	Artificial Intelligence
4IR	Fourth Industrial Revolution
ICTs	Information and Communication Technologies
IoT	Internet of Things
ITU	International Telecommunications Union
PPP	Public Private Partnerships

Author details

Ben Mkalama* and Bitange Ndemo
University of Nairobi, Nairobi, Kenya

*Address all correspondence to: ben@mkalama.co.ke

References

[1] UNDP. *Human Development Report 2019: Human Development for Everyone*. New York: United Nations Development Programme; 2019

[2] Ng'weno A, Porteous D. *Let's Be real: The informal sector and the gig economy are the future, and the present, of work in Africa*. In: Centre for Global Development, Study Group on Technology, Comparative Advantage, and Development Prospects. Centre for: Global Development; 2018

[3] Anwar, M. A., & Graham, M. (2020). Between a rock and a hard place: Freedom, flexibility, precarity and vulnerability in the gig economy in Africa. Competition & Change, *0*(0), 1-22.

[4] Deloitte. (2018). *Africa Is Ready to Leapfrog the Competition through Smart Cities Technology*. Deloitte.

[5] Al Nuaimi, E., Al Neyadi, H., Mohamed, N., & Al-Jaroodi, J. (2015). Applications of Big Data to Smart Cities. *Journal of Internet Services and Applications*.

[6] Onyima J, Ojiagu C. Digital technology and formalization of informal businesses: A case of African traditional spiritualists. International Journal of Academic Research in Business and Social Sciences. 2017;*7*(11)

[7] Dawson, M. (2011). Theory of Scarcity Versus Theory of Abundance. *The Internet Writing Journal*.

[8] UNCTAD. *Smart Cities and Infrastructure.* Economic and Social Council, Commission on Science and Technology for Development. Geneva: United Nations Publications; 2016

[9] Nam, T., & Pardo, T. A. (2011). Conceptualising Smart City with dimensions of technology, people, and Instutions. *The Proceedings of the 12th Annual International Conference on Digital Government Research.*

[10] Sarkar, A. (2020). *Global Business Models of Smart and Sustainable Cities.* Retrieved from Smart Cities Journal: https://www.thesmartcityjournal.com/en/articles/859-global-business-models-of-smart-and-sustainable-cities.

[11] Future Trends Trust. (2018). *Disruptive Business Models.* Future Trends Trust: https://www.fundactionbankiter.org/documents/20183/156075/Business+Model+Disruption+Report+EN+2.pdf/e238c273-e45c-4ccd-bd3a-802926094203.

[12] ILO. Resolution concerning decent work and the informal economy. In: *The General Conference of the International Labour Organization Meeting in its 90th Session*. International Labour: Organization; 2002 https://www.ilo.org/public/english/standards/relm/ilc/ilc90/pdf/pr-25res.pdf

[13] McKinsey Global Institute. (2011). *Internet Matters: The Net's Sweeping Impact on Growth, Jobs and Prosperity.* McKinsey: https://www.mckinsey.com/insights/high tech telecoms internet/internet matters.

[14] Muegge, S. (2012). Business Model Discovery by Technology Entrepreneurs. *Technology Innovation Management Review*.

[15] Ndemo, B. (2020). Slum digitisation, its opponents and allies in developing smart cities: The case of Kibera, Nairobi. In S. Hawken, H. Han, & C. Pettit, *Open Cities Open Data.* Singapore: Palgrave Macmillan.

[16] Vanags, A., & Gravelis, L. (2015). The benefits of 4G: Sweden and Estonia - European leaders in 4G.

TeliaSonera Institute Discussion Paper No.16. https://www.biceps.org//assets/docs/petijumu-serija/TSIDP16.

[17] Chesborough H. *Open Innovation Results*. In: *Going beyond the Hype, and Getting Down to Business*. Oxford University: Press; 2020. DOI: 10.1093/oso/9780198841906.001.0001

[18] Dungy TN, Ndofor HA. A dialogue on the informal economy in Africa. Africa Journal of Management. 2019;*5*(4):401-407

[19] GSMA. *2019 State of the Industry Report on Mobile Money*. GSMA: Mobile Money Programme; 2019

[20] ITU. (2019). *International Telecommunications Union.* Retrieved July 2020, from Global and Regional ICT Data: www.itu.int/itu-d/itu-d ict statistics.

[21] Genesis Analytics Limited. *Towards a Digital Workforce: Understanding the Building Blocks of Kenya's Gig Economy*. Nairobi: Mercy Corps' Youth Impact Labs; 2019

[22] Onkokame, M., Schoentgen, A., & Gillwald, A. (2018). *What is the state of microwork in Africa? A View from seven countries.* Retrieved July 2019, from research ICT Africa: https://researchictafrica.net/after-access_the-state-of-microwork-in-africa/

[23] Smit, H., Johnson, C., Hunter, R., Dunn, M., & van Vuuren, P. F. (2019). *Africa's Digital Platforms and Financial Services: An Eight-Country Overview.* Insight2impact. Johannesburg.: Insight2impact.

[24] Wood A, Graham M, Lehdonvirta V, Hjorth I. Good gig, bad gig: Autonomy and algorithmic control in the global gig economy. Work, Employment and Society. 2019;*33*(1):56-75

[25] Afrilabs and Briter Bridges. (2019). *Building a Conducive Setting for Innovators to Thrive: A Qualitative and Quantitative Study of a Hundred Hubs across Africa.* Afrilabs and Briter Bridges.

[26] Postman, N. (1993). *Technopoly: The Surrender of Culture to Technology.* New York: Vintage Books: ISBN 978-0-679-74540-2.

[27] African Development Bank. *Potential of the fourth industrial revolution in Africa*. In: African Development Bank; Technopolis; Research ICT Africa. Tambourine Innovation: Ventures. African Development Bank; 2019

[28] World Bank. *World Development Report 2016: Digital Dividends*. The World Bank: Washington D.C; 2016. DOI: 10.1596/978-1-4648-0671-1

Chapter 3

Does Smart City Development Promote Urbanization in India?

Sabyasachi Tripathi

Abstract

The recent explosion of urbanization is mainly driven by the developing countries in the world. Therefore, urban planners in less developed countries face huge pressure to create planned urbanization which includes the higher provision of infrastructure and basic public services. The part of this planned urbanization 'smart city' development is one of the important initiatives taken by many countries and India is one of them. In terms of the size of the urban population through India ranked the second position in the world but in terms of the percentage of the urban population, it ranks very low. Therefore, to promote the urbanization Government of India (GoI) has taken 'Smart Cities Mission' initiatives for 100 cities in 2015. In this context, the present chapter quantitatively assesses the impact of smart city development on the urbanization in India. Urbanization is measured by the size, density, and growth rate of the population of the smart cities. On the other hand, we use factor analysis to create infrastructure index by considering city level total road length, number of latrines, water supply capacities, number of electricity connections, hospitals, schools, colleges, universities, banks, and credit societies. OLS regression analysis suggests that infrastructure has a strong positive effect on urbanization. Therefore, the smart city mission is very much essential for the promotion of urbanization in India. Finally, we suggest that we need to have more smart cities in the future so that a higher rate of urbanization promotes higher and sustainable economic growth.

Keywords: urbanization, infrastructure, smart city mission, India

1. Introduction

The United Nations World Urbanization Prospects shows that in 2017, 4.1 billion people were living in urban areas. This indicates that more than half the world population (55%) lived in urban areas. In this context, India's urbanization is much slower than many developing countries and even its peers such as China, Brazil, and Russia. The latest Census data shows that the percentage of India's urbanization was 31.15% in 2011. On the other hand, China (or Brazil or Russia) has experienced about 49.2% (or 84.3% or 73.7%) urbanization rate in 2010. The reluctant urbanization in India can be because of a lack of governmental supportive policies or challenges in managing the urban dynamics [1]. On the other hand, China's urban policies are focused on integrated urban and rural development, the creation of city clusters to spread the benefits of urbanization, and the promotion of sustainable urban development. Though China's urbanization is more policy-induced, India's urbanization is more market-determined. Therefore, appropriate urban policies in India are required for proper design and implementation.

No country has ever reached middle-income status without a significant increase in urbanization [2]. Urbanization has contributed not only to higher income but also it has improved people's lives [3, 4]. Therefore, the promotion of urbanization is very important for many developing countries such as India. Currently, India is facing numerous challenges as a result of enormous urban dwellers. India has now two challenges; first, it has to speed up the urbanization rate, and secondly, it has to make proper or planned urbanization so that the maximum benefits of urbanization are achievable. Urbanization use resources such as excess labor and land more productively and becomes the engine of economic growth.

To achieve planned urbanization for higher and sustainable economic growth Government of India (GoI) has taken Smart Cities Mission initiatives. In June 2015, the Ministry of Urban Development (MoUD) released a mission statement and guidelines for the Smart Cities Mission. This program replaced the previous major central government's flagship program Jawaharlal Nehru National Urban Renewal Mission (JNNURM) and wanted to move India's cities forward under Prime Minister Narendra Modi's leadership. Under the guidelines, several strategies are sketched by which an applicant entity can apply to achieve smart city designation [5].

Though there is no universally accepted definition of 'smart city', India's smart city development mission is meant to invest more on the core infrastructure elements such as water, electricity, sanitation, solid waste management, public transport, e-governance, etc. GoI also has proposed eight features of comprehensive development for smart cities. This includes promotion of mixed land use, housing and inclusiveness, creation of walkable localities, preservation and development of open spaces, promotion a variety of transport options, making governance citizen-friendly and cost effective, giving an identity to the city, applying Smart Solutions to infrastructure and services in area-based development to make them better.

In this context, the present chapter assesses the impact of higher infrastructure availability on the population size of the smart cities in India. For the analysis, we consider only 85 smart cities in India that have populations more than 1 lakh (class I cities). The relevance for consideration of these cities that the class I cities accommodate about 70.2% of the total urban population in 2011. This indicates that India's urbanization is concentrated in and around the class I cities. Therefore, it is important to investigate whether a further increase in the infrastructure of class I cities escalate population or not. It is very important to increase urbanization in India as it is having a slower rate of urbanization.

2. Review of literature

There are very few following studies which explore the impact, structure, and implementation strategies of smart cities program in India. Russell et al. [6] argued that the Smart Cities Mission marks a continued shift for urban development policy in India away from direct government intervention. They argued that the cities nominated for the Smart Cities Mission have adequate levels of public services, a lower percentage of slums, and are bigger. Therefore, providing basic infrastructure to these cities is against the smart city ideas and concepts.

Praharaj et al. [7] indicated that Indian cities need synergy across urban policies for better results. They also stated that smart city plans lack integration and have a conflict with statutory master plans. Praharaj et al. [8] explored the relationship between active civic engagement and the availability of basic digital infrastructure and socio-economic standards in Indian cities. They provide important lessons for building future smart and connected cities as well as promoting healthy urban relationships and welfare, in the emerging economies of the world. Aijaz and Hoelscher [9] argued that to make the 'smart city mission' more equitable and sustainable the fair engagement of citizens and all stakeholders need to involve. Praharaj and Han [10] stated that the Indian smarty city discourse predominantly corporate-driven and technology-focused. Therefore, smart cities should engage with sustainability and community issues. Randhawa and Kumar [11] argued that smart city development policies lack concerns towards the natural environment which is an important dimension of sustainable development of a city.

Rana et al. [1] found that that 'Governance' is the most significant category of barriers for smart city development followed by 'Economic; 'Technology'; 'Social'; 'Environmental' and 'Legal and Ethical' in India. Hoelscher [12] stated that the smart cities agenda in India appears to be characterized by a failure to conceptualize and develop an integrated set of policies, and while a clearer (yet contested) concept is emerging, the prospects for success are uncertain. Praharaj and Han [13] found that the vast disparities remain across India's urban centers, located in different geographical regions, in terms of access to social capital and physical infrastructure. Their analysis suggests that education, health, and social services are important drivers in the urban typology building process. The small to medium sized cities in India are missing basic community infrastructure. This implies that smart city development strategy which considers one-size-fits-all by assuming importance of foundational infrastructure has the shortcomings. Tripathi [14] argued that smart cities in India should ensure smart distribution of benefits of urban economic growth to the poorer section of urban dwellers for future development of urban India.

Adapa [15] presents a comprehensive review of the existing smart city frameworks and cleaner production initiatives in the Indian context. Aijaz [16] argued that the negative effect of India's urbanization includes informal-growth of peri-urban areas, escalating water crises, social exclusions, an extension of slums, and mismanagement of solid waste. The author argued that the success of smart city development only possible if civic institutions correctly understand the city's social, economic, and physical requirements and its diversity. At the same time, citizens should show a greater sense of civic responsibility.

The brief review of the literature mainly suggests that how smart city development initiatives can be more effective if properly implemented. In other words, what are the important dimensions of India's urbanization that have to be considered for the successful implementation of smart city development strategies which will lead to successful urban development in India? However, these studies have missed important dimensions of India's urban development policy which is how to increase the urbanization rate which is essential for economic development. Therefore, the present study attempts to fulfill this gape for better urban development in the future.

3. Empirical analysis

To estimate the impact of infrastructure on smart cities we consider the following econometric model:

$$Urbanization = \alpha_0 + \sum_{i=1}^{10} \beta_i \boldsymbol{Infrastructure}_i + e_i \tag{1}$$

where e_i represents well-behaved error term and α_0 stands as constant. Ordinary least squares (OLS) method is used to analyze the impact of infrastructure on urbanization in India. Based on Tripathi [17, 18] city population size, city population density, and city population growth rate are considered to measure the urbanization in this paper. On the other hand, city-level availability of infrastructure is measured by considering city level total road length, number of latrines, water supply capacities, number of electricity connections, hospitals, schools, colleges, universities, banks, and credit societies.

In the context of the positive impact of infrastructure on urbanization, Tiebout [19] indicated that accessibility and superiority of public facilities such as parking facilities, police protection, roads, parks, and municipal golf courses are very important for choosing a municipality. Therefore, consumer voters would migrate to a city that satisfied their demand for infrastructure. Harris and Todaro's [20] model explained that rural to urban migration depends on expected rural–urban income differential rather than rural–urban wages. This indicates that urban condition is better with higher infrastructure facilities which attract more rural people [18].

In the context of India, several studies (e.g., [21–24]) argued that India's urban areas lack adequate infrastructure which requires urgent attention. Pradhan [25] investigated the impact of infrastructure on urbanization in India, using a composite infrastructure development index based on three sub-indexes: physical infrastructure, social infrastructure, and financial infrastructure. Using multivariate principal component analysis, the study confirmed that infrastructure has a significant positive impact on urbanization in India. On the contrast, Tripathi [18] argued that the improvement of infrastructure in large cities may not increase population concentration, but it will improve the living conditions and business activities that increase economic growth potential. Based on these studies we expect a positive or negative effect of infrastructure on urbanization driven by smart city development.

Details about the variable measurement and data sources are provided in Appendix A. **Table 1** presents the summary statistics of each variable used in the analysis. The coefficient of variation (CV) measures the dispersions of data points in a data series. Log of the city population, city population density, and city-wise total number of colleges have lower values of a coefficient of variation (CV) which indicates that there are little differences in their means, implying a more symmetrical distribution. However, it is not the case for the city-wise total number of credit societies, city-wise total water supply capacity, the city-wise total number of banks, and the city-wise total number of latrines.

Table 2 shows the raw correlation of the variables. The results show that the log of the city population is positively associated with all the infrastructure variables. Most importantly, the log of city population highly correlated with city-wise road length, the city-wise total number of latrines, city-wise total number of electricity connection, and city wise total number of schools. On the other hand, the correlation between city population densities and infrastructure variables is not strong. Similar results are obtained for the correlation between city population growth rate and infrastructure variables.

We now investigate the impact of infrastructure on the urbanization. Based on Tripathi [17, 18], we consider the city population, density, and growth rate for the measurement of urbanization. We consider a total of 10 variables to measure the infrastructure and stand as interdependent variables. **Table 1** shows that there are considerable variations between the minimum and maximum values of the variables. The correlation coefficients show that data are more correlated as the values increase. Hence, factor analysis is considered to reduce the number of independent variables to obtain appropriate estimation.

Variable	Mean	Standard deviation	Minimum	Maximum	Coefficient of variation
Log of city population (v1)	13.52	0.91	11.59	15.96	6.75
City population growth rate (v2)	19.02	22.54	−60.00	111.00	118.46
City population density (v3)	9084.07	5974.53	679	32622.00	65.77
City-wise total road length (v4)	1160.34	1536.60	9.00	11812.00	132.43
City-wise total number of latrines (v5)	198560.00	300578.80	1114	2063946.00	151.38
City wise total water supply capacity (kilo liter) (v6)	115657.90	231942.40	0.00	1200000.00	200.54
City wise total number of electricity connection (v7)	330136.00	449066.80	25500.00	2700000.00	136.02
City-wise total number of hospital (v8)	198.67	310.98	5.00	1706.00	156.53
City-wise total number of schools (v9)	859.68	1174.64	9.00	8397.00	136.64
City-wise total number of colleges (v10)	67.98	75.29	1.00	532.00	110.76
City-wise total number of universities (v11)	2.18	3.10	0.00	18.00	142.35
City-wise total number of banks (v12)	185.51	366.68	2.00	2247.00	197.66
City-wise total number of credit societies (v13)	272.95	806.75	0.00	5193.00	295.56

Note: calculations are based on 85 observations. Source: Author.

Table 1.
Description of data used for the analysis.

To ensure the validity of data, Kaiser-Meyer-Olkin (KMO) and Bartlett's Test of Sphericity are used. The KMO test is performed by using STATA version 14.1. The estimated results in **Table 3** show that factor analysis is highly recommended as the KMO value is 0.851. The probability of Bartlett's test of Sphericity is very significant ($0.000 < 0.01$). Thus, factor analysis is desirable.

The initial eigenvalues (i.e., a variance of the factor) are presented in **Table 4**. The most variance is presented by the first factor, the next maximum amount of variance is considered by the second factor, and so on. The negative eigenvalues indicate that the matrix is not full rank suggesting six factors for the analysis can be considered at most. On the other hand, the KMO criterion recommends that factors with Eigenvalues ≥ 1 should be considered for the analyses. Therefore, the only first factor is relevant for the study that accounts for about 86% of the variance in the solution.

The factor loadings (pattern matrix) according to the uniqueness i.e., a variance is exclusive to the variable and not contributed by other variables is presented in **Table 5**. The bigger values of uniqueness indicate that variables are not properly explained by the factors. For instance, 93.3% of the variance in 'total credit society'

	V1	V2	V3	V4	V5	V6	V7	V8	V9	V10	V11	V12	V13
V1	1.00												
V2	0.13	1.00											
V3	0.23	−0.11	1.00										
V4	0.68	−0.09	0.16	1.00									
V5	0.70	0.04	0.22	0.80	1.00								
V6	0.42	0.04	−0.04	0.44	0.51	1.00							
V7	0.75	0.05	0.27	0.81	0.91	0.41	1.00						
V8	0.38	−0.12	0.08	0.35	0.24	0.37	0.30	1.00					
V9	0.68	−0.04	0.17	0.87	0.81	0.43	0.82	0.25	1.00				
V10	0.64	0.01	−0.05	0.78	0.67	0.53	0.63	0.34	0.82	1.00			
V11	0.56	0.04	0.10	0.55	0.69	0.35	0.68	0.37	0.68	0.56	1.00		
V12	0.59	−0.04	0.27	0.69	0.79	0.32	0.81	0.13	0.71	0.53	0.52	1.00	
V13	0.32	0.04	0.05	0.23	0.20	−0.02	0.23	0.04	0.25	0.31	0.18	0.25	1.00

Note: see ***Table 1*** *for variable definitions. The calculation is based on 85 observations. Source: Author.*

Table 2.
Correlation coefficient of the variables used for the analysis.

KMO measure of sampling adequacy		0.851
Bartlett's test of sphericity	Approximate chi-square	707.040
	Df	45
	Sig.	0.000

KMO, Kaiser-Meyer-Olkin. Source: Author's calculation.

Table 3.
KMO and Bartlett's test.

Factor	Eigenvalue	Difference	Proportion	Cumulative
Factor1	5.61362	5.01985	0.8569	0.8569
Factor2	0.59377	0.23801	0.0906	0.9475
Factor3	0.35576	0.14509	0.0543	1.0019
Factor4	0.21067	0.11021	0.0322	1.034
Factor5	0.10046	0.03567	0.0153	1.0493
Factor6	0.0648	0.09602	0.0099	1.0592
Factor7	−0.03122	0.03265	−0.0048	1.0545
Factor8	−0.06387	0.06123	−0.0097	1.0447
Factor9	−0.1251	0.04279	−0.0191	1.0256
Factor10	−0.16789	.	−0.0256	1

Source: Author's calculation.

Table 4.
Explanation of total variance.

Variable	Factor1	Uniqueness
Total road length	0.8937	0.2013
Total number of latrines	0.9241	0.146
Total water supply	0.5232	0.7263
Total number of electricity connections	0.9179	0.1575
Total number of schools	0.923	0.1481
Total number of hospital	0.3619	0.869
Total number of colleges	0.8166	0.3332
Total number of universities	0.721	0.4802
Total number of banks	0.7801	0.3914
Total number of credit societies	0.2583	0.9333

Source: Author's calculation.

Table 5.
Factor loadings (pattern matrix) and unique variances for one factor model.

is not contributed by the other variables in the overall factor model. On the contrary, the 'total number of latrines' that has very low variance (14.6%) is not shared by other variables. As the values of factor loading for approximately all variables are higher (>0.3), we can conclude that factor 1 is defined by all six variables that are considered to produce an infrastructure index. Quite importantly, factor1 is mostly related to the city-wise number of electricity connections and city-wise number of latrines. It is also important to note that as we are using one factor only, factor rotation which helps to see the underlying dimensions (scales) more clearly is not suitable as there's nothing to rotate.

The linear regression analysis is used to investigate the impact of infrastructure on urbanization in India. **Table 6** presents the results of the regression analysis. The factor score values for the one selected factor is considered as the independent variable. Regression models 1–5 present the estimated results for three dependents variables i.e., size, growth, and density of city populations. To control the heteroscedasticity problem we estimate the robust standard errors.

Regression 1 shows that the infrastructure index has a positive and statistically significant effect on the smart city population in 2011. A 10% increase in infrastructure index increases the smart city population by 7.1%. This indicates that higher infrastructure investment increases the population of smart cities. On the other hand, a higher level of infrastructure also increases the population density of the smart cities in regression 4. The coefficient 0.123 indicates that a 10% increase in infrastructure index increases smart city density by 1.2%. However, infrastructure may not increase the growth rate of the city population as it has a statistically insignificant effect on it in regression 5. This is quite evident as most of the large cities considered for smart city development experienced a negative growth rate. For example, Thiruvananthapuram experienced a 14% negative population growth rate from the period of 2001 to 2011. Therefore, smart city development does not increase the population growth rate of smart cities.

To estimate the robustness of the results we consider smart city population data for 2020 and 2025 from World Urbanization Prospects (WUP): The 2018 Revision [26]. The WUP provides a data population of urban agglomerations with 300,000 inhabitants or more in 2018. On the other hand, though 11% of the total proposed work under the smart city mission completed in 2019, still we have to wait for 2021 (i.e., next Census data) for the evaluation of the impact of infrastructure on the population of smart cities. As some of the smart cities that are considered for our survey have a population less than 3 lakh we could collect data only 77 smart cities. The regression results 2 and 3 show that available

	Dependent variable				
	Log of city population 2011	**Log of city population 2020**	**Log of city population 2025**	**Log of city population density 2011**	**Log of city population growth rate 2011**
	(1)	(2)	(3)	(4)	(5)
Infrastructure index	0.711***	0.628***	0.632***	0.123*	0.218
	(0.165)	(0.147)	(0.145)	(0.074)	(0.183)
Constant	13.516***	7.142***	7.256***	8.961***	2.987***
	(0.064)	(0.079)	(0.079)	(0.065)	(0.096)
F Statistics	18.44***	18.33***	18.95***	2.80*	1.42
R^2	0.5886	0.4656	0.4648	0.0415	0.0343
Observations	85	77	77	83	67

Robust standard errors in parentheses. Source: Estimated using Eq. (1).
$^{*}p < 0.1$.
$^{***}p < 0.01$.

Table 6.
Results of regression analysis.

infrastructure in 2011 has a positive and statistically significant effect on the log of the smart city population in 2020 and 2025. This indicates that infrastructure has a big role in the promotion of urbanization in India and smart city mission is very important for that.

4. Conclusions

The present chapter assesses the impact of infrastructure on the urbanization by smart cities in India. Smart city urbanization is measured by population, density, and growth rate of the population of the 85 smart cities in India. On the other hand, smart city-wise availability of infrastructure is measured by the considering city level total road length, number of latrines, water supply capacities, number of electricity connections, hospitals, schools, colleges, universities, banks, and credit societies.

The factor analysis is used to create an infrastructure index by considering all the infrastructure variables. The OLS regression analysis is used to measure the impact of infrastructure on urbanization. The OLS regression results suggest that the availability of infrastructure has a positive and statistically significant effect on the urbanization measured by the smart city population and densities of the smart city population. This indicates that the smart city mission promotes India's urbanization.

India's cities and towns are having a serious lack of adequate infrastructure facilities. The Report on Indian Urban Infrastructure and Services [21] urged that urban India severely faces deficiency in the provision of urban public services such as street lights, solid waste management, roads, sewerage, and drinking water. The report estimated that Rs 39.2 lakh crores at 2009–10 prices are required over a 20-year period to achieve this growth. The outlay on urban roads accounts for Rs 17.3 lakh crore (or 44%) of this amount. In this perspective, the smart cities mission is appropriate for the promotion of urbanization in India by huge investment in infrastructure. It is very much important to indicate that India had a total of 7935 cities and towns in 2011. Therefore, smart cities initiatives only for 100 cities may not fulfill the urbanization dream for India. In the coming years, India should have more smart cities to explore the benefits of urbanization for higher economic growth.

Appendix A: Measurement of variables and data sources

City population, density and growth: City population data is collected from Census of India, 2011. Website: https://www.census2011.co.in/urbanagglomeration.php

Total road length: Both Kachcha road length and Pucca road length are considered for the measurement of total road length of a city. Source: Town amenities, District Census Hand Book, Census of India 2011. Website: http://censusindia.gov.in/2011census/dchb/DCHB.

Number of Latrines: Total number of pit, flush/pour, services, and other latrines. Source: Town amenities, District Census Hand Book, Census of India 2011.

Total water supply: Total protected water supply in city. Source: Town amenities, District Census Hand Book, Census of India 2011.

Electricity connection: Total number of electricity connections in domestic, industrial, commercial, road lighting, electricity, and other connections. Source: Town amenities, District Census Hand Book, Census of India 2011.

Total hospitals: It includes allopathic hospitals, alternative medicine hospitals, dispensary/health Centers, family welfare centers, maternity and child welfare centers, maternity homes, TB hospitals/ clinic, and nursing homes Source: Town amenities, District Census Hand Book, Census of India 2011.

Total number of schools, colleges, and universities: It includes all the private and governments' school, colleges and universities of a city. Source: Town amenities, District Census Hand Book, Census of India 2011.

Total number of banks: It includes nationalized banks, private commercial banks, and cooperative banks.

Total number of credit societies: Total number agricultural and non-agricultural credit societies.

Author details

Sabyasachi Tripathi[1,2]

1 Institute for Statistical Studies and Economics of Knowledge, National Research University Higher School of Economics, Moscow, Russia

2 Adamas University, Kolkata, India

*Address all correspondence to: sabya.tripathi@gmail.com

References

[1] Rana, N. P., Luthra, S., Mangla, S. K., Islam, R., Roderick, S., Dwivedi, Y. K., 2018, Barriers to the development of smart cities in Indian context, Information Systems Frontiers 21, 503-525.

[2] Annez, P. C., Buckley, R., 2009. Urbanization and growth: Setting the context. In Spence, M., Annez, P.C. and Buckley, R.M. (eds.). Urbanization and growth. Commission on growth and development. Washington, DC: The World Bank. pp. 1-45.

[3] Jacobs, J., 1984. Cities and the Wealth of Nations: Principles of Economic Life, New York: Random House.

[4] Hall, P., 1998. Cities in civilisation: Culture, innovation and urban order, London: Weidenfeld and Nicolson.

[5] Ministry of Urban Development (MoUD), 2015. Smart Cities Mission Guidelines. Retrieved online 15th February 2017 from. www.moud.gov.in/.

[6] Russell, M. S., Pathak, P. A., Agrawal, G., 2019. India's "smart" cities mission: A preliminary examination into India's newest urban development policy, Journal of Urban Affairs 41:4, 518-534.

[7] Praharaj, S., Han, J.H., Hawken, S., 2018. Urban innovation through policy integration: Critical perspectives from 100 smart cities mission in India, City, Culture and Society 12 (1), 35-43

[8] Praharaj , S. , Hoon , J. , Hawken , S., 2017 . Innovative civic engagement and digital urban infrastructure: Lessons from 100 smart cities mission in India. Procedia Engineering 180 , 1423-1432 .

[9] Aijaz, R., Kristian Hoelscher, K., 2015. India's smart cities mission: An assessment, ORF Issue Brief, Issue no. 124, December.

[10] Praharaj, S., Han, H., 2019a. Cutting through the clutter of smart city definitions: A reading into the smart city perceptions in India, City Culture and Society 18, 1-10,

[11] Randhawa, A., Kumar, A., 2017. Exploring sustainability of smart development initiatives in India, International Journal of Sustainable Built Environment 6 (2), 701-710.

[12] Hoelscher, K., 2016. The evolution of the smart cities agenda in India, International Area Studies Review 19(1): 28-44.

[13] Praharaj, S., Han, H., 2019b. Building a typology of the 100 smart cities in India, Smart and Sustainable Built Environment 8(5), 400-414.

[14] Tripathi. S., 2017. How to develop an equitable distribution of urban GDP by smart city development in India", Regional Science Inquiry 9 (2), 131-146.

[15] Adapa, S., 2018. Indian smart cities and cleaner production initiatives – Integrated framework and recommendations, Journal of Cleaner Production 172, 3351-3366.

[16] Aijaz, R., 2016. Challenge of making smart cities in India. Asie. Visions, No. 87, Ifri.

[17] Tripathi, S., 2013. Do Large agglomerations lead to economic growth? Evidence from urban India", Review of Urban and Regional Development Studies 25 (3)(2013), 176-200,

[18] Tripathi S., 2018. Does a higher level of infrastructure increase population in large agglomerations? Evidence from India", Review of Urban and Regional Development Studies, 30 (2), 145-168.

[19] Tiebout, C.M. 1956. A Pure Theory of Local Expenditures. The Journal of Political Economy 64(5), 416-424.

[20] Harris, J.R., Todaro, M.P., 1970. Migration, unemployment and development: A two-sector analysis. American Economic Review 60(1), 126-142.

[21] Government of India. 2011. High Powered Expert Committee Report on Urban Infrastructure and Services. [Last accessed 9th February, 2019] Available from URL: http://icrier.org/pdf/FinalReport-hpec.pdf

[22] Infrastructure Development Finance Company. (2011). India Infrastructure Report 2011. Water: Policy and Performance for Sustainable Development. Oxford University Press, New Delhi, India. [Last accessed] Available from URL: http://www.idfc.com/pdf/report/IIR-2011.pd

[23] Nair, V., 2012. Infrastructure Development in India, Dr. Marri Channa Reddy Human Resource Development Institute, Hyderabad. [Last accessed on 19th January 2018] Available from URL: http://www.mcrhrdi.gov. in/Group12012/introductiontoppp.pdf

[24] Toutain, O., Gopiprasad, S., 2006. India Infrastructure Report 2006. Oxford University Press, New Delhi. [Last accessed] Available from URL: http://www.idfc.com/pdf/report/IIR-2006.pdf

[25] Pradhan, R., 2007. Does infrastructure play role in urbanization: Evidence from India. Indian Journal of Economics and Business 6(1), 81-92.

[26] United Nations, 2018. Department of Economic and Social Affairs, World Urbanization Prospects: The 2018 Revision, Online Edition.

Section 2

Smart Building and Architecture

Chapter 4

Smart Buildings: A Model Approach for Institutional Buildings

Kumar Avinash Chandra

Abstract

Smart Buildings should be seen from a multi-industrial standpoint, involving the right combination of architecture, structure, information technology, automation, environment and energy, services and facility management such as to minimize life-cycle costs, maximize comfort and adapt properly to cultural stimuli. Intelligent architecture concerns with intelligent design to meet cultural and contextual requirements, with proper use of IT and smart technology, as well as with optimal building exploitation and cost-effective maintenance over its life-time. This might also include intelligent and responsive facades. Facility management looks for the best financial management for maintenance, rebuild and renovation, for the best space utilization, for the best daily operational services and for maximizing user satisfaction.

Keywords: internet of things, sensors, smart city, smart buildings, smart governance, testbed, urban development, sensor networks

1. Introduction

Today, like the pattern in developing nations across the globe there is major shift of the population for rural to urban areas. As per a survey, our nation India, also perceives brisk for population shifting in urban areas by huge figures. The accumulating trend of swerving populace to civic living been discerned. The group of researchers stated that approximately 55% percent of population across the globe resides in the urban area, and is believed the percentage to hike up to 70% by the year 2050 [1]. The facts states that 31.2% (approximately 377 millions) of increase in urban population in 2011, the result also predict the numbers to rise to 40% by the year 2030 and up to 60% of the country's total population would move to the urban living in the nation.

The nimble relocation of populace in civic is usually confronted by service delivery and infrastructure management, are most asserted among all others challenges offered due to population explosion. The local management responsible for urban management should always have the smarter means for the cop-up with any confrontations and any related affairs due to emerging population relocation as in health care management, congestion in traffic management, infrastructure development, waste management, energy demand, pollution, etc. The concept for smart city engages for sophisticated civic modus, which within self has various sophisticated system fir daily requirements and challenges faced by the habitat of a area. The concept of Smart City coined being a blueprint for tackling with all these challenges mentioned. An intelligent and smart game plan for, manages components as in for all the challenges is provided in within Smart City.

With the growth and expansion in the city , new agile, shrewd and ingenious approach is required for the advancement in operational competence, enhancing productivity and as well as diminishing the managerial expenses [2]. Gradually, there has been increase in the IoT appliances such as smart boxes, TV sets, etc. by the dwellers. Even in the sectors of chattels real the appositeness of akin gadgets has upturned as in for Smart locks, thermostats, smart alarms, intelligent voice assistant and many more such gadgets. The neoteric augmentation in the field of digital automation has made the smart cities slicker than antecedent version of self. A smart or intelligent metropolis is rigged with the sensors as in for commutation, state-of-the-art cameras on the streets for influx management on the streets and for the purpose of cognizance as well, sensors at parking for monitoring the vacant slots (if any), etc.The eloquent amelioration in the permissive appliances tech, as in NFC, ingrained actuators, RFID tags, etc. Alongside materialization pertinent utility and appliances the IoT been lauded as abut dominating development to the contemporary hooked and ambulatory hobnob infrastructure. The recent prognosis as envisages that IoT would be imperative chunk of FI, as its akin appliances might surmount the total numbers of mobile and computer devices been accessed by the individuals. For such sequential events unfurls in the impending time frame, deduction of the schema and architecture delineation of FI be dependent on the staunchly be swayed by stipulation of IoT.

A framework of connectivity is catered by the recent turmoil in FI by which plebeians, society can annex with each other and as well as the devices as well. A study conducted which states, the total number of gadgets which are annexed with each other is much greater in number than total of humans on the planet [3]. The technical advancements and elucidations for scientific know-how for conceiving Smart Cities are sprouting and are surfacing up. **Figure 1** shows the inter-linkage of individual commodity as in terms of IoT. Multifarious facets of an entity can be stirred by IoT as in healthiness, commutation salvation etc. As in for governance it could play vital job as in to provide better efficiency, policy making, close and obscure monitoring, energy policy, pollution measurements, etc.

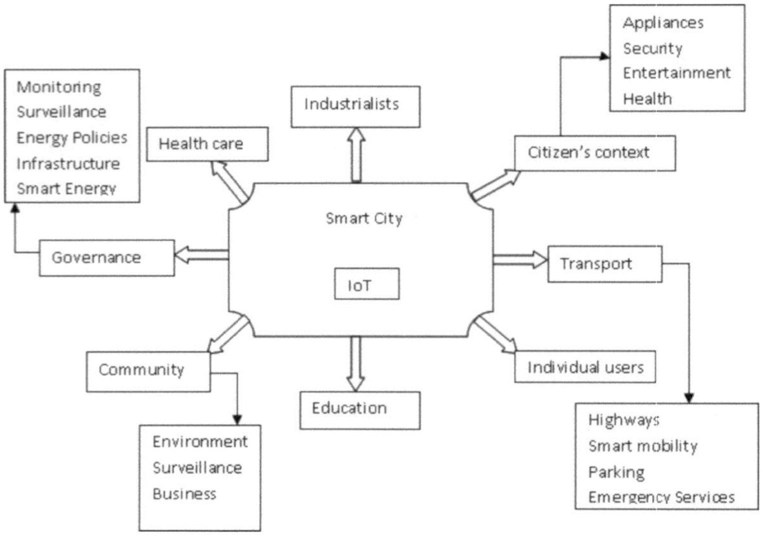

Figure 1.
Main aspects of smart city and inter-linkages based on IoT.

2. Indian perspective of smart cities

Government of India (GoI), in its' election manifesto for 2014 proposed development of 100 smart cities, which in later stage transformed to brown city from green city. In other words, GoI which earlier planning for developing 100 new cities as Smart Cities later planned to develop the existing cities into smart cities. And for this purpose, SCM and AMRUT a completely different wing under Ministry of Urban Development (MoUD) was setup, which was considered a compelling stride for encyclopedic enactment for Smart Cities (**Figures 2** and **3**). The implementation

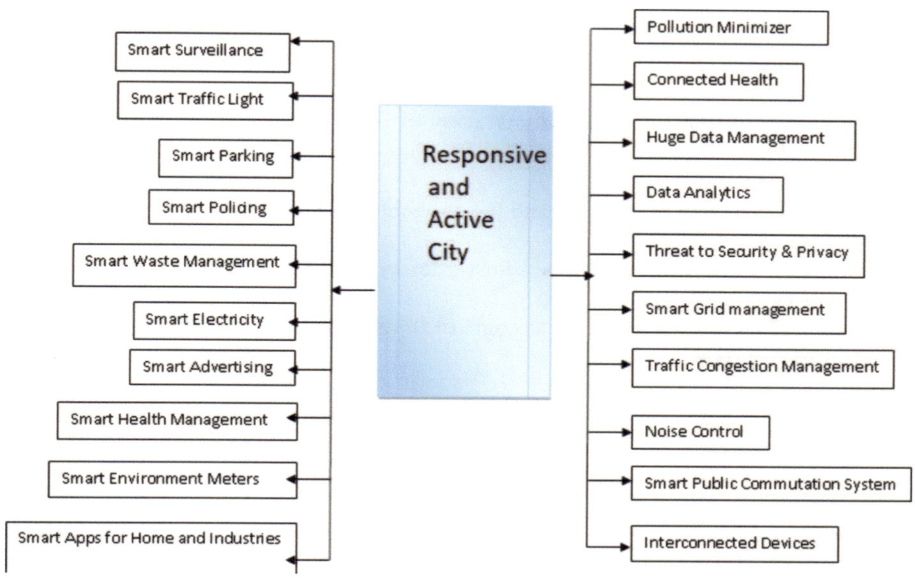

Figure 2.
Smart cities: Available technologies (right) and challenges (left).

Figure 3.
Example of prevailing smart building technologies (Image source: https://thegibraltarmagazine.com).

of SCM be annexed as the contingency plan to knuckle down to defiance of securing the intent of urbanization as per nationwide domestic development plan. Sectors where SCM needs to cynosure are:

i. Development of competent infrastructure for Civic Establishments and provincial governments.

ii. Development of competent Civic Administrative Organization.

iii. Enacting upon decentralization policy.

iv. Curtailing disagreement in civic domain.

v. Developing permissive plight for decent and broad urbanization.

The clamant materialization for gestating of Smart Cities along Indian lexicon be enunciated as adhere to:

i. A civic is obligated to be viable and imperishable.

ii. Fundamentals for the reliable administration be cherished.

iii. Abiding foresight, technology, strength of the governance and supportive administration and schema.

iv. Adequacy of the Civic administration to enact the above.

v. The nation is required to erect its own allusion for ontogenesis of Smart Cities.

Smart City be defined as in Indian lexicon as "A Smart City would be the one which plans judiciously to meet its aspirations and challenges in a sustainable manner while fostering principles of good governance. These are achieved in a Smart City by utilizing the enhanced power of technology, engaging with a more aware and informed citizenry and creating a more competent and capacitated set of people working within an accountable framework."

The schema for regional area augmentation has been designed by SCM and MoUD with intent to revamping fiscal development and aspect of living. The schema has basically trilateral factors [4]: a) Area – Based Development (ABD) responsible for uplifting of the regional extant inclusive of the blighted areas into advanced and planned ones; b) Green – field Projects which would develop new provinces into state-of-the-art centre so as to facilitate the exploding populace; c) Pan –city Development (PAN) which shall anticipate the appositeness of the elicited smart and intelligent elucidation to prolonging city framework.

3. Technologies for Smart Cities

The hefty fortification of IoT is playing pivotal guise in the administering of Smart City ventures. The constant advancements in the technology are enabling facilitating Smart Cities across the globe. The commodity by individuals be in service on routine basis are rigged with digital and computerized gears, mechanism and covenants so as to make them pertinent and associated with other linked and connected devices with Internet Protocol. The competence of surveillance and supervise of obscure and secluded area as well is ease with help of IoT. Apart from that one could administer remotely. The important physiognomy of the Smart City is the enormous heterogeneous data from the various sensors and devices deployed within for administrative purposes. The super meteoric accretion of smart cities and the IoT coaxed various challenges for all researchers and industries as well for designing of a conducive and impeccable smart city.

With the use of Standard Web Protocols for communication [5, 6], IoT enact as Broadband Network having Internet at its concenter. For the employment of IoT demand for the communication standards that operates placidly amidst the numerous commodity whichever be computed, implicated and can hatch variance in purlieus. Among all the technologies pertinent to IoT are confabulated in brief as follows:

3.1 RFID

The arrangement comprises of a chip or a tag along a chip to read the tag. This advancement in technology can be used for registering any individual or an object for the intent to self recognized by the system. Each of the interlinked objects or gadgets accredited with diacritic identity [7].

3.2 Addressing

The prevailing fad in the fields of IoT could facilitate kinship of the individual associated gadgets and equipments so as to entrench smart and intelligent purlieus. The individual identity of the associated gadgets and equipments is must for in IoT.

3.3 Wireless sensor networks (WSNs)

With the help of WSNs data from different sources be collected easily and then be used in for various sectors as in Traffic Management, wellness program, pollution control, etc. It could also be tagged with some other sensors as in RFID to infer much accurate details about the individual object (**Table 1**).

Network type	**NFC**	**WPAN**	**WPAN**	**WPAN**	**WLAN**	**WLAN**
Year	2011	2002/2005	2003	2007	2012	2009
Network Size	—	7	245	65,535		30
Bit Rate	424 Kbps	3Mbps	55Mbps	250Kbps	> 7Gbps	248Mbps
Frequency	13.56 MHz	2.4 GHz	2.4 GHz	868–915 MHz/ 2.4GHz	2.4/5/60 GHz	2.4/5 GHz
Range	0.2 m	100 m	100 m	75 m	5 m	50 m

Table 1.
Important communication standards within IoT.

4. Smart buildings

About 30–40% of total power usage and discharge of CO2 is occurring at edifice [8]. The government is trying to clinch to reduce the energy consumption in new and as well as extant infrastructure as well so as to secure sustainability of the environment. Government through its stake holders, real-estate developers, land owner, proprietors, tenants and customers is trying to reforms in the sector to reducing the carbon emission in the buildings. The further energy effectualness can be abated to achieve the objective set by IPCC [9]. Many researchers and industries are working in this regard making the building smart so as to minimize the consumption of energy within the infra-structure.

Over last few decades, there has been rigorous research and advancement in the over Smart Intelligent buildings. Though theoretically, in disquisitions and also in technical communiqué the phrase 'Smart' is being cited more often in last few years. With the advancement in new and boost in the technologies, the smart buildings have secured much enthusiasm from researchers and industries as well [10]. Computerization and automation has been so much part of modern days living standard required to chasten our living. Nowadays, everyone is longing to manage and administer the gadgets and devices installed remotely and effortlessly.

The buildings are elementary and fundamental fragment of the society. The buildings in-houses the inhabitancy, plaza, emporiums, office area, deli, residentiary and market complexes, etc. Designing the smart city is intrinsic stride for the ontogenesis towards the Smart City. The Smart Buildings or Intelligent buildings, in defiance of diversified interpretation of the IB from the various Industrials, researchers be observed from the perspective of industrial and technical crux, implicating amalgam construction design, framework, power, utility, technological advancements, environment and amenity administration in a manner so as to aggrandize the assuages and also to curtail the circuition price [11, 12], by means of aid from advanced state-of-the-art technology and information technology advancements. The Smart Intelligent Buildings are also well efficient of curtailing the intramural energy dilapidation.

The ambit of automation and modern IT industry, the modern Smart Intelligent Buildings be confabulated based on technical elegance and assimilation in assorted multitudinous folds [13]. Facilities like control over aegis, avenue, luminosity, elevators, data, Infobahn etc. falls as in basal or crux of the folds of the Smart Buildings. The assimilation of various functionality of the basal fold forms the next in hierarchy of the layers. Assimilated communication system forms the next in line. The comprehensive grid structure of all the Smart buildings forms the apogee for the folds or layers of assimilation. The advancement in IT sector administers the crucial and omphalic aspect here in Smart Buildings, explicitly in cognizance of the subsequent:

- Imposing for energy and policies codes and protocols, Building bylaws.
- Assimilating with the nearby power/smart grids
- Self guided and intelligent uninterrupted building responds.
- Visualization of diminishing carbon emission and energy savings.

5. System architecture

The new modern advancement in wireless system and sensor grid network can provide provides convenience in structuring for the Smart Buildings as in to administer the smart and complaisant abutment for the occupants. The continuous and unceasing monitoring and administering of the various factors in and out of a building is the necessity to diminish the energy consumption of a building. Sensors and actuators installed at the proper locations and also be attainable at any moment over the grid, is very important for the same purpose of conducive management and automation. As there been development in IoT technology and Smart

DOI: http://dx.doi.org/10.5772/intechopen.94943

buildings in recent years, the dominion protocols, building acts and standards for industrial regulations is also been changes accordingly with due course of time, as in Consumer Electronic Bus (CEBus), Local Control Network (LCN). Suite of Internet Protocol (IP) turned to be new and paramount inclination for amalgamation of various services. Smart modern gateways are deployed at the network edge which then equipped with the access to protocols based on IP. The technology based on IEEE 802.154.4 helps in sending IPv6 packets efficiently through 6LoWPAN technique for header compression [14]. The following figure (**Figure 4**) depicts the rough architecture for the system.

The various sensors and actuators installed in a building combine to be the elementary or fundamental cause for data or information procreation. The initial level or Level one for the system be the raw data acquired from the sensor nodule, which then by communication service act upon farther processing of the data acquired. The architecture for the Smart Building is described with as mentioned principles of design.

- Information Assemblage. At the primary stage at this level, the raw data from the various sensors and actuators implanted at the building is collected and is stored for processing at later stage.

- Dossier Processing. Here at this level, all the data or information compiled in the previous stage is put together and processed so as all the data is stored in common format as in Resource Description Framework (RDF). Resource Description Framework (RDF) is most trivial approach for data castling over the web. Pre-refined data at this level will then be used for morphological knowledge and ambivalent inference at the preceding level.

- Dossier Assimilation and Inference. The exploitation of domain dependent distinct data is allowed through morphological web technology. Here all the information from the previous level is assorted and classified in as classes. Later the data collected is categorized into two base Data or Object Property and association of all the data is to defined with the either of two property based on Web Ontology Language (OWL).

  SPARQL is a query language subsidiary of RDF which is adopted to salvage and beguile the records hoarded in the RDF form. This stage galvanize towards amalgamation of low-level database.

- Gadget Control and Admonition. All the data processed in the previous stage of the architectural hierarchy is now ready to be used by various applications installed for smart activity and administration to ease of human.

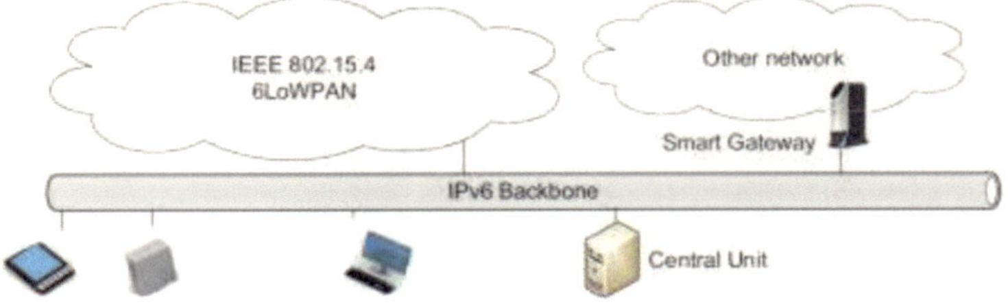

Figure 4.
Architecture for administrating and controlling over the system.

6. System application in buildings

To many individuals in the field of building automation continuance praxis, IoT plugging might look alike a jargon for gray wont. Might be that be total untrue. The coronation of sensors and imbrute praxis, angling to comprehend benediction associated with it as in crouched obligation to continuance of the building and viable competence with ameliorated superintendency and crouched corps outlay. The appositeness of IoT overture aggrandized prospects as in worth, viable efficacy, affinity through enhanced valise and liquidity management [15].

6.1 Architectonics

Apprehending the statistics on real time basis and cloud dependent dossier capacitate the enterprises in optimization of potency and curb operational squeezes as well. Annexation of the machinery and heirloom appurtenances could large amount of extra debts. Monitoring of those appurtenances installed do provides inestimable insights for the management for lapse, delinquency and to reckon regime bent as well [16–19].

6.2 Surveillance

Modern art-of-the-state designed IoT dossier with agitation apprehension sensors are the next generation system which could be wireless and also be quite setup for temporary setup as well, easily manageable remotely. These monitoring systems are battery operated and unlike conventional CCTV setup compared to are cost effective as well [20].

The present-day advanced system can also be used for securing the electric panels, which could trigger an effective alarm which when detects any unplanned apprehension in the proximity. If GPS be attached for advanced feature can also be triggered to activate with the alarm so as to manage theft if in case it occurs.

IoT based hazard control alarm system could prove very effective in the case of emergency. Fire or smoke alarm which when connected with the cloud dossier can automatically contact nearby fire station and police in case of any emergency. This could curtail ample amount of man hour for maintenance of the alarm and time-honored gratuitous auditing of the system.

6.3 Automation management

The system equipped modern sensors could be instated easily and effectively to manage the equipment as in light power, HVAC, fire, security, etc. and cloud-based dossier helps in supervise and oversight easily and efficiently [21–23]. Parameters as in temperature, automated door operation, humidity, air quality and pressure, etc. Apart from managing machinery as lift, escalators they also be prognosticate in case of disruption and be prompted for abrupt alleviative alacrity.

The system equipped with sensors once triggered could easily transfer statistics to cloud for further processing and record. It helps in control and optimization and also eliminate the long man-hour for data collection as well also increase the effectiveness with cost diminishing.

7. Conclusion

In years to come the art-of-the-state real estate development can be visualized globally. The trend of IoT, be then conceptualized in practical manner by next few years. The assemblage of IoT and updated sensors certainly do aggrandize the efficacy, performance, wherewithal, unlimitedness and also curb the outlay over the building. Buildings are the one large power consumer, fact to the government across the globe have their focus now on them, regulations and mandates are updated regarding buildings are updated regularly for carbon footprint mandate.

Apart from management of power the IoT controlled system helps in diminishing carbon emission. With technicality point of view the development of such buildings with proper architecture and standardized codes would not only interoperability be salubrious but cost amiable as well. The bottom-line discussion of this paper is modern sensor based IoT equipped buildings be a necessity in coming years for a healthy and future-secured environment and be cost effective at the same time when making our life a lot simpler and easier.

Acknowledgements

I would like to express my sincere gratitude and thanks to my be loving parents for their everlasting support and love.

Conflict of interest

The authors declare no conflict of interest.

Abbreviations

IoT	internet of things
RFID	radio frequency identification
NFC	near field communication
FI	future internet
SCM	smart city mission
AMRUT	Atal Mission for Rejuvenation and Urban Transformation
IB	intelligent buildings

Author details

Kumar Avinash Chandra
Department of Electrical Engineering, Dr. A.P.J.A.K. Women's Institute of Technology, Darbhanga, Bihar, India

*Address all correspondence to: energy.avinash@gmail.com

References

[1] United Nations, "World Urbanization Prospects", 2014.

[2] Khatoun, R, Zeadally, S: 'Smart Cities: concepts, architecture, research opportunities'. Commun ACM, 2016,59(8) pp 46-57.

[3] J. Cubbi, R. Buyya, S. Marusic, and M. Palaniswami, "Internet of things (IoT): A Vision, architectural elements and future directions,". Future Gener. Comput Sys, vol 29, pp 1645-1660,2013.

[4] MoUD, "Smart City mission and guidelines", 2015.

[5] I. Atzori, A.Lera and G. Morabita, "The Internet ofthings: A Survey" Commut. Netw, vol 54, pp 2787-2805, 2010

[6] *A. bassi*, and G. Horn "Internet of Things in 2020: A roadmap for the Future", European Commission Information Society and Media, 2003.

[7] A.K. Evangelos, D.T. Nikolakos, and C.B. Anthony, "Integrating RFIDs and smart objects into a Unified Internet of Things architecture", Advances in Internet of Things, Vol 1, pp. 5-12, 2011.

[8] Building Energy Data Book, 2007, US Department of Energy.

[9] IPCC Forth Assessment report, Climate Change, 2007, Mitigation of Climate Change.

[10] Transforming the market: Energy Efficiency in Buildings, World Business Council for Sustainable Development (WBCSD), 2009.

[11] A. Dunkels and J. Vasseur. IP for Smart Objects. IPSO Alliance White Paper #1, Sept, 2008

[12] Communication from the European Commission, Energy Efficiency: Delivering the 20% target (772), Nov, 2008

[13] M.L. Wild, "A Little give or takeon ELectricity", New York Times, April 29, 2009.

[14] C.Y. Yong, B. Qiao, D.J. Wilson, M.Wu, "Coordinated management of Intelligent pervasive spaces", 5th IEEE Int'l Conference on Industrial Informatics, Vol1, pp 529-534, June 2007

[15] https://www2.deloitte.com/content/dam/Deloitte/nl/Documents/real-estate/deloitte-nl-fsi-real-estate-smart-buildings-how-iot-technology-aims-to-add-value-for-real-estate-companies.pdf

[16] IOT Security Foundation. Accessed on Jan. 13, 2017. [Online].

[17] A. Al-Fuqaha, M. Guizani, and M. Ayyash, "Internet of Things: A survey on enabling technologies, protocols and applications," IEEE commun. Surveys. Tuts., vol 17, no. 114, pp2347-2376, 4th quart., 2015.

[18] J. Granjal, E. Monteiro and J. Sa Silva, "Security for the Internet of Things: A survey of existing protocols and open research issues," IEEE Commun. Surveys. Tuts., vol 17, no. 13, pp. 1234-1312, 3rd Quad. 2015.

[19] C. Lai, R. lu and X. Shen, "Towards secure large scale machine to machine communication in 3GPP networks: Challenges and solutions," IEEE Commun. Mag., vol 53, no. 12, pp. 12-13, Dec, 2015

[20] Z. Li, Z. Tang abnd Y. Yang, "Research on architecture of security video surveillance n/w casade system with big data," World J. Eng., vol. 13, no. 1, pp77-81. [online].

[21] P. Bellagente, P. Ferrari, A. Flammini and S. Rinaldi, "AdoptingIoT framework for the energy management of smart buildings: A real test case," in Proc. IEEE 1st Int. Forum Res. Technol. Soc. Ind. Levaraging Better Tomorrow (RTSI), Turin, Italy, Sep. 2015, pp 138-143.

[22] F. Tao, Y. Wang, Y. Zuo, H. Yang and M. Zhanga, "Internet of Things in Product life-cycle energy management," J. Ind. Inf. Integer., vol. 1, pp-26-39, mar. 2016, doi: 10.1016/j.jii.2016.03.001.

[23] Y. Sun, T. Y. Wu, G. Zhao and M. Guizani, "Efficient Rule engine for smart building systems," IEEE Trans. Comput., vol. 64, no. 6, pp. 1658-1669, jun. 2015, doi:10.1109/ TC.2014.2345385.

Chapter 5

Standard Elevator Information Schema: Its Origins, Features and Example Applications

Jonathan Beebe and Ahmad Hammoudeh

Abstract

A generational change is taking place in building transportation systems as manufacturers and maintenance companies begin to integrate their products and services with the technologies of smart buildings and smart cities. Frequently this integration relies on the Internet of Things and cloud services. The diverse and heterogeneous nature of such collaborations requires a common shared semantic understanding of the complex and dense information that may be generated by transportation systems in buildings. The Standard Elevator Information Schema (SEIS) provides this in a format which is both machine and human readable. The role of the schema is to provide the 'vocabulary' for these collaborations. At the same time the schema specifies the properties, relationships and validation rules that define the information model, which could form the foundation upon which all elements of building transportation control and monitoring functions are constructed. SEIS is published under the Collective Commons licence and is free to download and incorporate into any product with the objective of reaching the broadest audience. This chapter discusses the origins and features of SEIS and provides a varied set of example applications. Consideration is also given to the issues of cyber security and data protection.

Keywords: lift, elevator, XML, XSD, schema, information model, IoT, REST

1. Introduction

The end of the second decade of the twenty-first century has witnessed a growing diversity of devices and information processing which smart buildings and cities are introducing into vertical (now also 'two-dimensional', c.f. thyssenkrupp elevator MULTI [1], and maybe 'three-dimensional' in the future [2]) building transportation systems. A new generation of lift systems has been developed offering increased levels of personal response particularly as a result of supporting destination call registration. This has been accompanied by increasing sophistication and variety of human interfaces, for users in many roles. Also there has been a rapid expansion of cloud-based analysis services, using Artificial Intelligence, that are capable of managing and processing vast amounts of data. The Internet of Things (IoT) is a key enabler of this technology evolution, so a section is included providing an overview of the IoT with the further objective of clarifying this somewhat misused appellation.

The variety of information domains and related equipment and service suppliers involved in these developments makes it increasingly important to conform to an agreed standard vocabulary and associated set of rules for interchange of information – a schema – that is specific to the application domain i.e. transportation systems in buildings.

Indeed the design and configuration of such systems would be enhanced considerably if such a schema was to be implemented internally as a foundation information model as well as for external communications. The continuing practice adopted by manufacturers of implementing proprietary communications, while promoting vendor lock-in, is no longer viable when travellers, administrators, owners, technicians, etc. experience many points of contact with such a variety of systems from disparate suppliers – For example, users cannot be expected to select the application of the lift manufacturer or maintenance company in whose lifts they wish to travel to register a call from their personal mobile device! However, the need for a common shared information model extends far beyond this trivial example and is relevant to most aspects of the operation of building transportation systems, particularly when machines are communicating directly with other machines without human interpretation of the exchanged information.

The Standard Elevator Information Schema (and complimentary Escalator Schema) which defines the information model is a formal declaration of the elements, their properties, interrelationships and validation rules and is directly meaningful to both humans and machines ('things').

2. Lift system communications: current status and potential developments

The richness of information and the potential diversity of interactions with external systems is much greater for passenger lift systems than for escalators and walkways and so this section refers mainly to lifts. However, a truly smart building will integrate communications with all transportation systems wherever practicable – probably a subset of what is possible for the lifts. In this section we look at what is currently available (discussed in greater detail in CIBSE Guide-D [3]) whilst casting an eye to potential developments that are within sight.

2.1 Current state of development

At the time of writing, a new generation of lift technology is appearing that embodies capabilities to communicate beyond the lift motor room through an expanding range of applications and interfaces.[1] Previously the lift system user interface has been specifically aimed at the primary users, i.e. the passengers, via simple electrical devices such as call buttons, floor position indicators, hall direction indicators and gongs, etc. The introduction of computer technology and networked communication between devices has lead to an increasing complexity of the hardware of lift control equipment and also its associated passenger controls (e.g. destination call stations). At the same time, the new technology has promoted more sophisticated software in the form of control algorithms as well as a proliferation

[1] Inclusion (or omission) of references to manufacturers' products is purely for example and does not represent endorsement (nor rejection) of those products.

of external systems which collect, process and distribute the resulting complex and rich information.

Lift manufacturers [4–8] and third-party suppliers [9–14], sometimes in collaboration with major companies providing software and services (such as Microsoft and IBM), are developing new products and services which interconnect many lift installations concurrently thereby considerably extending the types of user that can interact in some way with a lift installation. The following subsections briefly discuss the different classes of internal and external systems that may be interconnected, identify the user roles that could benefit as a result and outline how these benefits might be delivered to those users.

2.2 Data logging

Data logging is in essence the capability to create and record a stream of data describing the activity of the lifts and their current status. As such, it should not be considered independently of external applications which then analyse and present the data. From the earliest days of data logging, and particularly since it became viable to install logging as a permanent feature rather than a commitment of expensive equipment only justifiable for a few days, it was acknowledged that the volume and diversity of the logged data necessitated a repeatable and automated (i.e. software-based) analysis.

Most modern lift control systems include some form of data logging, although this is usually in a proprietary format and available only to the staff of the manufacturer for the purposes of test, installation and maintenance. For older installations, a number of generic gateway devices are available from independent suppliers, which sample commonly available electrical signals and buses and in some cases additional sensor devices (e.g. accelerometer) to generate a stream of logged data, but again in a proprietary format. However, as building owners are increasingly demanding access to logged data of their lift activity (since they are in any case the designated owners of that data under Data Protection legislation) it is becoming common practice to include a requirement in the specification of new installations and refurbishments for some form of gateway that allows the connection of third-party logging systems [15] where the owner specifies the data format.

Data logging operates at several different levels, where access is restricted to persons possessing different skill sets and purposes:

Maintenance staff on site – to interrogate and alter system parameters and systems technicians able to update software and trouble-shoot. This data might include door timing, parking and priority floors and zone definitions, etc. More recently, device-specific operational 'signatures' (e.g. temperature, vibration, door operation counts, etc.) have been logged, which are used to identify abnormalities and drift due to wear and failure in specific components. There is an emerging trend of linking to advanced on-line web services that offer automated analysis and 'learning by example', enabling lift manufacturers to evolve their service offering and to drive preemptive maintenance scheduling. This analysis is of particular benefit to centralised maintenance managers and technicians who arrive on site better informed and provisioned with spares.

Owners/operators – to access performance data allowing analysis of how well the lift installation is managing the passenger demands, and duration of periods when lifts are unavailable etc. This data might include the number of stops, number of landing calls and their origins, car calls, system response times and an analysis of the demand patterns in terms of landing calls. Interest is growing in monitoring and reporting the energy consumption of building transportation systems. Remote display (possibly in the lobby or building manager's office of the relevant building) of centrally held information is often also offered.

Maintenance management off-site – to review the operation of the lift, check any irregular behaviour and to plan a service schedule matching the duty cycle of each lift. Also to provide a 'forensic record', helping to establish the reason for a lift failure. Such data would include: fault and warning indications such as the safety circuit being interrupted prematurely, doors failing to open, abnormal journey duration, excessive re-levelling operations, door lock failures, machine and machine-space temperatures, etc. A number of major manufacturers [5–8] and a growing number of independent suppliers [9–14] offer remote monitoring services and display of lift status with real-time visual display of car and door activity plus registered calls.

Architects, transportation system designers and researchers – may use logged data to specify new installations or refurbishments for planning and subsequent assessment of traffic handling capacity and suitability of control policies matching the particular demands of a building. The data to support this would include the number and timing of landing calls (particularly in the case of destination calls), car calls, floor position, direction commitments and door operations of lifts, passenger loading, the allocation of lifts to respond to calls, etc. If data collection is fully comprehensive, it may be possible to replay lift operation into a traffic simulation program, closing the loop between traffic simulation for planning and what is happening in the actual installation [16]. A less data-intensive approach is discussed later in this chapter (Section 7, Example 4).

2.3 Remote control

Remote control concerns a very carefully controlled flow of data in the opposite direction to data-logging (i.e. from the external environment to the lift controllers and associated displays). The safety of passengers must first be thoroughly assessed for risks in the context in which the control is to be used – for example simply including a remote control to return all lifts to an entry floor of a hospital might put patient lives at risk.

Maintenance management off-site – A car or landing call may be entered to check that a lift is in service. No form of direct remote control, other than that available locally to passengers, is allowed as this could jeopardise passenger safety, for example if the security of the system were to be compromised through a cyber attack.

Intending passengers – With the increasing use of destination control traffic control systems a more personalised lift service may be provided to passengers. Call-stations are no longer restricted to lobby locations. For example, applications have been developed to allow calls to be registered from personal mobile devices enabling improved planning of call allocations and reducing queues for call registration, these are often linked to building access control [17, 18]. Requests need not be limited to human passenger traffic – a recent press-release [19] announced the development of a robot delivery agent that could distribute supplies and accompany visitors within a building (e.g. hospital, hotel, office), which is able to register calls for lift travel between floors.

Travelling passengers – Public information displays and announcements may be provided to passengers [20–22] notifying both waiting and travelling passengers of the services, facilities and events on the floors which they may select as a destination, or to which they are already travelling. A second type of user interface may be required which provides a management system allowing both general and targeted information (e.g. based on destination floors, etc.) to be prepared and updated before presentation to one or more groups of users (e.g. in a multi-tenanted building).

Machine to machine – In future it is likely that remote applications will be able to update, within strictly defined limits, a limited range of high-level parameters (e.g. parking and priority floors, zones, energy consumption targets, etc.) in the lift controllers thereby linking smart cities and buildings to influence lift performance. Such remote applications would use artificial intelligence to 'learn' the optimal parameter settings based on external conditions (e.g. weather, calendar, traffic, etc.). It is even possible that other more general sources (e.g. anonymised social media discussions) could be interpreted automatically, enabling members of the public freely to provide information which could lead to modification of the operation of the services that they use, though probably without the deliberate intention of doing so.

3. Overview of IoT

Similar to many booming computer technologies, the concept of the Internet of Things (IoT) has attracted a lot of hype and when it comes to what IoT is, stakeholders have a variety of interpretations. Such interpretations often are biased towards the interests of those stakeholders who wish to emphasise their own assets of IoT [23]. In the same vein, lift manufacturers are announcing products which claim to use IoT technology. However, IoT is more than a network of internet-connected devices. It entails compliance with a set of essential features. So what is IoT?

The International Telecommunication Union [24] defined IoT as:

'A global infrastructure for the information society, enabling advanced services by interconnecting (physical and virtual) things based on existing and evolving interoperable information and communication technologies'.

For a general overview, Atzori et al. [25] groups the defining features of IoT into three main categories called 'visions'. IoT is the intersection of those three visions:

- 'Things-oriented' vision – focuses on the things' identity and functionality
- 'Internet-oriented' vision – emphasises the role of the network infrastructure
- 'Semantics-oriented' vision – focuses on systematic approaches towards representing, organising and storing, searching and exchanging the things-generated information.

Taking a slightly more abstract approach and distancing the network infrastructure, Raggett [26] of the W3C organisation looks at the challenges and the risk of fragmentation of the IoT and proposes a 'Web of Things'

- Things standing for physical and abstract entities
- Applications decoupled from underlying protocols
- Shared semantics and rich metadata

The 'Applications' here will be specific to a business domain or discipline and the 'Shared Semantics' need to be described by a formal definition, such as a UML domain model [27].

In a detailed discussion of IoT in the building transportation industry [28], the key conclusion was that the successful integration of building transportation systems (particularly lifts) into the IoT requires a standard definition of the semantics of any information exchanged between inter-operating devices (the 'things'). It is the third, 'Semantics-oriented' vision where the greatest value can be added that is specific to the domain of lift operation because doing so is key to fulfilling the requirement of the IoT in 'enabling advanced services'.

We can conclude that integrating lifts into the IoT is not simply a matter of establishing communications between devices over the internet, nor is it achieved just by using Internet messaging protocols. It is both of these characteristics plus the ability to run applications that exchange information in a way that is meaningful to all the things involved in the exchange i.e. there is a shared definition of the semantics.

A feature of the current 'human Internet', which is now taken for granted and which needs to be replicated for the IoT, is that of search and discovery facilities. In support of this, services in the IoT must be discoverable so that devices which have never communicated with them before can use these services and vice-versa that services can discover devices. The significance of this ability is illustrated by imagining the ensuing disruption if it were necessary, when installing a new lift system, to make manual updates to all the systems with which the lifts might eventually communicate.

4. Standards

4.1 Generic standards

Successful interoperation of systems from different suppliers is dependent on the faithful adherence to standards at all levels. In particular, the IoT relies on many standards and standard services e.g. Wi-Fi, IPv6 addressing (required to address every individual connected thing uniquely), http and web-service protocols, time-of-day services, load-balancing, cloud computing, etc.

As ETSI says on its website [29].

'Smart objects produce large volumes of data. This data needs to be managed, processed, transferred and stored securely.

The use of standards

- ensures interoperable and cost-effective solutions
- opens up opportunities in new areas
- allows the market to reach its full potential

The more things are connected, the greater the security risk. So security standards are also needed to protect the individuals, businesses and governments which will use the IoT'.

4.2 Domain-specific standards

Interoperating applications that are specific to a business domain must be able to understand the domain-specific information contained in any data that is shared via the network. This is because there is a machine (rather than a human who can make inferences and interpretations) in both the generating and consuming roles. A lexicon or dictionary of the relevant terms and the rules for their use, usually in the form of a schema, is required to support such applications.

4.2.1 Existing standards specific to the lift systems domain

Some standards specific to lift systems do exist, and continue to be enhanced, describe –

the active state of the dynamic components of a group of lifts:

- CANopen-Lift [30] and

- BACnet, although at least two different implementations of this exist:

  1) BSR/ASHRAE 2016 [31] and

  2) National Standards Committee of the People's Republic of China [15].

and the component parts:

- BIM (proposed, not published)

However, there is a real need for an agreed and adopted formal schema to describe the day-to-day operational characteristics of a system of lifts, which is independent of the underlying electrical connectivity and network protocols. Such independence is a key requirement for successful integration of lifts into the IoT.

4.2.2 Standard elevator information schema

The Standard Elevator Information Schema (SEIS) [32] (and complimentary Escalator schema) was developed in 2003 [33], from earlier work, and uses standard software development tools and methods to model the information and operation of a lift system. Whilst it certainly describes the data normally associated with remote monitoring (car and landing call registration and cancellation, car movement, floor-position, and door state, etc.) the SEIS is much more comprehensive, including elements of processed information such as demand profiles, car journey plans and energy consumption. This level of detail enables the development of sophisticated applications and services, against a common standard, which therefore do not have to concern themselves with, for example, the intricacies of individual car control. The current state of development of IoT applications for lifts makes it thoroughly appropriate that such a standard should now be widely adopted throughout the Building Services industry.

5. Origins and features of the standard elevator information schema

5.1 Origins

The origins of SEIS can be traced back to a refurbishment project in 1980 of several groups of lifts in a classical (11 storey) building in central London – Bush House. The traffic demand was non-standard and continued throughout 24 hours of every day of the week. At the time there were virtually no computer controlled lifts anywhere in the world, and this provided the opportunity to design computer software from first principles for both individual car and group supervisory controllers.

5.1.1 Interface abstraction

Working from the available published literature [34], the Bush House central controller design was developed around an abstract model of the primary elements of information that are essential to the control of a group of lifts. Each control element is linked to its external environment – electrical sensors, relays, indicators and other control computers – through an interface software layer which communicates via very short (less than 6 characters) textual event messages. Even delay timers are implemented as messages that the controller sends to itself, which are held for a specified period before being delivered. This architecture has a number of significant advantages:

1. The interface messages are directly intelligible to both humans and computers.

2. It was very easy to simulate messages in order to test the operation of the central control software thoroughly before it was installed with real equipment.

3. External devices such as a dynamic graphical status display and systems for performance monitoring, problem detection and data-logging could be driven directly from the interface messages which were exactly the events the controller itself was seeing. The project delivered all of these external devices.

The great benefit of the interface abstraction layer is that the controller is not predicated on a specific configuration of equipment nor a set of electrical connections to the hardware of the lifts. The design made it very straight forward to port the controller to a different building where components from other lift manufacturers might be used. An example of this abstraction is the use of a 64-bit absolute position encoder providing sub-millimetre resolution of the lift position in the shaft. From the point of view of the lift drive control, it is important to know very accurately the position of landing levels, slow-down points, etc., the thousands of positions in between however, are ignored. On the other hand, from the perspective of the logical control element of the lift, it is most important to know the floor position of the lift, including when the lift is travelling between floors and taking into account the direction of travel when a slow-down point is reached – hence the 'Next Possible Stopping Floor' of the lift was defined (now `<CarDynamicData><Floor>` in SEIS).

5.1.2 Event driven

A controller only receives a message when a change of state occurs, so it is truly 'event driven' and it interprets each received event in the context of the source of the event message and the current state of other known information. The source of the message needs to know nothing about the functionality that lies behind the interface, allowing great flexibility for the controller designers to evolve and improve their design. This 'encapsulation' is a fundamental characteristic of the object orientated programming paradigm [35].

5.2 Publication of a standard schema

However, it was not until the introduction of functionality into Internet web pages, supported by tools and technologies based on XML, that it became practicable to publish a formal statement of the full set of terms and relationships necessary to define lift operation. The format of this publication is a single XML schema file written in XML Schema Definition (XSD) language that is at once readable by humans and a wide range of computer operating environments. It contains specifications of:

- simple types, which can include validation rules for permissible numeric value ranges and text formatting
- complex types as heterogeneous collections of simple types
- aggregation relationships with optionality and multiplicity restriction rules, and every definition can be elaborated with textual documentation for the benefit of human readers.

So it was that in 2003 the knowledge from the earlier work was translated into an XSD – the Standard Elevator Information Schema – by Beebe and made publicly available [32] under the Collective Commons licence. It is hoped that through this mode of publication, discussion of the relative merits of any alternative schemas, should they be developed, would be limited to technical concerns thereby avoiding commercial and proprietorial considerations. The reference works of the day [36, 37] were analysed to identify significant nouns and their definitions that might form the atomic elements of the schema to ensure a wide-reaching scope. At the same time terms that were not directly relevant to lift operation were carefully excluded in order to avoid duplication or conflicts with schemas for other domains, currently available or potentially in the future. Based on the foundation elements and a more general survey of research across the field of lift traffic analysis, monitoring and simulation, further useful complex elements were identified and added to the schema e.g. demand profile, travel plan, door cycle, etc.

Two possible interaction modes were envisaged for SEIS at its inception (though a further mode was introduced with the application of REST transactions and more may be discovered as experience is gained):

- **Aggregated** –all the currently known information that is held under a specific node is communicated in a single instance message as an aggregation of elements. For example, this mode is used to drive a simple graphical status display example in this chapter (Section 7, Example 1).
- **Event** – only information which has changed is communicated as a time-stamped event message. This mode is appropriate when it is preferable to use communication bandwidth economically and the current state is maintained by the recipient of the message – see Event Data Log Example on the SEIS website [38].

This was the starting point for the root element types in the schema:[2]

5.3 Aggregated information

Aggregated information is subdivided, again to improve the efficiency of communications, into:

- **StaticData** – describing the unchanging parametric information of the lift systems.
- **DynamicData** – describing the current state of operational lifts.

[2] NB the text of the following subsections is graphically illustrated on the SEIS website – SEIS element names are in Courier New typeface [39]

5.3.1 Static data

For a lift car (`CarStaticData`) – the list of floors served by the specific lift, the number of car decks and door operators, the reference speed profile(s), door timing and energy profile(s).

For a group (`GroupStaticData`) – information such as the floors served by the group, named floors, pre designated priority floors, floors included in fixed zones plus a selected dispatcher algorithm name. Additionally, the group static data contains the collection of car static data for the lifts in that group.

For a building (`LocationStaticData`) – the location name along with the collection of group static data and/or car static data for the groups and any single cars at the location.

5.3.2 Dynamic data

`DynamicData` describes the current state of individual lifts and each group as a whole. The root element of each aggregation of dynamic data elements is time-stamped to specify the instant that it is valid in order to avoid inaccuracies due to possible communication delays. It should be noted that many of the dynamic data elements are specified as optional (multiplicity 0...1) to provide as much flexibility as possible to accommodate different installation configurations. Again, the dynamic data is separated into car-specific, group-specific and location-specific aggregations of elements of information:

`CarDynamicData` – includes floor position, committed (or next) direction of travel, door state, load, registered car and allocated landing calls. An optional additional element defines the `Route` which the car will follow in responding to its list of currently registered calls, which can be used for example to drive passenger information displays or to assist development and test activities.

`GroupDynamicData` – includes registered passenger landing calls (directional and/or destination) plus the collection of car dynamic data for the lifts in that group. Additionally, an optional `DemandProfile` for the group covering a number of data collection periods.

`LocationDynamicData` – is currently, simply the location name along with the collection of group dynamic data and/or car dynamic data for the groups and any single cars at the location. Additionally, an optional `DemandProfile` for the location covering a number of data collection periods.

5.4 Event information

Every event is time-stamped and includes the identifiers of the lift and/or group which generated the event. Optionally, the event may also include a destination identity which specifies the target object of the event (i.e. rather than the destination of the event message). Events may be communicated individually as a stream of updates or batched, providing flexible and efficient options for the design of data logging and status reporting systems. The time-stamp ensures that the sequence of events and the time they occurred are retained even if a communication link is temporarily lost or deliberately terminated.

A predefined set of event types, discovered from a thorough review of available literature plus experience of designing controller and management systems, is included in SEIS and this set may be enhanced as further experience is gained. It is therefore a requirement of SEIS that it is backwards compatible. Each predefined event type may have special properties, derived from SEIS type definitions, which are specific to that event.

`ExceptionEventType` – (describing a specific event instance) may contain other events thereby creating a context for the root event. This may represent the occurrence of an error, warning or simply some noteworthy information.

Recent developments of applications, which enable pre-emptive or predictive maintenance, use operating 'signatures' of devices to identify deviations from normal behaviour as an early indication of potential failure. Currently, the addition of an optional Signature property to the `LogEventType` is under consideration. Such a Signature data-type would need to be flexible and include the possibility of textual representation of a sequence of binary or analogue data values and an associated sampling interval.

5.5 Message formats

We have already noted that SEIS is written in XML Schema Definition language (XSD) so XML might seem a logical choice for formatting messages that conform to the schema. The benefit of this is that received XML messages can be validated automatically against the schema using a number of third-party libraries and there are several editor tools that can generate validation classes directly from the schema in a variety of programming languages. Such tools and libraries can relieve software developers from the burden of significant code development and continuing maintenance effort.

However, an XML message format is not a requirement and similar auto-generation of validation code can be found for other formats, for example JSON. The advantage of JSON is that it is a much more efficient representation and removes a great deal of the bulk of tags that are required for XML. Even if XML is used, more compact messages can be achieved by translating SEIS's long tag names into short acronyms. This could even be achieved in real time by using an XSL style-sheet (this is illustrated in Section 7. Example 3).

5.6 REST

SEIS is a schema that defines an information model – it does not include any specification of functionality and so there are no supporting function calls specific to the operation of passenger lifts. Instead, the same set of standard generic functions can be requested from any supported node of the model and these functions are not at all related to building transportation systems. Thus every supported node in the model must offer the same four functions:

- GET
- PUT
- POST
- DELETE

which might be implemented, for example, as a 'RESTful' [40] interface in a variety of programming technologies, or as a relational database with stored procedures, or similar. This underlines the technology-independent nature of SEIS. Nodes which have a multiplicity greater than 1 (i.e. lists) must support queries using the properties and referenced links of the node (this is illustrated in Section 7. Example 3). SEIS is independent of any network topology (e.g. proxies, gateways, firewalls, etc) so is scalable, which is another characteristic of REST.

6. Security

Threats to networked and other inter-connected systems are continually evolving. The adoption and growth of the IoT throughout the building services domain serves to increase the vulnerability of autonomous networked devices to attack from external hazards and malign enterprises. To ensure that remote systems and applications continue to be updated with protection against new threats as they appear and with minimum risk of introducing vulnerabilities, security measures should adhere to published standards and be developed using third-party tools, which are developed and maintained by experts. The SEIS schema itself does not include cyber-security measures, which is correct since it is concerned only with describing building transportation systems.

Connection to any external network or communication channel, whether confined to the building perimeter or extending beyond it raises critical cyber security issues and potential data-protection concerns, which are discussed in CIBSE Guide-D [3].

Securing the integrity of systems from both

- outside interference or control by unauthorised external agents and

- unauthorised access to information gathered and/or held by the system and the proprietary software code of the system itself

should be addressed directly and must be continually reviewed against newly emerging threats. This responsibility extends throughout all roles in a business and all its processes (i.e. staff induction, in-service training and departure; product design philosophy, customer/client interaction, service procedures, etc., etc.). It cannot simply be addressed by technology-based solutions.

7. Examples of potential SEIS applications

7.1 Example 1: dynamic graphical status display

A lift group controller supplies a SEIS conformant XML document, on demand, to a standard web browser. The browser invokes a locally held XML Style-sheet (XSL) transform to render the raw XML as an HTML graphical representation of the current status of the group (so, if desirable, the presentation might be rendered differently in each browser instance):

A static version of the display can be accessed from the SEIS website [41]. The displayed graphic (i.e. right-hand panel of **Figure 1**) is interactive so that car and landing calls may be registered by clicking on the 'button' images where calls are not currently registered (hovering over the button normally displays the text of the request that would be generated in the browser's status bar).

This example demonstrates how SEIS supports and promotes the Model-View-Controller (MVC) [42] software design pattern – SEIS defines the Model. The lift data is made independent of its presentation, so it is the choice of the client application (in this case a simple web browser) as to how to represent the data. If the client is a data logger then it may be that no transform is needed. The example graphic display is very crude and not easy to read, but a more sophisticated transform might produce a very attractive result. Of course, it is not necessary to display the XML source and XSL Stylesheet frames as they are simply there to illustrate this example.

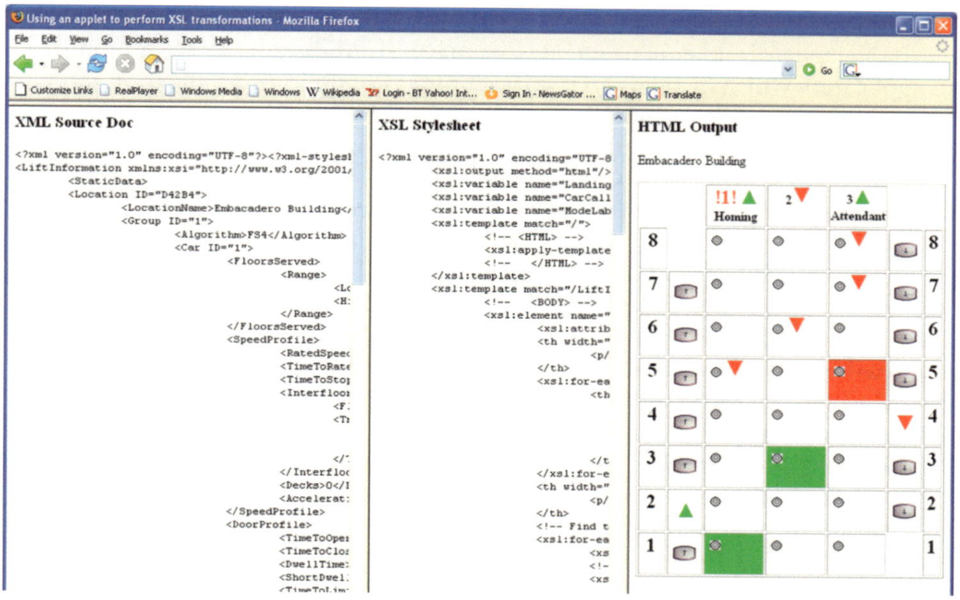

Figure 1.
XML source (left) is transformed by XSL (centre) renders HTML dynamic graphic status display (right).

7.2 Example 2: auto validation

This example shows how simply by referencing the SEIS schema (**Figure 2**, line 3 highlighted), an XML document [43] is validated. An error has been introduced (**Figure 2**, line 107 highlighted) into the document, assigning an invalid value to the `Direction` of Car 1's `CarDynamicData`, which is only valid if selected from the enumeration defined in the schema. The error is generated by the common class libraries that support XML. It has then been picked up and displayed to the user (**Figure 2**, bottom left circled), by the XML editor program in which the XML is being viewed. The resulting action will differ depending on the type of application which is receiving the erroneous XML, but the error detection is common.

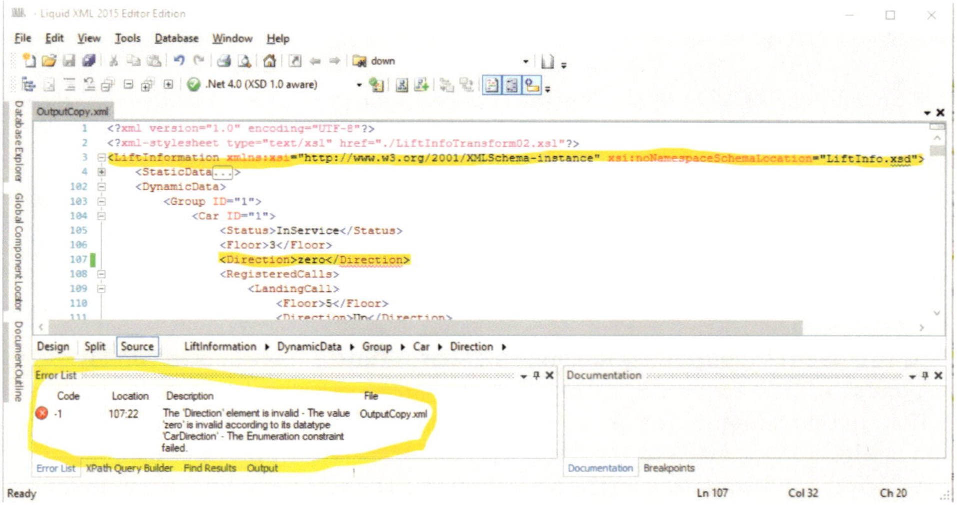

Figure 2.
XML document containing enumeration value error fails validation against SEIS schema.

This example demonstrates the support, which the schema provides for its own inevitable updates and enhancements that may be introduced over the lifetime of an application which depends on it. Maintaining the application validation software is greatly simplified (so is much less costly) because it is unlikely that each installed instance of the application will need to be upgraded on site.

7.3 Example 3: dispatcher interface

This example illustrates a group dispatcher published as a RESTful interface, conforming to SEIS and supporting the 'publish and subscribe' mode of interaction. The role of a dispatcher is to allocate passenger calls to the optimal lift in a group, as evaluated by an internal algorithm. The published interface is independent of the algorithm and it is the intention of the design that the dispatcher interface may be implemented in any installation configuration.

The aim of the example is to demonstrate how, by reference to a schema (i.e. SEIS) the dispatcher can operate simply through queries to and updates of the information without the need for a specific 'allocate this call' function. The benefits of this approach are that:

1. No assumption is made about the type of algorithm used by the dispatcher and therefore the inevitable debate is avoided about which parameters must be passed in such a call. The interface is compatible with any algorithm technique from dynamic sectoring to neural networks based on cost functions. In REST this is referred to as 'separation of concerns'.

2. Furthermore, a simple 'allocate' function call is not appropriate in this application because any resulting allocation(s) must be communicated to several subscribed agents/devices, not just the initiator of the call. In fact there is not even a restriction on the number of dispatchers which may be operating concurrently for the same group of lifts.

Because the computing power and network bandwidth available to such an application, probably running in the lift motor room, are likely to be 'constrained' this example was implemented for the CoAP [44] protocol, which is specifically designed to minimise processing and communication demands. CoAP messages have a similar format to HTTP messages used by web-browsers to access a web server – each request includes:

- an 'address' (URI) which identifies the exact location of a particular resource that is the object of the request,

- optionally a 'query' string which acts as a filter for the data of the resource and

- optionally a data 'payload' which can carry data to or from the requested resource.

CoAP devices (endpoints) include a standard address '.well-known' which responds to a GET request by listing all the resources that the device supports so that, in this example, any lift can discover what nodes of the SEIS information model are supported by the dispatcher under the `<GroupDynamicData>` node as well as information on the algorithm and the group configuration under `<GroupStaticData>`. In this example we concentrate on the `<RegisteredCalls><LandingCalls>` sub-nodes within both the `<GroupDynamicData>`

and `<CarDynamicData>` nodes of a SEIS-conformant information model – these are published as CoAP resources.

Subscribing to a node is equivalent to making a GET request (with an 'Observe' flag) where the response containing the requested information is delayed until a change of state occurs in the node, and a response is re-sent with every subsequent change. Call stations (and mobile devices, etc. via secured apps on the Internet) subscribe to the `<GroupDynamicData><RegisteredCalls><LandingCalls>` node. Lift car controllers subscribe to the same node.

In this example messages are formatted in JSON, replacing the long XML element tag-names with short two-letter acronyms, to reduce message size. So for example `LandingCallType` data is represented as:

the query – {"Fr=01","Dr="UP""} and

the payload:

`{"Ss":0,"Df":0,"St":0.0,"Dn":0.0,"Ct":0.0,"Rk":0,"La":0,"Id":12345678}`

where:

Floor→Fr; Direction→Dr; Status→Ss; Destination→Df; StartTime→St; Duration→Dn; AssignedTo→La; CallID→Id; etc.

The transformation to JSON and auto-validation (as shown in Example 2) can still be achieved using standard software development tools to generate code automatically, by making reference to a schema definition.

The following sequence describes the process of allocating landing calls to lifts:

1. When a passenger registers a landing call, a POST message containing `<Call-Registration>` data is directed to the `<GroupDynamicData><RegisteredCalls><LandingCalls>` node (**Figure 3**):

Address of node: `"GroupDynamicData/RegisteredCalls/LandingCall"`
Node query: `{["Fr=01","Dr="UP""]}`
Payload: `{"Ss":0,"Df":10,"St":0.0,"Dn":0.0,"Ct":0.0,"Rk":0,"La":0}`

2. The dispatcher responds simply with a positive acknowledgement (`ACK`) to show that the POST request has been accepted. The response includes a unique reference (`CallID` - **Figure 3** dashed ellipse), so that the new data can be accessed in future, and which will be used to cross-reference future events associated with this landing call:

Response: `Code=2.01, Type=ACK, MID=142, Token=03: {"Content-Format":"text/plain"} Payload: 8 Bytes 34350805`

3. The dispatcher then performs what it is uniquely designed to do – finding the optimal lift to allocate to unallocated landing calls, at the same time possibly

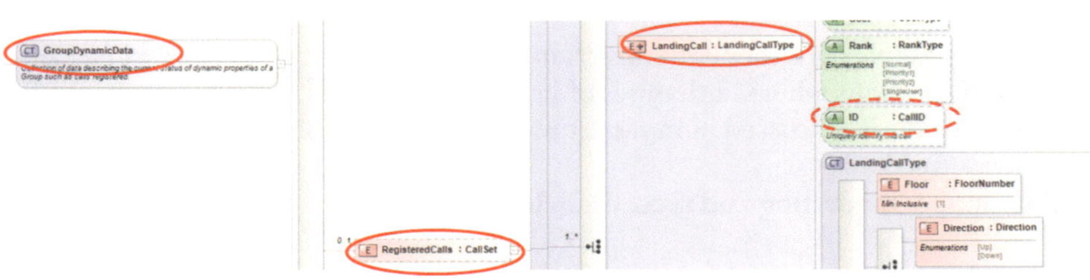

Figure 3.
<GroupDynamicData><RegisteredCalls><LandingCall>.

re-allocating other calls. The resulting allocations will be updated in the `<GroupDynamicData><RegisteredCalls><LandingCalls>` node.

4. The update will trigger the dispatcher interface to issue GET response `CallAssignment` messages to any car that has subscribed to the node and also to the subscribed call stations (possibly via a cloud-based app in the case of a mobile device). The message includes the `CallID` reference so that if necessary the call station can be notified to inform the passenger that their call has been registered (`"Ss":1`) and, in the case of a destination call, which lift has been allocated to their call:

Response payload: {"Ss":1,"Df":10,"St":40578.6,"Dn":0.0, "Ct":0.0,"Rk":0, "La":2,"Id":34350805}

At the same time the resulting allocation will be notified to any subscribed lift and the allocated lift car will update its own `<CarDynamicData><RegisteredCalls>` `<LandingCalls>` node, which signals acceptance of the allocation (see **Figure 4**).

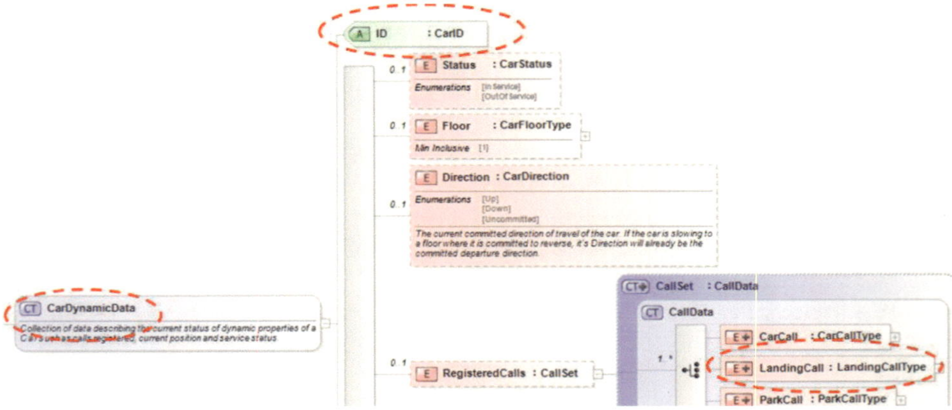

Figure 4.
<CarDynamicData><RegisteredCalls><LandingCalls>.

5. When the allocated lift arrives, to cancel the call it sends a `CallCancellation` (`"Ss":2`) UPDATE message to the dispatcher interface (in this case the system response time – Dn – was 20.5 s):

Address of node: `"GroupDynamicData/RegisteredCalls/LandingCall"`

Node query: `[""Fr=01","Dr="UP"","Id=34350805"]`

Payload: {"Ss":2,"Df":10,"St":40578.6,"Dn":20.5,"La":2," Id":34350805"}

This process works either with direction calls or destination calls and in fact supports configurations where both types of landing call station are used concurrently.

This example is discussed in more detail in a separate publication [45].

7.4 Example 4: modelling and recording demand profiles

An architect designing a 10 storey office building wishes to simulate the operation of the lifts, particularly with respect to the expected traffic to and from the restaurant on the top floor during the lunch period of a working day. The building has two entry/exit floors at the lowest levels – garage (floor 1) and a main terminal lobby (floor 2) for pedestrian entry/exit. The CIBSE Office lunch peak traffic template [46] is selected to define the traffic demand profile to drive the evaluation simulation. This template describes traffic in each of 12 five-minute intervals over the lunch hour in terms of the travelling population (actually the number of passenger arrivals) during the interval as a percentage of the total building population. The travelling population is divided by percentage into each of three categories – Incoming, Outgoing and Interfloor.

7.4.1 Selection of suitable traffic template

Even for a building with just a single main terminal floor, the lunch time template is considered to be very intense since it presents the features of both start and end of day peak flows, but in the presence of concurrent interfloor traffic. However, the building which is the subject of this design is not simple since it has two entry/exit levels. For the purposes of lift traffic analysis it is assumed that there is no resident population on the garage and lobby floors and furthermore, since the staff of the restaurant are not expected to travel during the busy lunch period, all three floors can be treated as if they are exits/entries of the building.

So, the specification by the template of traffic in terms of 'incoming' and 'outgoing' flows is convenient, as this example demonstrates, because the restaurant can be treated as a pseudo entry/exit floor. Consequently, for the building in question 'outgoing' traffic involves both up and down travelling passengers, and similarly for 'incoming' traffic.

During the lunch hour, the busiest period for the restaurant floor in this example, it is assumed that the "outgoing/incoming" traffic will be split 50% to/from the restaurant and 50% through the two real entry/exit floors where 10% passes through the garage and the remaining 40% via the lobby.

7.4.2 Populating the SEIS DemandProfile

With the parameters of the traffic template established, it is now possible to calculate the arrival rates, for each 5-minute interval, of passengers at the populated floors heading for an exit floor (outgoing) and similarly the arrival rates at the entry floors heading for the populated floors (incoming), lastly between populated floors (interfloor). We also have the means of estimating the destination floor of each arriving passenger and so the number of passengers arriving at a floor and sharing a destination floor is derived according to the above percentages for each type of traffic flow.

The result is a three-dimensional matrix of per-interval arrival and destination rates that is most easily generated via the passenger data editor of a lift traffic simulator [47], which then will be responsible for generating individual passenger journeys, based on its own internal statistical formula. Several simulation runs should be generated to ensure a valid sample size. Subsequently, the passenger data matrix should be loaded into a spreadsheet for easier manipulation. The matrix is most concisely and consistently expressed as a SEIS `DemandProfile`, (see **Figure 5**), which provides a crucial mechanism for linking the building under design with the reality of its operation.

The data in the spreadsheet can then be exported as XML in the form a `DemandProfile` as a permanent record of the design and analysis that has been conducted. A key element of the `DemandProfile` is the `ResponseQuality` (red ellipse in **Figure 5**), which links demand rate to response, thereby capturing a measure of the quality of service during the sample period. The particular value of this link is illustrated in this example.

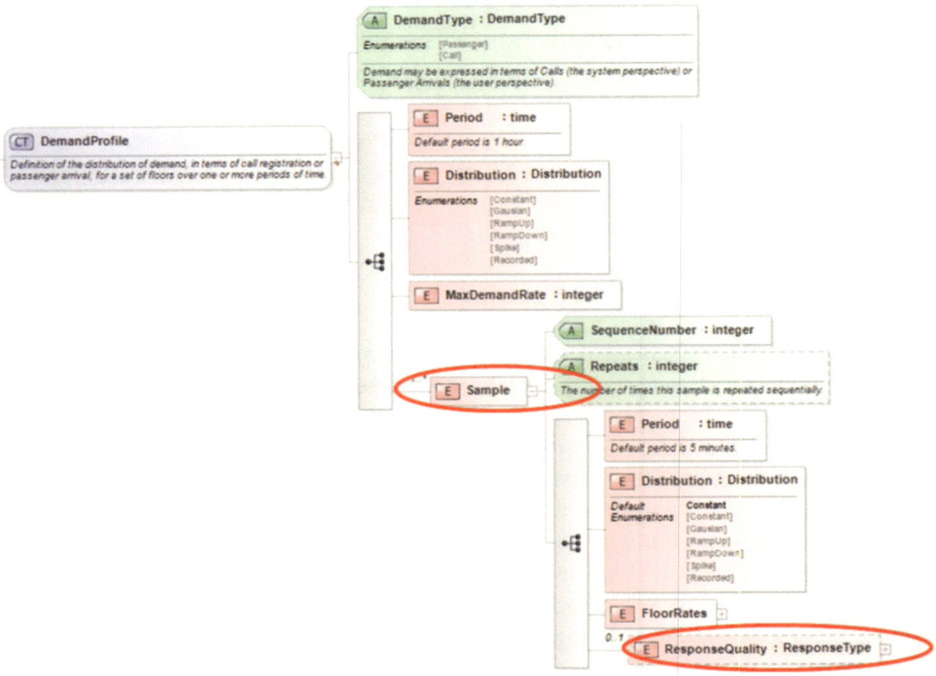

Figure 5.
Definition of a SEIS DemandProfile element.

7.4.3 Using demand profiles throughout the lifecycle of the building

During the building design phase the demand profile can be used to configure a lift traffic simulation, where the simulator will be responsible for generating individual passenger journeys, based on its own internal statistical formula, over each period of the demand profile. These simulations can then be used to assess the suitability of the proposed lift system – its handling capacity and its anticipated quality of service.

After the building is constructed and inhabited by the user population, data logging should be set up to record the real `DemandProfiles` as they occur, which should be compared with those of the design phase to ensure that a good match has been achieved. A match of design against recorded is more readily achieved for destination calls than for direction calls (where some interpretation of the data is required) but the demand profile is compatible with both (even a mix of) call types. Many demand profiles should be recorded to ensure a statistically valid set of results is used and the records should continue over the lifetime of the building to ensure that the characteristics of passenger demand have not changed and the ability of lifts to satisfy it has not deteriorated. If the match is not demonstrated the recorded results should be used to generate further simulations where the parameters of the lift system may be adjusted in the search for a more appropriate configuration. All demand profiles (planned and recorded) should be referenced and co-located in the Building Information Model (BIM level-3) [48] – once it has become standard practice to maintain such a resource for the management of a building and its assets throughout its operational lifecycle.

8. Conclusion

The origins and features of SEIS have been discussed in relation to the increasing interconnection and interaction of building transportation systems with the buildings and cities which they support. The importance of conforming to a standard definition of the operation of building transportation systems cannot be overstressed and SEIS provides this in an open and accessible format while being compatible with a properly secured environment, allowing commercial confidence and intellectual property to be protected. The examples presented have illustrated how wide and varied is the range of potential applications for SEIS in this domain. In particular the example of a group dispatcher indicates the role that SEIS could play in supporting the interoperation of devices (potentially the 'things' in the Internet of Things) where different autonomous elements, which may not have interacted previously, must collaborate.

Author details

Jonathan Beebe* and Ahmad Hammoudeh

1 Jonathan Beebe Ltd, Bury, Lancashire, UK

2 Princess Sumaya University for Technology, Jordan

*Address all correspondence to: jonathan@jonathanbeebe.com

References

[1] thyssenkrupp MULTI. Cable-Less Elevator That Moves Not Only Vertically But Also Horizontally [Internet]. Available from: https://multi.thyssenkrupp-elevator.com/ [Accessed: 05 January 2020]

[2] So A, Al-Sharif L, Hammoudeh A. Traffic analysis of a three-dimensional elevator system. Building Services Engineering Research and Technology. 2018;**39**(1):5-20

[3] CIBSE. CIBSE Guide D. Transportation Systems in Buildings. CIBSE Pub. London. 2015/20; ISBN: 978-1-906846-64-0 (book); ISBN: 978-1-906846-64-7 (PDF); Revision expected 2020

[4] thyssenkrupp. MAX Product Brochure. Available from: https://max.thyssenkrupp-elevator.com/assets/pdf/TK-Elevator-MAX-Brochure_EN.pdf [Accessed: 05 January 2020]

[5] KONE. KONE e-Link Product Information [Internet]. Available from: https://www.kone.co.uk/new-buildings/kone-people-flow-intelligence/monitoring-solutions/ [Accessed: 30 March 2019]

[6] Schindler. Schindler Ahead Product Information [Internet]. 2016. Available from: https://www.zdnet.com/article/huawei-talks-up-cloud-ambition-rides-iot-elevators-with-schindler/ [Accessed: 30 March 2019]

[7] OTIS. OTIS EMS Panorama Product Factsheet. Available from: http://otusazottcdaw1.cloudapp.net/site/ca-eng/OT_DL_Documents/OT_DL_DownloadCenter/Product%20Information%20EMS/EMS%20fact%20sheet.pdf [Accessed: 30 June 2019]

[8] Mitsubishi. Mitsubishi Electric- Mel Eye Product Brochure. Available from: http://www.mitsubishielectric.com/elevator/products/basic/elevators/control_system/pdf/catalog.pdf [Accessed: 30 March 2019]

[9] Kollmorgen. Kollmorgen Visual Lift Product Brochure. Available from: http://www.kollmorgen.de/uploads/tx_fwdownloads/VisualLift_EN.pdf [Accessed: 30 March 2019]

[10] Escalade. iCON Product Information [Internet]. Available from: http://www.iescalade.com/icon/ [Accessed: 30 December 2019]

[11] Peters R, Nickerson J. Lift and escalator management systems: Requirements and implementation. In: Proceedings of the 10th Symposium on Lift & Escalator Technologies; September 2019. Northampton; 2019. pp. 103-112

[12] Data Hoist. Analytics and IoT for All Elevators [Internet]. Available from: http://datahoist.com/index.php/about-us/ [Accessed: 30 September 2019]

[13] LiftNet-USA. Liftnet Online Brochure [Internet]. Available from: https://www.liftnet.com/about/liftnet-online-brochure/ [Accessed: 30 December 2019]

[14] LiftNet-Germany. Product Datasheet. Available from: https://liftnet.org/fileadmin/liftnet1/Docs/LIFTnet/LIFTnet2.0-Datenblatt_2018.1_E.pdf [Accessed: 04 January 2020]

[15] PRC National Standards Committee of People's Republic of China. GB/T 24476-2017 - Specification for Internet of Things for Lifts, Escalators and Moving Walks. 2018. Available from: https://www.chinesestandard.net/PDF/English.aspx/GBT24476-2017 [Accessed: 04 January 2020]

[16] Peters R, Smith R. Analysis of elevator performance and passenger demand with destination control. Elevator Technology. 2008;**17**. Available as reprint from: https://www.peters-research.com/index.php/support/articles-and-papers/41-analysis-of-elevator-performance-and-passenger-demand-with-destination-control [Accessed: 06 May 2020]

[17] Schindler. Schindler myPORT Product Information [Internet]. Available from: https://www.schindler.com/us/internet/en/mobility-solutions/products/destination-technology/myport-technology.html [Accessed: 30 March 2019]

[18] KONE. KONE Access Solutions Product Information [Internet]. Available from: https://www.kone.co.uk/new-buildings/kone-people-flow-intelligence/access-solutions/ [Accessed: 30 March 2019]

[19] thyssenkrupp. thyssenkrupp Launches Elevator Interface to Allow Seamless Multi-Level Robot Movement Throughout Hotels, Hospitals and Other Building [Internet]. Available from: https://storage.thyssenkruppelevator.com/assets/downloads/news/pr/tkE-RoboticsIntervaceinTKE-10172019.pdf [Accessed: 30 October 2019]

[20] Gateway. The Magic Mirror which Opens Up the Lift Car to the Outside World Product Webpage [Internet]. Available from: http://gateway.luve.it/?page_id=2 [Accessed: 30 March 2019]

[21] KONE. KONE Infoscreen Product Factsheet [Internet]. Available from: https://chad.kone.com/Images/factsheet-kone-infoscreen_tcm162-18655.pdf [Accessed: 30 March 2019]

[22] OTIS. OTIS eView Factsheet [Internet]. 2016. Available from: https://files.otis.com/otis/en/au/contentimages/otiseViewFactSheet.pdf [Accessed: 30 March 2019]

[23] Minerva R, Biru A, Rotondi D. Towards a definition of the internet of things (IoT). IEEE Internet Initiative. 2015;**1**:1-86

[24] International Telecommunication Union. Overview of the Internet of Things. 2012. Available from: https://www.itu.int/rec/dologin_pub.asp?lang=e&id=T-REC-Y.2060-201206-I!!PDF-E&type=items [Accessed: 17 January 2020]

[25] Atzori L, Iera A, Morabito G. The internet of things: A survey. Computer Networks. 2010;**54**(15):2787-2805

[26] Raggett D. Using Semantics and Rich Metadata to Bridge IoT Silos—W3C's Work on the Web of Things. 2015. Available from: https://www.w3.org/2015/12/09-wot-m2m.pdf [Accessed: 29 December 2019]

[27] Rumbaugh J, Jacobson I, Booch G. The Unified Modeling Language Reference Manual. Reading, England: AddisonWesley; 1999

[28] Beebe J. Integration of lift systems into the Internet of Things and the need for an open standard information model. In: Proceedings of the 6th Symposium on Lift & Escalator Technologies; 21-22 September 2016. Northampton; 2016. pp. 41-52

[29] ETSI website. Internet of Things [Internet]. Available from: http://www.etsi.org/technologies-clusters/technologies/internet-of-things [Accessed: 29 December 2019]

[30] CAN in Automation. CiA® 417 Series: CANopen Application Profile for Lift Control Systems [Internet]. 2018. Available from: https://www.can-cia.org/can-knowledge/canopen/cia417/ [Accessed: 10 February 2019]

[31] BSR/ASHRAE. Proposed Addendum to Standard 135-2016, BACnet® — A Data Communication Protocol for Building Automation and Control

Networks (Second Public Review Draft). 2016. Available from: http://www.bacnet.org/Addenda/Add-135-2016bs-PPR2-ISC-Draft-3_chair_approved.pdf

[32] Beebe J. Standard Elevator Information Schema [Internet]. Available from: http://www.std4lift.info/ [Accessed: 29 December 2019]

[33] Beebe J. An XML-based standard for communicating elevator information. In: Proceedings of the International Congress on Vertical Transportation Technologies (ELEVCON 2005). Beijing; 2005

[34] Barney GC, Dos Santos SM. Lift Traffic Analysis, Design and Control. Stevenage, England: Peter Peregrinus; 1977

[35] Wikipedia. Object Oriented Programming, Encapsulation [Internet]. Available from: https://en.wikipedia.org/wiki/Object-oriented_programming#Encapsulation [Accessed: 11 January 2020]

[36] Barney G, Cooper D, Inglis J. Elevator & Escalator Micropedia. Incorporating microGuide-D. London: GBA Pub; 2009

[37] Barney G, Al-Sharif L. Elevator Traffic Handbook: Theory and Practice. 2nd ed. Abingdon, England: Routledge; 2016

[38] Beebe J. Standard Elevator Information Schema, Example Application - Event Data Log [Internet]. Available from: http://www.std4lift.info/LiftInfoExampleEventDataLog.pdf [Accessed: 10 January 2020]

[39] Beebe J. Standard Elevator Information Schema - Lift Information Type [Internet]. Available from: http://www.std4lift.info/HTMLSchema/index.html?url=liftinformationtype.html [Accessed: 10 January 2020]

[40] Wikipedia. Representational State Transfer [Internet]. Available from: https://en.wikipedia.org/wiki/Representational_state_transfer [Accessed: 11 January 2020]

[41] Beebe J. Standard Elevator Information Schema - Simple Display [Internet]. Available from: http://www.std4lift.info/SimpleDisplay/Output.xml [Accessed: 14 January 2020]

[42] Wikipedia. Model-View-Controller [Internet]. Available from: https://en.wikipedia.org/wiki/Model-view-controller [Accessed: 02 February 2020]

[43] Beebe J. Standard Elevator Information Schema - Validation Failure [Internet]. Available from: http://www.std4lift.info/SimpleDisplay/Output.error.xml [Accessed: 14 January 2020]

[44] CoAP. Constrained Application Protocol RFC 7252. Available from: https://tools.ietf.org/html/rfc7252 [Accessed: 05 January 2020]

[45] Beebe J, Towards A. Global traffic control (dispatcher) algorithm - requirements analysis. Transportation Systems In Buildings. 2018;**2**:1-18. DOI: 10.14234/tsib.v2i1.147

[46] CIBSE. CIBSE Guide D. Transportation systems in buildings, Sec 4.7.3. CIBSE Pub. London. 2015/20; ISBN: 978-1-906846-64-0 (book); ISBN: 978-1-906846-64-7 (PDF)

[47] Peters Research. Elevate™ Traffic Analysis and Simulation Software [Internet]. Available from: https://www.peters-research.com/index.php/elevate/about-elevate [Accessed: 14 January 2020]

[48] Innovate UK and Infrastructure and Projects Authority. Creating a Digital Built Britain: What You Need to Know – BIM Maturity Levels [Internet]. Available from: https://www.gov.uk/guidance/creating-a-digital-built-britain-what-you-need-to-know [Accessed: 21 January 2020]

Chapter 6

A Universal Methodology for Generating Elevator Passenger Origin-Destination Pairs for Calculation and Simulation

Lutfi Al-Sharif

Abstract

The origin-destination matrix is a two-dimensional matrix that describes the probability of a passenger travelling from one floor in the building to another. It is a two-dimensional square matrix. The row index denotes the origin floor and the row index denotes the destination floor for the passenger journey. A previous chapter described the methodology for constructing the origin-destination matrix (OD matrix) from the user requirements. However, that chapter placed the restriction that any floor must either be assigned as an entrance floor or an occupant floor, but not both. This chapter relaxes this restriction and shows a method for developing the origin-destination matrix that allows any floor to either be an entrance floor; an occupant floor; or both. The origin destination matrix can be compiled using three sets of parameters: the mix of traffic (incoming traffic, outgoing traffic, inter-floor traffic; and inter-entrance traffic); the floor populations; and the entrance percentage bias (i.e., the relative strength of the arrivals at the entrance floors). The origin-destination matrix can be used for the generation of random passenger origin-destination pairs (which is necessary when using the Monte Carlo Simulation (MCS) method to calculate the round-trip time or in elevator traffic software).

Keywords: elevator, lift, round trip time, incoming traffic, outgoing traffic, inter-floor traffic, inter-entrance traffic, general traffic conditions, Monte Carlo simulation, origin-destination matrix

1. Introduction

In an elevator traffic system within a building, the origin-destination matrix is a compact-concise tool that is used to clearly describe the probability of a passenger travelling from one floor in the building to another. It is a two-dimensional square matrix. The diagonal elements of the matrix are equal to zero, since rational passenger behaviour is assumed (i.e. no passengers will travel from a floor to the same floor). Moreover, the sum of all the elements within the matrix is equal to one (representing the universal set in probability). The row index of the matrix denotes the origin floor of the passenger's journey, whilst the column row denotes the destination floor of the passenger's journey.

It is worth noting that all the events in the *OD* matrix are mutually exclusive (i.e., if one of them takes place in a round trip, the others cannot take place in the same round trip). As an example, if a passenger chooses to go from the third floor to the seventh floor in a round trip, he/she cannot also go from the eighth floor to the second floor in the same round trip.

The general format for an *OD* matrix is shown below. All the diagonal elements are equal to zero, as it is assumed that passengers are rational and would not travel from a floor to the same floor.

$$OD_{adj} = \begin{bmatrix} 0 & p_{1\cdot 2} & \cdot\cdot & \cdot\cdot & p_{1\cdot N-1} & p_{1\cdot N} \\ p_{2\cdot 1} & 0 & \cdot\cdot & \cdot\cdot & p_{2\cdot N-1} & p_{2\cdot N} \\ \cdot\cdot & \cdot\cdot & \cdot\cdot & \cdot\cdot & \cdot\cdot & \cdot\cdot \\ \cdot\cdot & \cdot\cdot & \cdot\cdot & \cdot\cdot & \cdot\cdot & \cdot\cdot \\ p_{N-1\cdot 1} & p_{N-1\cdot 2} & \cdot\cdot & \cdot\cdot & 0 & p_{N-1\cdot N} \\ p_{N\cdot 1} & p_{N\cdot 2} & \cdot\cdot & \cdot\cdot & p_{N\cdot N-1} & 0 \end{bmatrix}$$

The origin-destination matrix is critical for the elevator traffic design process, specifically for the following two functions:

1. The origin-destination matrix is used to generate passengers with origin and destination pairs to be used in finding the round-trip time using Monte Carlo simulation. This can be done for both calculation and simulation [1].

2. The origin–destination matrix can also be used to derive the probabilities of any type of event taking place in a round trip (e.g., the probability of a journey taking place between the third and sixth floors without stopping at the fourth and fifth floors). Deriving these probabilities is critical to deriving equations for evaluating the round-trip time.

The final origin-destination matrix must obey a number of rules, listed below:

1. The sum of all the elements in the *OD* matrix should add up to 1.

2. All the elements of the diagonal of the matrix should be equal to 0.

It is worth noting that work has been carried out in trying to estimate the origin-destination traffic from elevator movements in the building [2–5]. The outcome can be used in a number of ways, including generating virtual passenger traffic [6] or deciding on the suitable group control algorithm to adopt during that period of time [7].

Previous work has shown how the origin destination matrix can be derived [8, 9]. However, those two pieces of work have assumed that any floor must either be an entrance floor or an occupant floor, but not both. The work presented in this paper relaxes this requirement and allows any floor to be both an entrance floor and an occupant floor.

Examples of traffic mix conditions that are believed to be representative of the lunch-time peak traffic conditions in many modern office buildings include: 40%:40%:20% [10]; 45%:45%:10% [11]; 42%, 42%, 16% [12]; incoming, outgoing and inter-floor traffic, respectively.

In addition, it is possible to calculate the round-trip time for general traffic conditions under the Poisson arrival process assumption [13, 14] or the plentiful-passenger-supply assumption [15, 16].

This framework can then be combined with other design methodologies in order to be used as a comprehensive universal elevator traffic system design tool, whereby the average passenger waiting time and the average passenger travelling times can also be added as user requirements, and outputs such as the car capacity, the elevator speeds can be provided as outputs of the design.

Existing software packages [17] employ a slightly different method for generating passenger origin-destination pairs, which leads to marginally different results.

Section 2 describes the two types of floors. Section 3 describes the possible modes of traffic in a building. Section 4 describes how the traffic can be described in a building. Section 5 shows how the universal origin destination matrix can be derived. A numerical example is given in Section 6. Section 7 discusses the issue of verification to ensure the correctness of the final OD matrix. Conclusions are drawn in Section 8.

2. Types of floors

This section looks at the classification of floors. Any floor in a building can either be described as an entrance floor or an occupant floor (or both as presented in this paper). An entrance floor (referred to in short as *ent* floor) is a floor through which passengers can either enter or exit the building. An entrance floor can also be referred to an entrance/exit floor, as it becomes an exit floor when the traffic is out-going.

An occupant floor is a floor through which passengers cannot enter or exit the building, but where they will reside during their stay in the building (referred to in short as an *occ* floor). Based on the assumption above, all floors in the building would be denoted as either entrance floors or occupant floors (or both as is allowed in this paper). In some buildings, a floor could simultaneously be an entrance floor and an occupant floor (e.g., ground floor that has a population on it).

An example of a building that is represented in this form is shown below in **Table 1**. Each floor is either an entrance/exit floor or an occupant floor, although one floor is designated with both functions (entrance and occupant) which is the ground floor. The percentage of passengers entering the building via an entrance floor is denoted as the percentage arrival rate $[Pr_{arr}(i)]$. The population of a floor expressed as a percentage of the total building population is denoted as the percentage population $[(U(i)/U)]$.

Where a floor is an entrance floor, it has a nonzero value for its percentage arrival rate and a zero-percentage population. Where a floor is an occupant floor, it has a nonzero percentage population but a zero-percentage arrival rate. Where it has mixed designation, it can have both.

Floor notation	Floor name	Percentage arrival rate	Population	Population percentage
N	*L5*	0	50	0.10
N-1	*L4*	0	50	0.10
...	*L3*	0	75	0.15
...	*L2*	0	75	0.15
...	*L1*	0	100	0.20
3	*G*	0.7	150	0.30
2	*B1*	0.2	0	0.00
1	*B2*	0.1	0	0.00

Table 1.
Representation of traffic in a building.

It is customary for the entrance floors to be contiguous and for the occupant floors to be contiguous, but this is not necessary. An example of a noncontiguous floor is where a restaurant is located on the top floor of a building and is classified as an entrance/exit floor (as it is functionally external to the building albeit not physically external). The design methodology presented later in this paper can cope with the general case where the entrance floors are noncontiguous and the occupant floors are noncontiguous.

Table 1 shows a building with three entrances with unequal percentage arrival rates (0.7 ground floor denoted as G; 0.2 from the basement denoted as B1; 0.1 from the basement denoted as B2). There are six occupant floors (G and L1 to L5). They have unequal population percentages.

The lowest floor in the building is denoted as floor 1 and the topmost floor as floor N. The following convention is followed in describing the values of arrival percentages and populations for the floors:

$Pr_{arr}(i)$ is used to denote the arrival percentage of the ith floor, where i runs from 1 to N.

$U(i)/U$ is used to denote the percentage population of the ith floor, where i runs from 1 to N, where $U(i)$ is the population of the ith floor and U is the total building population.

The representation of traffic in a building as shown in **Table 1** is the default format and represents pure incoming traffic into the building. Under such a traffic condition, all passengers would be entering the building form the entrance floors and heading to the occupant floors.

A general format for representing the traffic in a building is shown in **Table 2** below. As a generalisation, the term arrival can be extended to arrival/departure to cover both passengers entering the building under incoming traffic conditions and passengers leaving the building under out-going traffic conditions. By setting a population percentage for a floor to zero, it is an indication that it is an entrance floor; by setting an arrival percentage for a floor to zero, it is an indication that it is an occupant floor. By setting a value to the population percentage and a value to the percentage arrival rate, it is an indication that it has dual function.

It is worth noting that the summation of the percentage arrival/departure rates is 1 and the summation of all the building percentage populations is 1 as shown in Eqs. (1) and (2) below.

$$\sum_{i=1}^{N} \Pr_{arr}(i) = 1 \tag{1}$$

Floor	Percentage arrival/Departure	Percentage population
N	$Pr_{arr}(N)$	$(U(N)/U)$
N-1	$Pr_{arr}(N-1)$	$(U(N-1)/U)$
…	…	…
…	…	…
…	…	…
2	$Pr_{arr}(2)$	$(U(2)/U)$
1	$Pr_{arr}(1)$	$(U(1)/U)$

Table 2.
General representation format for a building.

Start floor (origin)	End floor (destination)	Type of traffic	Description
Entrance/exit	Occupant	Incoming traffic	Passengers arriving at the building
Occupant	Entrance/exit	Outgoing traffic	Passengers leaving the building
Occupant	Occupant	Inter-floor traffic	Passengers moving within the building (restaurants, meeting rooms)
Entrance/exit	Entrance/exit	Inter-entrance traffic	Usually very small

Table 3.
Types of traffic.

$$\sum_{i=1}^{N}\left(\frac{U(i)}{U}\right)=1 \tag{2}$$

Two row vectors (arrays) can be developed based on the arrival percentages and the floor population percentages, as shown below. The first row vector is the percentage arrivals, denoted as Pr_{arr}:

$$\boldsymbol{Pr_{arr}} = [Pr_{arr}(1)\ \ Pr_{arr}(2)\ \ \dots\ \ \dots\ \ \dots\ \ Pr_{arr}(N-1)\ \ Pr_{arr}(N)] \tag{3}$$

The convention that will be used in this paper is that the indexing will start from the lowest floor in the building and increase upwards. The percentage population can be also be organised into a row vector:

$$\boldsymbol{U_{nor}} = [\,U(1)/U\quad U(2)/U\quad \dots\quad \dots\quad \dots\quad U(N-1)/U\quad U(N)/U\,] \tag{4}$$

The "nor" subscript stands for "normalised".

3. Types of traffic

This section classifies the possible types of journeys and thus the possible types of traffic. Every passenger journey must logically have an origin and a destination. Considering that any floor can either be an entrance/exit floor or an occupant floor, there can exist in theory four types of journeys depending on the classification of the origin and destination floors for each journey.

A journey that starts from an entrance/exit floor and terminates at an occupant floor is denoted as an incoming traffic journey. A journey that starts from an occupant floor and terminates at an entrance/exit floor is denoted as an outgoing traffic journey. A journey that starts from an occupant floor and terminates at an occupant floor is denoted as an inter-floor journey. A journey that starts from an entrance/exit floor and terminates at an entrance/exit floor is denoted as an inter-entrance journey. These four types of traffic are listed in **Table 3** below.

4. Description of the traffic in a building

It has become customary to describe the prevailing traffic in a building at any one point in time as a mixture of the four types of traffic described in the previous section.

The percentage of the traffic that is incoming at any one point in time is denoted as *ic*; the percentage of the traffic that is outgoing at any one point in time is denoted as *og*; and the traffic that is inter-floor at any one point in time is denoted as *if*; and the traffic that is inter-entrance is denoted as *ie*. The combination of these four numbers can be used to describe the traffic mix as shown below (where any of these parameters can vary between 0 and 1):

$$ic : og : if : ie$$

As expected, the sum of all four numbers should add up to 1 as shown in Eq. (5) below. Thus, assigning values for three of these numbers automatically sets the value of the fourth parameter.

$$ic + og + if + ie = 1 \tag{5}$$

As an example, one suggested composition of the lunchtime traffic conditions can be described by the following representation [11]:

$$ic : og : if : ie \text{ as } 0.45 : 0.45 : 0.10 : 0.0 \text{ respectively.}$$

5. The origin-destination (*OD*) matrix

At the end of Section 2, the percentage floor populations were compiled in a concise normalised format in the shape of a row vector denoted as $\boldsymbol{U_{nor}}$. In addition, the percentage arrivals were compiled in a concise format in the shape of a row vector denoted as ***Prarr***.

In order to compile the overall final origin destination, it is first necessary to develop the four origin-destination matrices that contain the contribution from the four types of traffic (incoming, outgoing, inter-floor and inter-entrance). Once these four matrices have been fully developed, the final overall origin-destination matrix is found by adding these four matrices.

5.1 Finding the initial values of the four OD matrices

The first matrix (the incoming traffic matrix) is obtained by multiplying the transpose of the percentage arrival vector ($\boldsymbol{Pr_{arr}}$) by the normalised population vector ($\boldsymbol{U_{nor}}$). As the dimensions of the percentage arrival row vector is 1 by N, when it is transposed, it dimensions become N by 1. The dimensions of the population normalised row vector is 1 by N. When an N by 1 vector is multiplied by a 1 by N vector, this results in an N by N matrix.

$$OD_{ic\_ini} = Pr_{arr}{}^{T} \bullet U_{nor} \tag{6}$$

The second matrix (the outgoing traffic matrix) is obtained by multiplying the transpose of the normalised population vector ($\boldsymbol{U_{nor}}$) by the percentage arrival vector ($\boldsymbol{Pr_{arr}}$) as shown in Eq. (7) below.

$$OD_{og\_ini} = U_{nor}{}^{T} \bullet Pr_{arr} \tag{7}$$

The third matrix (the inter-floor traffic matrix) is obtained by multiplying the transpose of the normalised population vector ($\boldsymbol{U_{nor}}$) by the normalised population vector ($\boldsymbol{U_{nor}}$) as shown in Eq. (8) below.

$$OD_{if\_ini} = U_{nor}{}^{T} \bullet U_{nor} \tag{8}$$

The fourth matrix (the inter-entrance traffic matrix) is obtained by multiplying the transpose of the percentage arrival vector ($\boldsymbol{Pr_{arr}}$) by the percentage arrival vector ($\boldsymbol{Pr_{arr}}$) as shown in Eq. (7) below.

$$OD_{ie\_ini} = Pr_{arr}{}^{T} \bullet Pr_{arr} \tag{9}$$

where the subscript T denotes the transpose of a matrix.

It is worth noting that each of these matrices that result from the multiplication have a total sum of all the elements that is equal to 1.

5.2 Adjusting the four OD matrices

Once the four initial Origin-Destination matrices have been produced by multiplying the relevant vectors, the next step is to remove the irrational traffic (travelling from a floor back to the same floor). This is done in three steps as discussed below:

a. Finding the value of the adjusting factor, M, for each matrix. The value of M is necessary in order to re-adjust the sum of the matrix back to 1, later on. M is calculated as shown below:

$$M = 1 - \left(\sum_{i=1}^{N} (pr_{ii}) \right) \tag{10}$$

b. The second step is to zero all the diagonal elements of the four matrices. This removes the irrational behaviour of a passenger going from a floor back to the same floor.

$$p_{ii} = 0 \quad for \quad i = j \tag{11}$$

c. But as the diagonal items have been zeroed, the sum of all the elements in the matrix no longer adds up to 1. Thus, the third step is to re-adjust the remaining nonzero elements of the matrices in order to restore the sum of all elements in the matrix back to 1. This is done by dividing each element in the matrix by the value of M, found in Eq. (10) above.

Once these three steps have been carried out, the resulting four OD matrices become the final matrices.

$$OD_{ic\_fin}, OD_{og\_fin}, OD_{if\_fin}, OD_{ie\_fin}$$

5.3 Finding the final OD matrix from the four OD matrices

The final origin–destination matrix can now be calculated by adding the weighted sum of the four final matrices. Each one is multiplied by the percentage of the traffic that is represents. This is shown in Eq. (12) below.

$$OD_{fin} = ic \bullet OD_{ic\_fin} + og \bullet OD_{og\_fin} + if \bullet OD_{if\_fin} + ie \bullet OD_{ie_{fin}} \tag{12}$$

This matrix can also be referred to as the "*normalised* origin-destination matrix". It is referred to as *normalised* as it only depends on the arrival percentages of the floors, the population percentages of the floors and the mix of traffic. It does not depend on the passenger arrival intensity. The matrix in this form can now be used in order to generate random passenger origin–destination pairs for evaluating the round-trip time using the Monte Carlo Simulation (*MCS*) method. It can also be used to evaluate the round-trip time using formula by calculation. The sum of all the elements of the matrix is equal to 1. The diagonal element of the matrix must be equal to zero.

6. Numerical example on evaluating the universal origin-destination matrix and generating passenger origin-destination pairs from it

This section presents a practical numerical example on developing the universal original-destination matrix.

6.1 The arrangement

A building has three entrance floors (B2, B1 and G) and six occupant floors (G, L1, L2, L3, L4 and L5). The ground floor (G) has a dual function, hence that is why is appears in both lists. The percentage arrival rates from the three entrances are 0.1, 0.2 and 0.7 from B2, B1 and G respectively. The percentage populations of the occupant floors are: 30%, 20%, 15%, 15%, 10% and 10% for the occupant floors G, L1, L2, L3, L4 and L5, respectively. All the building details are shown in **Table 4**. It is required that the overall origin–destination matrix be developed based on a traffic mix of 40%:30%:20%:10% of incoming, outgoing, inter-floor and inter-entrance traffic, respectively.

It is worth noting that the ground floor in this case is a dual function floor: it is an entrance floor and an occupant floor simultaneously. This is the sort of example that requires a universal OD matrix, whereby any floor can be an entrance and an occupant floor at the same time.

The next step is to create the initial values of the four distinct matrices: the incoming traffic matrix, the outgoing traffic matrix, the inter-floor matrix and the inter-entrance. These are shown below:

#	Percentage arrival	Population	Percentage population
L5	0	50	0.10
L4	0	50	0.10
L3	0	75	0.15
L2	0	75	0.15
L1	0	100	0.2
G	0.7	150	0.3
B1	0.2	0	0
B2	0.1	0	0

Table 4.
Generalised representation of traffic for a building depending on the traffic mix.

6.2 The incoming traffic matrix

The first matrix to be developed is the incoming traffic matrix, as shown below.

OD$_{\textbf{ic\_ini}}$

		0	0	0.3	0.2	0.15	0.15	0.1	0.1
		B2	B1	G	L1	L2	L3	L4	L5
0.1	B2	0	0	0.03	0.02	0.015	0.015	0.01	0.01
0.2	B1	0	0	0.06	0.04	0.03	0.03	0.02	0.02
0.7	G	0	0	0.21	0.14	0.105	0.105	0.07	0.07
0	L1	0	0	0	0	0	0	0	0
0	L2	0	0	0	0	0	0	0	0
0	L3	0	0	0	0	0	0	0	0
0	L4	0	0	0	0	0	0	0	0
0	L5	0	0	0	0	0	0	0	0

The value of M for the incoming traffic matrix is calculated as shown below:

$$M = 1 - \left(\sum_{i=1}^{N} (pr_{ii}) \right) = 0.79 \tag{13}$$

Zeroing the diagonal and then dividing all the elements by M gives the final incoming traffic matrix, shown below.

OD$_{\textbf{ic\_fin}}$

0	B2	B1	G	L1	L2	L3	L4	L5
B2	0	0	0.01519	0.010127	0.007595	0.007595	0.005063	0.005063
B1	0	0	0.03038	0.020253	0.01519	0.01519	0.010127	0.010127
G	0	0	0	0.070886	0.053165	0.053165	0.035443	0.035443
L1	0	0	0	0	0	0	0	0
L2	0	0	0	0	0	0	0	0
L3	0	0	0	0	0	0	0	0
L4	0	0	0	0	0	0	0	0
L5	0	0	0	0	0	0	0	0

6.3 The outgoing traffic matrix

The same process is applied in order to calculate outgoing traffic matrix.

OD$_{\textbf{og\_ini}}$

		0	0	0.3	0.2	0.15	0.15	0.1	0.1
		B2	B1	G	L1	L2	L3	L4	L5
0.1	B2	0	0	0	0	0	0	0	0

0.2	**B1**	0	0	0	0	0	0	0	0
0.7	**G**	0.03	0.06	0.21	0	0	0	0	0
0	**L1**	0.02	0.04	0.14	0	0	0	0	0
0	**L2**	0.015	0.03	0.105	0	0	0	0	0
0	**L3**	0.015	0.03	0.105	0	0	0	0	0
0	**L4**	0.01	0.02	0.07	0	0	0	0	0
0	**L5**	0.01	0.02	0.07	0	0	0	0	0

The value of M for the incoming traffic matrix is calculated as shown below:

$$M = 1 - \left(\sum_{i=1}^{N} (pr_{ii}) \right) = 0.79 \tag{14}$$

Zeroing the diagonal and then dividing all the elements by M gives the final incoming traffic matrix, shown below.

OD$_{\mathbf{og_fin}}$

0	**B2**	**B1**	**G**	**L1**	**L2**	**L3**	**L4**	**L5**
B2	0	0	0	0	0	0	0	0
B1	0	0	0	0	0	0	0	0
G	0.011392405	0.02278481	0	0	0	0	0	0
L1	0.007594937	0.015189873	0.053164557	0	0	0	0	0
L2	0.005696203	0.011392405	0.039873418	0	0	0	0	0
L3	0.005696203	0.011392405	0.039873418	0	0	0	0	0
L4	0.003797468	0.007594937	0.026582278	0	0	0	0	0
L5	0.003797468	0.007594937	0.026582278	0	0	0	0	0

6.4 The inter-floor traffic matrix

The same process is applied in order to calculate inter-floor traffic matrix.

OD$_{\mathbf{if_ini}}$

		0	**0**	**0.3**	**0.2**	**0.15**	**0.15**	**0.1**	**0.1**
		B2	**B1**	**G**	**L1**	**L2**	**L3**	**L4**	**L5**
0.1	**B2**	0	0	0	0	0	0	0	0
0.2	**B1**	0	0	0	0	0	0	0	0
0.7	**G**	0	0	0.09	0.06	0.045	0.045	0.03	0.03
0	**L1**	0	0	0.06	0.04	0.03	0.03	0.02	0.02
0	**L2**	0	0	0.045	0.03	0.0225	0.0225	0.015	0.015
0	**L3**	0	0	0.045	0.03	0.0225	0.0225	0.015	0.015
0	**L4**	0	0	0.03	0.02	0.015	0.015	0.01	0.01
0	**L5**	0	0	0.03	0.02	0.015	0.015	0.01	0.01

The value of M for the incoming traffic matrix is calculated as shown below:

$$M = 1 - \left(\sum_{i=1}^{N} (pr_{ii}) \right) = 0.805 \tag{15}$$

Zeroing the diagonal and then dividing all the elements by M gives the final incoming traffic matrix, shown below.

OD$_{\mathbf{if_fin}}$

0	B2	B1	G	L1	L2	L3	L4	L5
B2	0	0	0	0	0	0	0	0
B1	0	0	0	0	0	0	0	0
G	0	0	0	0.014907	0.01118	0.01118	0.007453	0.007453
L1	0	0	0.014907	0	0.007453	0.007453	0.004969	0.004969
L2	0	0	0.01118	0.007453	0	0.00559	0.003727	0.003727
L3	0	0	0.01118	0.007453	0.00559	0	0.003727	0.003727
L4	0	0	0.007453	0.004969	0.003727	0.003727	0	0.002484
L5	0	0	0.007453	0.004969	0.003727	0.003727	0.002484	0

6.5 The inter-entrance traffic matrix

The same process is applied in order to calculate inter-entrance traffic matrix.

OD$_{\mathbf{if_ini}}$

		0	**0**	**0.3**	**0.2**	**0.15**	**0.15**	**0.1**	**0.1**
		B2	**B1**	**G**	**L1**	**L2**	**L3**	**L4**	**L5**
0.1	**B2**	0.01	0.02	0.07	0	0	0	0	0
0.2	**B1**	0.02	0.04	0.14	0	0	0	0	0
0.7	**G**	0.07	0.14	0.49	0	0	0	0	0
0	**L1**	0	0	0	0	0	0	0	0
0	**L2**	0	0	0	0	0	0	0	0
0	**L3**	0	0	0	0	0	0	0	0
0	**L4**	0	0	0	0	0	0	0	0
0	**L5**	0	0	0	0	0	0	0	0

The value of M for the incoming traffic matrix is calculated as shown below:

$$M = 1 - \left(\sum_{i=1}^{N} (pr_{ii}) \right) = 0.46 \tag{16}$$

Zeroing the diagonal and then dividing all the elements by M gives the final incoming traffic matrix, shown below.

OD_{if_fin}

0	B2	B1	G	L1	L2	L3	L4	L5
B2	0	0.00434783	0.015217	0	0	0	0	0
B1	0.004348	0	0.030435	0	0	0	0	0
G	0.015217	0.03043478	0	0	0	0	0	0
L1	0	0	0	0	0	0	0	0
L2	0	0	0	0	0	0	0	0
L3	0	0	0	0	0	0	0	0
L4	0	0	0	0	0	0	0	0
L5	0	0	0	0	0	0	0	0

6.6 Finding the final overall OD matrix

The final step is to combine all the four matrices by multiplying each matrix by the associated traffic percentage mix and adding them up. This gives the final OD matrix shown below.

OD_{fin}

	B2	B1	G	L1	L2	L3	L4	L5
B2	0	0.004348	0.030407	0.010127	0.007595	0.007595	0.005063	0.005063
B1	0.004348	0	0.060815	0.020253	0.01519	0.01519	0.010127	0.010127
G	0.02661	0.05322	0	0.085793	0.064345	0.064345	0.042896	0.042896
L1	0.007595	0.01519	0.068071	0	0.007453	0.007453	0.004969	0.004969
L2	0.005696	0.011392	0.051054	0.007453	0	0.00559	0.003727	0.003727
L3	0.005696	0.011392	0.051054	0.007453	0.00559	0	0.003727	0.003727
L4	0.003797	0.007595	0.034036	0.004969	0.003727	0.003727	0	0.002484
L5	0.003797	0.007595	0.034036	0.004969	0.003727	0.003727	0.002484	0

As expected, the diagonal of the matrix has zero elements and the sum of all the elements in the matrix is 1.

6.7 Converting the PDF to a CDF

The probability density function (PDF) is then converter to a cumulative distribution function (CDF) by integration. The integration order is run along the row, then the second row, then the third row, etc. As expected, the first element of the CDF array is zero and the last two elements are equal to 1. This is in recognition of the fact that the diagonal elements in the PDF are always zeros.

	B2	B1	G	L1	L2	L3	L4	L5
B2	0	0.004348	0.034755	0.044882	0.052477	0.060072	0.065135	0.070198
B1	0.074546	0.074546	0.135361	0.155614	0.170804	0.185994	0.196121	0.206248
G	0.232858	0.286078	0.286078	0.371871	0.436216	0.500561	0.543457	0.586353
L1	0.593948	0.609138	0.677209	0.677209	0.684662	0.692115	0.697084	0.702053

L2	0.707749	0.719141	0.770195	0.777648	0.777648	0.783238	0.786965	0.790692
L3	0.796388	0.80778	0.858834	0.866287	0.871877	0.871877	0.875604	0.879331
L4	0.883128	0.890723	0.924759	0.929728	0.933455	0.937182	0.937182	0.939666
L5	0.943463	0.951058	0.985094	0.990063	0.99379	0.997517	1	1

6.8 Generating a number of passenger origin-destination pairs from the CDF

In this sub-section, five examples on the random sampling of passenger origin-destination pairs.

This is carried out by generating a random number uniformly distributed between 0 and 1. In the examples shown below, they are shown with 3 decimal places for simplicity. In each case, the number is to find the first number in the CDF that is larger than the random number generated. Once this is found, the row of the element represents the origin floor of the passenger origin-destination pair, and the column index represents the destination floor.

Five examples:

- Random number: 0.772; next *smallest* number in the table larger than this number is: 0.777648, which has a row index of L2 and a column index of L1. So, the origin–destination pair is L2 to L1.

- Random number: 0.591; next *smallest* number in the table larger than this number is: 0.593948, which has a row index of L1 and a column index of B2. So, the origin–destination pair is L1 to B2.

- Random number: 0.001; next *smallest* number in the table larger than this number is: 0.004348, which has a row index of B2 and a column index of B1. So, the origin–destination pair is B2 to B1.

- Random number: 0.998; next *smallest* number in the table larger than this number is: 1, which has a row index of L5 and a column index of L4. So, the origin–destination pair is L5 to L4.

- Random number: 0.862; next *smallest* number in the table larger than this number is: 0.866287, which has a row index of L3 and a column index of L1. So, the origin–destination pair is L3 to L1.

7. Verification

In order to check the correctness of the final origin destination matrix, it is possible to carry out a validation process as presented in the steps discussed below:

1. The OD matrix is set up with dimensions of N by N and set to 0.

2. The number of trials is set to a large number (e.g., 100,000 trials).

3. The trial numbers are split into four groups in proportion to the percentage mix of traffic. For example, if the traffic mix is 0.4:0.3:0.2:0.1 for incoming, outgoing, inter-floor and inter-entrance traffic, then 40,000 trials are set aside

for incoming traffic passenger, 30,000 for outgoing traffic passengers, 20,000 for inter-floor passengers and 10,000 for inter-entrance passengers.

4. Starting with the 40,000 passenger trials (representing incoming traffic), random sampling is carried out on the origin floor for the passenger by using $\mathbf{Pr_{arr}}$, and random sampling is carried out on the destination floor using $\mathbf{U_{nor}}$. This gives an origin-destination pair: the corresponding entry in the OD matrix is incremented. If the origin floor and destination floors are equal the result is discarded, and the process repeated.

5. For the 30,000 passenger trials (representing outgoing traffic), random sampling is carried out on the origin floor for the passenger by using $\mathbf{U_{nor}}$ and random sampling is carried out on the destination floor for the passenger by using $\mathbf{Pr_{arr}}$. This gives an origin-destination pair: the corresponding entry in the OD matrix is incremented. If the origin floor and destination floors are equal the result is discarded, and the process repeated.

6. For the 20,000 passenger trials (representing inter-floor traffic), random sampling is carried out on the origin floor for the passenger by using $\mathbf{U_{nor}}$ and random sampling is carried out on the destination floor for the passenger by using $\mathbf{U_{nor}}$. This gives an origin-destination pair: the corresponding entry in the OD matrix is incremented. If the origin floor and destination floors are equal the result is discarded, and the process repeated.

7. For the 10,000 passenger trials (representing inter-entrance traffic), random sampling is carried out on the origin floor for the passenger by using $\mathbf{Pr_{arr}}$ and random sampling is carried out on the destination floor for the passenger by using $\mathbf{Pr_{arr}}$. This gives an origin–destination pair: the corresponding entry in the OD matrix is incremented. If the origin floor and destination floors are equal the result is discarded, and the process repeated.

8. At the end of the 100,000 trials, the OD matrix is divided by the number of trials (100000). The result should match the OD final overall matrix derived as shown in sections 5 and 6.

The source code in MATLAB that can be used to create the probability density function array, convert it to a cumulative distribution function and then generate random passenger origin destination pairs is listed in Appendix A. It comprises the main module and three functions.

8. Conclusions

This paper has presented a systematic methodology for converting the user requirement specification into an origin-destination matrix.

It converts the traffic conditions (i.e., traffic mix) and the floor percentages (the floor percentage arrivals and the floor percentage populations) into an origin destination matrix. The conversion process takes into consideration that irrational passenger behaviour must be disallowed (e.g., one passenger's journey cannot originate and terminate at the same floor). Thus, the diagonal of the matrix must be zeroed.

The origin destination matrix is a compact concise form for expressing the passenger movements within the building. It has been used for generating random

DOI: http://dx.doi.org/10.5772/intechopen.93332

passenger destinations for calculating the round-trip time using the Monte Carlo simulation method of within elevator traffic simulation software, an example of which is shown at the end of the paper.

Nomenclature

ic	percentage of incoming traffic under the traffic mix
if	the percentage of inter-floor traffic under the traffic mix
ie	the percentage of inter-entrance traffic under the traffic mix
N	the total number of floors in the building
og	the percentage of outgoing traffic under the traffic mix
$P_{arr}(i)$	the percentage arrival from the ith floor
p_{ij}	the passenger transition probability from the ith floor to the jth floor
U	the total building population
$U(i)$	the building population on the ith floor

A. Appendix A. Source Code in MATLAB

Generate_P_Passengers.m.

1. Generate Passenger Origin Destination Pairs Randomly

Author: Lutfi Al-Sharif Date: 27th April 2020 version: 1.0

```
% this MATLAB script generates passenger origin–destination pairs randomly
% based on three main parameters:
% the floor populations
% the entrance floors bias ratios
% the mix of traffic.
```

2. Set the Parameters

This building has 10 floors The floor index runs from 1 to 10. floors 1, 2 and 10 are entrance floors floor 10 is probably a restaurant.

```
% set the entrance bias
EB = [0.3 0.3 0 0 0 0 0 0 0 0.4];
% set the population on each floor
U = [0160160160100100100100100 20];
% set the traffic mix as percentages adding up to 1
% [i/c o/g i/f i/e]
% incoming: outgloing: interfloor: interentrance
Traffic = [0.45 0.45 0.1 0];
```

```
% set the number of required passengers
P = 12;
```

3. shuffle the seed to set it to a specific value

```
% in order to get the same sequence every run, set the seed to a specific
% value (e.g., 0 in this case).
sd = 0; % value of the seed
rng(sd)

% in order to get a different sequence everytime the programme runs, shuffle
the seed
rng('shuffle');
```

4. Create the PDF (probability density function)

call Create_PDF to create the PDF array.

```
PDF = Create_PDF(EB, U, Traffic).

PDF =
```

Columns 1 through 7

0	0.0229	0.0229	0.0229	0.0143	0.0143	0.0143
0.0229	0	0.0258	0.0258	0.0161	0.0161	0.0161
0.0229	0.0258	0	0.0029	0.0018	0.0018	0.0018
0.0229	0.0258	0.0029	0	0.0018	0.0018	0.0018
0.0143	0.0161	0.0018	0.0018	0	0.0011	0.0011
0.0143	0.0161	0.0018	0.0018	0.0011	0	0.0011
0.0143	0.0161	0.0018	0.0018	0.0011	0.0011	0
0.0143	0.0161	0.0018	0.0018	0.0011	0.0011	0.0011
0.0143	0.0161	0.0018	0.0018	0.0011	0.0011	0.0011
0.0029	0.0337	0.0309	0.0309	0.0193	0.0193	0.0193

Columns 8 through 10

0.0143	0.0143	0.0029
0.0161	0.0161	0.0337
0.0018	0.0018	0.0309
0.0018	0.0018	0.0309
0.0011	0.0011	0.0193
0.0011	0.0011	0.0193
0.0011	0.0011	0.0193
0	0.0011	0.0193
0.0011	0	0.0193
0.0193	0.0193	0

5. Convert the PDF to a CDF

CDF: Cumulative Distribution Function The CDF is ideal to use for the random sampling in order to randomly generate passenger origin–destination pairs.

```
CDF = PDF2CDF(PDF).

CDF =

Columns 1 through 7

     0    0.0229  0.0458  0.0686  0.0829  0.0972  0.1115
0.1659  0.1659  0.1917  0.2175  0.2337  0.2498  0.2659
0.3548  0.3806  0.3806  0.3836  0.3854  0.3872  0.3891
0.4465  0.4723  0.4752  0.4752  0.4771  0.4789  0.4807
0.5296  0.5457  0.5475  0.5494  0.5494  0.5505  0.5517
0.5875  0.6037  0.6055  0.6073  0.6085  0.6085  0.6096
0.6455  0.6617  0.6635  0.6653  0.6665  0.6676  0.6676
0.7035  0.7196  0.7215  0.7233  0.7245  0.7256  0.7267
0.7615  0.7776  0.7795  0.7813  0.7824  0.7836  0.7847
0.8080  0.8418  0.8726  0.9035  0.9228  0.9421  0.9614

Columns 8 through 10

0.1258  0.1401  0.1430
0.2821  0.2982  0.3319
0.3909  0.3927  0.4236
0.4826  0.4844  0.5153
0.5528  0.5539  0.5732
0.6108  0.6119  0.6312
0.6688  0.6699  0.6892
0.7267  0.7279  0.7472
0.7859  0.7859  0.8052
0.9807  1.0000  1.0000
```

6. Generate P passenger origin destination pairs using the CDF

call GenPass to generate P passenger origin–destination pairs usnig the CDF.

```
POD = GenPass(P,CDF)

% The resultant array POD is a two dimensional array that has dimensions of
% 2 rows by P columns. The first row contains the passenger origin floor;
% the second row contains the passenger destination floors.
% each column pertains to one passenger (a passenger origin–destination
% pair).
% the floor indexing runs from 1 to N.

% End.

POD =

  1  10  1  10  4  7  1  6   2  2  10  4
  7  8   4   3  8  2  9  10  3  8  9   2.
```

____.

```
function [PDF] = Create_PDF(EB, U, Traffic).

%This module, Create_PDF compiles the probability density function (PDF)
%for generating passenger origin–destination pairs.

% Author: Lutfi Al-Sharif
% Date: 27th April 2020% this function creates a square array that is referred to as a PDF
% PDF stands for probability density function (PDF).
% it is a square array that has dimensions of N rows by N columns
% where N is the total number of floors in the building.

% it is used in order to generate random passneger origin–destination-pairs.
% it depends on two important parameters: the relative percentages of the
% floor populations of the occupant floors, the relative entrance bias of
% entrance floors; and the mix of traffic (the percentage incoming traffic,
% outgoing traffic, interfloor traffic and interentrance traffic).
```

1. Background Theory and References

the methodology presented here in this function is based on a method of producing the origin destination matrix in the following papers.

```
% Al-Sharif L and Abu Alqumsan A M. An Integrated Framework for Elevator
% Traffic Design under General Traffic Conditions Using Origin Destination
% Matrices, Virtual Interval and the Monte Carlo Simulation Method.
% Building Services Engineering Research and Technology 2015; 36(6):
% 728–750.

%Al-Sharif L and Abu Alqumsan A M. Generating the Elevator
% Origin–Destination Matrix from the User Requirements Specification under
% General Traffic Conditions. Elevator Technology 2016; 21: 1–13.
% Proceedings of Elevcon 2016. Madrid/Spain: The International Association
% of Elevator Engineers.

%Al-Sharif L. Building the Origin–Destination
% Matrix under General Traffic Conditions and Using it to Generate
% Passenger Origin–Destination Pairs (METE XII). Lift Report 2016;
% 42(3):24–33.

% This paper introduces the methodology by which any floor could
% simultaneously be an entrance floor and an occupant floor

% Al-Sharif L. The Universal Origin–Destination-Matrix with Dual
% Designation Floors as Entrances and Occupant Floors. Lift Report 2018;
% 44(2):36–45.
```

2. extract the total number of floors and traffic mix from the variable Traffic

```
N = length(U);

% The variable Traffic is a one dimensional array, of size 1x4% the four
variables inside Traffic represent the ratio of incoming,
% outgoing, interfloor and interentrance respectively.
% they should all be less than or equal to 1% the sum of the four values must
add up to 1%
% the first element is the incoming traffic
% extract it and assign it to the variable i_c
i_c = Traffic(1);
% the second element is the outgoing traffic
% extract it and assign it to the variable o_g
o_g = Traffic(2);
% the third element variable is the interfloor traffic
% extract it and assign it to the variable i_f
i_f = Traffic(3);
% the fourth element in the Traffic Array element variable is the interfloor
traffic
% extract it and assign it to the variable i_f
i_e = Traffic(4);
Error using Create_PDF (line 49)
Not enough input arguments.
```

3. Generate the PDF

the next step is to prepare the probability density function (PDF) square array.

```
% find the total population U which is the sum of the individual floor
% populations
Utotal = sum(U);
% normalise it by dividing each element of the array by the total
% population.
% this will result in an array that has a total sum of 1
Unor = U/Utotal;
```

4. produce the first draft arrays for the four traffic modes

```
% the transpose of an array A is A'
% the tanspose is used here in order to produce a square array
% by multiplying the transpose of a row vector by another row vector of the
% same size.
% A square array is produced.

i_c_array = (EB'*Unor);
o_g_array = (Unor'*EB);
i_f_array = (Unor'*Unor);
i_e_array = (EB'*EB);
```

5. process them to make them ready for the final PDF

```
% zero the diagonals
% this assumes rational passenger behaviour whereby a passenger cannot
% travel from a floor back to the same floor.

for i = 1:N
  i_c_array(i,i) = 0;
  o_g_array(i,i) = 0;
  i_f_array(i,i) = 0;
  i_e_array(i,i) = 0;
end
% re-adjust to compensate for the loss of the value of the diagonal elements

% find the adjusting factors for all four arrays (which are the current sum
% of all the elements in the array (expected to be smaller than 1 due to
% teh fact that the diagonal was zeroed.

Mi_c = sum(sum(i_c_array)); % sum up all the elements of ic
Mo_g = sum(sum(o_g_array)); % sum up all the elements in og
Mi_f = sum(sum(i_f_array)); % sum up all the elements in if
Mi_e = sum(sum(i_e_array)); % sum up all the elements in ie

% then divide the array by the adjusting factor in order to restore the sum
% of its elements to 1
if Mi_c ~ = 0
i_c_array = i_c_array/Mi_c;
end

if Mo_g ~ = 0
o_g_array = o_g_array/Mo_g;
end

if Mi_f ~ = 0
i_f_array = i_f_array/Mi_f;
end

if Mi_e ~ = 0
i_e_array = i_e_array/Mi_e;
end

% Each of the four arrays now must sum up to 1.
% this can be checked by using the instruction sum(sum(A))
```

6. find the value of the PDF as the weighted sum of all the four traffic mode arrays

```
% now find the value fo the final PDF by doing a weight sum of the four
% arrays, multiplying each by its strength from the traffic array.

PDF = i_c*i_c_array+ o_g*o_g_array + i_f*i_f_array + i_e*i_e_array;

% the sum of all the element of the final PDF should also sum up to 1.
end
```

PDF2CDF.m.


```
function [CDFfromPDF] = PDF2CDF(PDF).
%This function converts the PDF to a CDF
% Author: Lutfi Al-Sharif
% Date: 27th April 2020
```

1. Extract the number of floors from the dimensions of the input argument, PDF and set to N

```
N=size(PDF,1); % as this is a square matrix, it does not matter if we take the
number of rows or the number of columns.

Error using PDF2CDF (line 7)
Not enough input arguments.
```

2. Convert the PDF to a CDF

the pdf is the probability density function. The cdf is the cumulative distribution function by integrating the PDF we can obtain the CDF.

```
CDFtemp = 0; % initialise a temporary variable that is hold a temporary value
during the conversion process.
CDF(N, N) = 0; % initialise all the values in the CDF to zero

for i = 1:N % run through the array one row at a time, so finish a complete row
then move to the next row and complete it.
% (i.e., rather than one column at a time).
for j = 1:N % move through column entres in the row being processed
    CDFtemp = CDFtemp+PDF(i,j); % keep track of the running sum in CDFtemp
    CDF(i,j) = CDFtemp;
end
end
% note that this CDF will always end with a 1, but might not start with a
% zero.
% it is important to be aware of this when carrying out random sampling on
% this CDF matrix.

CDFfromPDF=CDF;
end
```

GenPass.m.


```
function [ POD ] = GenPass(P,CDF)
```

1. Generate Passenger origin–destination pairs for P passengers

Author: Lutfi Al-Sharif Date: 27th April 2020

```
% this function will generate a number of passenger origin–destination
% pairs.
% The number of passengers is passed as a parameter P
% and the CDF (cumulative distribution function) is also passed to the
% function.
% it applies the principle of random sampling in order to find the
% origin–destination pair for a certain passenger.
% A random number that is uniformly distributed between 0 and 1 is
% generated using the function rand()
% this random number is then checked to see where it falls within the CDF
% array. Its position within the array decides the origin (the row) and
% the destination (the column) in the array.

% the output set of origin–destination pairs is compiled in the array
% called POD. it has dimensions of 2 x P.
% the first row contains the origins of the P passengers.
% the second row contains the correspinding destinations for these
% passengers.
```

2. Set up a Counter and apply it to the CDF to find the origin–destination pair

```
% the results will be placed in the POD with is a 2 by P array that
% contains all the passenger origins (in the first row) and their
% destinations (in the second row).
POD = 0; % zero the POD array.

% first set up a counter for the P passengers, using k as an index
for k = 1:P
% the ODfound is a Boolean variable that indicates that the value of
% the random variables has been matched to its position inside the CDF
% matrix.
% initialise it to false so that it can be set to true one found.
  ODfound = false; % the origin destination for this passenger has not been
found yet.
  tempCDF = 0; % tempCDF will hold that las indexed value of the elements in
the CDF.
  temp = rand();
%now run through the rows of the CDF one row at a time
for i = 1:length(CDF)
%now run through the elements of the row one column at a time
for j = 1:length(CDF)
% check if the random number is between two consecutive elements
% of the CDF array.
if ((temp>tempCDF) && (temp<=CDF(i,j)) && (ODfound == false))
          ODfound = true; % if found, then set temporary flag to indicate this
          POD(1,k) = i; % the row index is the origin for this kth passenger.
          POD(2,k) = j; % the column index is the destination for this kth
passenger
% the found origin and destination are placed in the kth column
% of the POD array. The origin in the first row and the
% destination in the second row.
```

```
end
end
end
end.

Error using GenPass (line 33)
Not enough input arguments.

end
```

Author details

Lutfi Al-Sharif
Mechatronics Engineering Department, The University of Jordan, Amman, Jordan

*Address all correspondence to: lutfi.alsharif@outlook.com

References

[1] Al-Sharif L, Al-Adem MD. The current practice of lift traffic design using calculation and simulation. Building Services Engineering Research and Technology. 2014;**35**(4):438-445

[2] Basagoiti R, Beamurgia M, Peters R, Kaczmarczyk S. Origin destination matrix estimation and prediction in vertical transportation. In: 2nd Symposium on Lift and Escalator Technologies. Northampton, U.K.: University of Northampton; 2012

[3] Basagoiti R, Beamurgia M, Peters R, Kaczmarczyk S. Passenger flow pattern learning based on trip counting in lift systems combined with on-line information. In: 3rd Symposium on Lift and Escalator Technologies. Northampton, U.K.: University of Northampton; 2013

[4] Kuusinen JM, Sorsa J, Siikonen ML. The elevator trip origin-destination matrix estimation problem. Transportation Science, articles in advance, INFORMS. 2014;**49**(3):433-719

[5] Kuusinen JM, Malapert A. The effect of randomisation on constraint based estimation of elevator trip origin-destination matrices. In: 4th Symposium on Lift and Escalator Technologies. Northampton, United Kingdom: University of Northampton; 2014

[6] Siikonen ML. Procedure for Controlling an Elevator Group Where Virtual Passenger Traffic Is Generated. US Patent Number 6 345 697 B1; 2002

[7] Kameli N. Predictor Elevator for Traffic During Peak Conditions. US Patent Number 5 276 295; 1994

[8] Al-Sharif L. Building the origin-destination matrix under general traffic conditions and using it to generate passenger origin-destination pairs (METE XII). Lift Report. 2016;**42**(3):24-33

[9] Al-Sharif L, Abu Alqumsan AM. An integrated framework for elevator traffic design under general traffic conditions using origin destination matrices, virtual interval and the Monte Carlo simulation method. Building Services Engineering Research and Technology. 2015;**36**(6):728-750

[10] Barney GC, Al-Sharif L. Elevator Traffic Handbook: Theory and Practice. 2nd ed. Routledge: Taylor & Francis, Abingdon-on-Thames, United Kingdom; 2016

[11] CIBSE. CIBSE Guide D: Transportation systems in buildings. Balham, London, United Kingdom: Chartered Institute of Building Services Engineers. 4th Edition 2010. pp. 2-4, Section 4.3.2

[12] British Council for Offices. BCO Guide to Specification 2009. London, United Kingdom: British Council for Offices; 2009

[13] Peters R. Lift traffic analysis: Formulae for the general case. Building Services Engineering Research and Technology. 1990;**11**(2):65-67

[14] Hakonen H, Lahdelma R. Calculation of Elevator Round-Trip Time for the Collective Control Algorithm in General Traffic Situations. Turku Centre for Computer Science, TUCS Technical Report No 671; 2005

[15] Al-Sharif L, Abu Alqumsan AM. Stepwise derivation and verification of a universal elevator round trip time formula for general traffic conditions. Building Services Engineering Research and Technology. 2015;**36**(3):311-330. DOI: 10.1177/0143624414542111

[16] Al-Sharif L. Calculating the elevator round trip time for the Most basic of cases (METE II). Lift Report 2014. 2014;**40**(5)

[17] Peters Research Ltd., Elevate. Available from: https://www.peters-research.com/index.php version 8.27

Section 3

Smart Energy in Smart Cities

Chapter 7

Data Compression Strategies for Use in Advanced Metering Infrastructure Networks

Chih-Wei Hsu and Sun-Yuan Hsieh

Abstract

Internet of Things technology has advanced rapidly. For example, numerous sensors can be deployed in a city to collect a variety of data, and such data can be used to monitor the city's situation. A possible application of such data is smart metering implemented by power suppliers for their consumers; smart metering involves installing a multiplicity of smart meters that, in conjunction with data centers, form a smart grid. Because a smart gird must collect and send data automatically, the establishment of advanced metering infrastructure (AMI) constitutes the primary step to establishing a smart grid. However, problems remain in smart metering: data traffic from smart meters flows rapidly at a huge volume, resulting in bandwidth bottlenecks. Thus, this chapter proposes some data compression technologies as well as a novel scheme for reducing the communication data load in AMI architectures.

Keywords: smart grid, smart meter, concentrator, compression, advanced metering infrastructure, data traffic

1. Introduction

The Internet of Things (IoT) connects a multiplicity of devices to make life convenient. Smart grids constitute one implementation of IoT, and many countries have launched smart grids to develop integrated energy supply systems. A smart grid incorporates automation, bidirectional communication, and advanced sensor measurement systems to streamline interactions between the client and power supplier [1–4]. In contrast to traditional grids, smart grids enable power suppliers to distribute power efficiently and to control the use of power during a given period [5]. Furthermore, the data collected from a smart grid enable automatic billing.

The analysis of energy consumption data has many advantages for both consumers and suppliers: Consumers can track their energy usage, particularly as it varies with the seasons. Power suppliers can monitor how power is utilized across the distribution grid, which can help them formulate power management and energy-saving measures [6–8]. The establishment of advanced metering infrastructure (AMI) constitutes the first step to constructing such an intelligent power gird.

AMI is central to a smart grid system and enables the system to automatically monitor usage. **Figure 1** illustrates the architecture of a typical AMI, indicating that it comprises several concentrators and smart meters that are connected to a meter data management system (MDMS) [9]. The smart meters send data to the

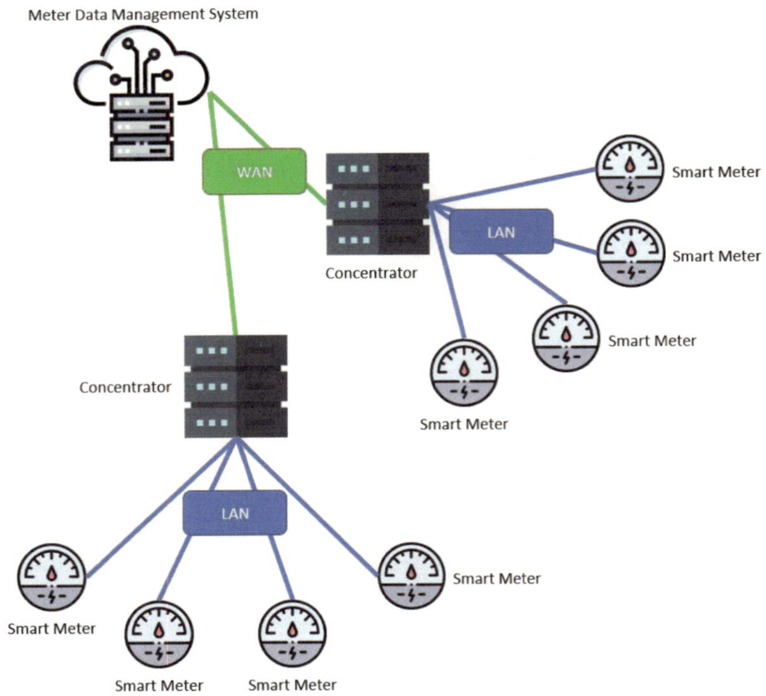

Figure 1.
AMI structure.

concentrators, and the concentrators send the data to the MDMS. The AMI architecture requires high-quality and high-speed networks for providing stable service and efficient monitoring, and it comprises two such networks, namely a local area network (LAN) and wide area network (WAN). The LAN connects the concentrators and smart meters and leverages cutting-edge communication technologies such as power-line communication and the Zigbee specification. The WAN connects the concentrators and MDMS and constitutes the most important part of the AMI system; this network supports a host of high-speed technologies such as broadband-over-power-line technology, 3G, and long-term evolution [10].

Nonetheless, smart meters have a very large data frequency—with a small packet being generated every 15–60 min—that will exceed the capacity of existing communication technologies [11, 12]; this makes data difficult to transmit. Furthermore, smart grids face challenges in storing these large volumes of data. To solve these two problems, data size in both communication and storage should be reduced. This chapter introduces some novel approaches to doing so.

2. Preliminaries

Smart grids have expanded considerably, and many organizations have noticed the advantages engendered by smart meters. Thus, research on smart grids has similarly expanded. Specifically, smart grids enable the automatic distribution of power on the part of suppliers and provide usage data, which can be used to develop various applications.

Smart meters constantly upload data to the MDMS. Such data pertain to, for example, interval data readings, meter remote disconnections, meter remote reconnections, and meter firmware patches. The volume of such data is considerably large, potentially causing network collapse. The reduction of packet size in a smart grid addresses this problem, and such a reduction has become a major topic of research. Thus, this section introduces some basic strategies that are currently adopted to solve this problem.

2.1 Basic method for data reduction in an AMI system

Among existing methods, the conventional approach is combining the messages sent by multiple meters to reduce either protocol overhead or the frequency of transmission. In this approach of message concatenation, energy consumption data, control signal messages, and firmware update packages are combined. This scheme entails a lower transmission volume because the packet header is reduced, where multiple messages can be sent in a header. However, one issue in this scheme is where messages ought to be concatenated.

2.1.1 Meter side

Message concatenation at the meter side results in a greater reduction in data size than does that at the concentrator side. Data size reduction is greater at the meter side primarily because a meter generates data within a certain period; thus, a longer interval between instances of meter-to-concentrator data transmission yields disproportionately large savings in data size. However, most meters contain low-performance hardware, hampering this method. Consequently, meters are either overloaded or unable to compress the data.

2.1.2 Concentrator side

Messages can also be concatenated at the concentrator side prior to transmission. This is the appropriate site for message concatenation [13, 14] because concentrators have more powerful hardware and process data more efficiently than meters do. In concentrator-side concatenation, meters send packages to the concentrator, the concentrator subsequently aggregates the data into one package, and the compressed message is finally sent to the MDMS. Concentrator-side concatenation not only reduces the volume of messages but also stabilizes the system.

However, despite the advantages of concentrator-side concatenation, existing devices on the market are incapable of supporting such concatenation. Currently, concentrators on the market are capable of executing only simple integrations, and these products all follow the PRIME standard of WAN communication [15].

3. Novel approaches

A message generated by a smart meter contains a 40–60-B header within the packet. Concentrators can collect messages sent from the meter side according to a Poisson process [16] before combining these messages to minimize possible traffic. The aforementioned procedure comprises two steps, namely combination and compaction, which are detailed as follows [17].

3.1 Combination

Combination entails combining messages into a larger packet to solve the problem of protocol overhead during communication. **Figure 2** illustrates how such combination reduces traffic. Combining meter messages on smart grids is efficient because each piece of data generated by a meter have a size of only a few bytes.

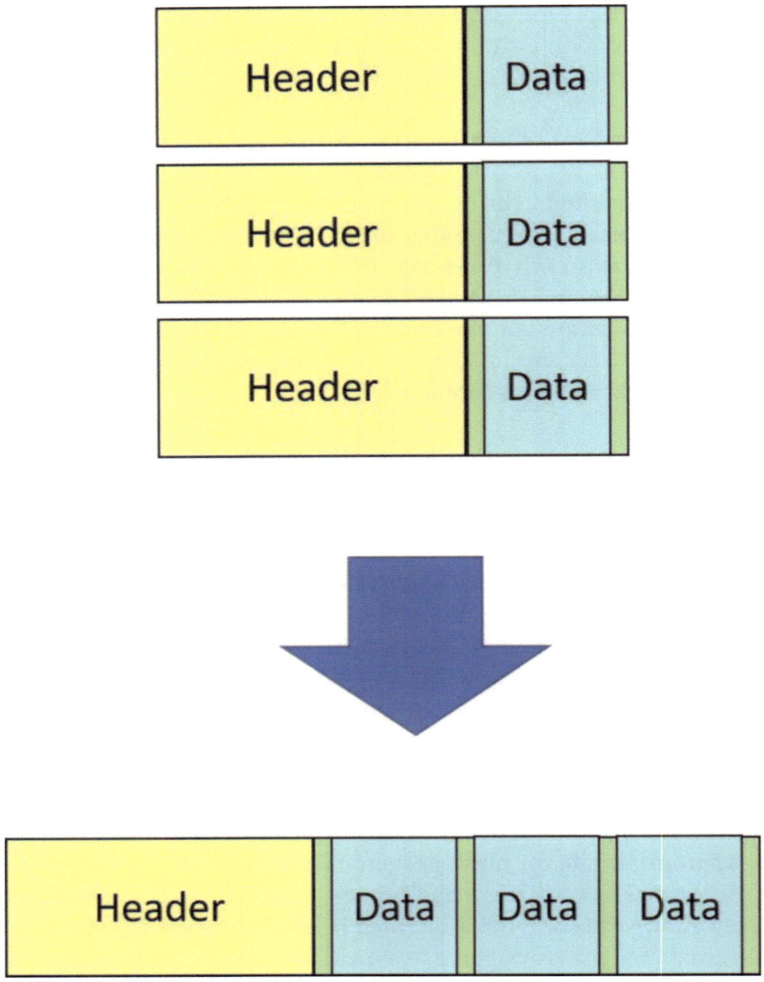

Figure 2.
Concept underlying the combination procedure.

Concentrators combine messages that have been sent by meters in a specified period, after which they compact the messages.

Figure 3 illustrates the operating procedures of a typical combination algorithm. Specifically, the concentrators receive meter messages, place the messages into a combination queue, combine the messages in the queue once some predetermined quantity of messages have been accumulated, and finally compact the messages into a package [17]. This method enables solving the protocol overhead problem. The data compaction scheme is detailed as follows.

3.2 Compaction

Compaction reduces the size of headers in large-scale AMI architecture, which is very helpful for reducing data volume. Messages can be compacted through either full compaction (FC) or loose compaction (LC) [17].

3.2.1 Full compaction

Figure 4 illustrates the operating procedures of an FC algorithm. Specifically, any piece of data received from the meter is stored in Q_c, and the usage in the

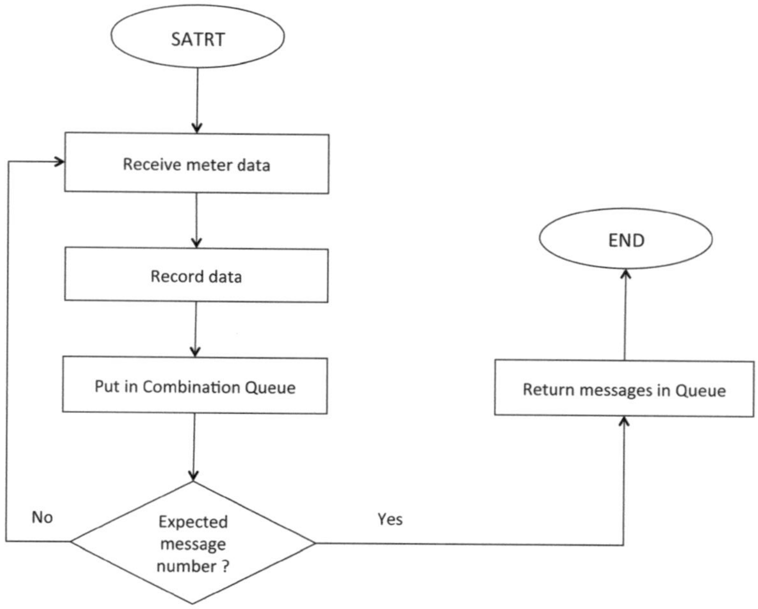

Figure 3.
Combination procedure.

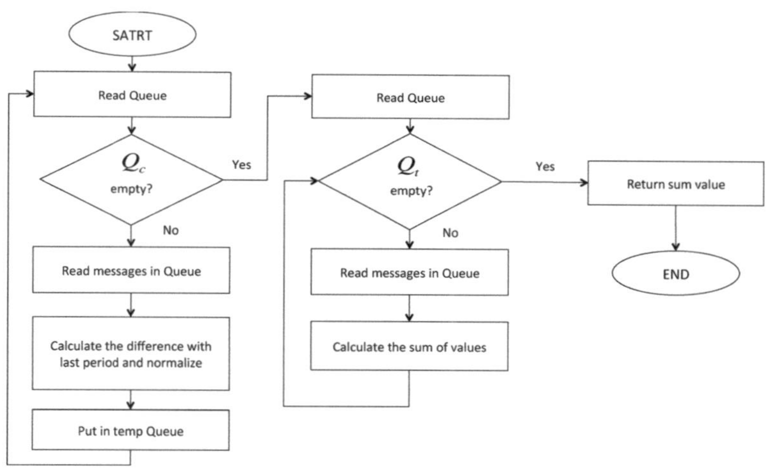

Figure 4.
FC procedure.

current period is calculated by determining the difference between the message in the current period and the message in the last period; this process is performed until Q_c is empty. Subsequently, before the usage data are stored in Q_t, the calculated value should be normalized; this can specifically be achieved by multiplying the value by 10^3 to convert any floating-point number into an integer. Once all messages are in Q_t, the concentrators can start calculating the sum of all values in Q_t. Once the FC procedure is complete, the concentrators obtain the total usage from the meter data, which helps power suppliers monitor the usage of an area to optimize power delivery.

To illustrate the FC procedure, **Table 1** lists some energy usage values in an area. First, all values are input into Q_c. Second, the difference between the previous usage and current usage is calculated for each period. Third, the values obtained from these calculations (212.880, 323.769, 388.112, 46.115, and 610.762 in this example)

Meter Number	Previous Usage	Current Usage	Difference
1	2153.232	2366.112	212.880
2	5698.564	6022.333	323.769
3	23154.003	23542.115	388.112
4	1542.121	1588.236	46.115
5	56213.225	56823.987	610.762

An Example of meter data.

Table 1.
Example meter data from an area.

are input into Q_t. Finally, these values in Q_t are summed to indicate the net difference in energy usage (1581.638 in this example).

3.2.2 Loose compaction

LC differs from FC in that LC is recoverable, whereas FC is not. **Figure 5** illustrates operating procedures of an LC algorithm. First, the concentrators obtain meter data by dequeuing Q_c and subtracting the meter value for the current period from the meter value for the previous period. Second, similar to second step of the FC algorithm, the value obtained in the first step is multiplied by 10^3 for normalization and subsequently input into Q_t. Third, LC sorts values in Q_t in descending order. Fourth, the positive difference between a value and its lower-valued neighbor is computed and then input into Q_l; the initial, lowest value in Q_t thus remains unchanged. For example, if $[a, b, c, d]$ is input into Q_t, then $[a, (b-a), (c-b), (d-c)]$ is input into Q_l. Finally, the values returned by the LC algorithm constitute the compacted data.

To illustrate the LC procedure, consider the values in **Table 1**. First, the LC algorithm takes the difference between past and present usages and inputs these differences into Q_c. In this example, these differences can be listed as follows: [212.880, 323.769, 388.112, 46.115, 610.762]. This list is then sorted in descending order as [46.112, 212.880, 323.769, 388.112, 610.762] and input into Q_t. The LC algorithm then dequeues each piece of data from Q_t and calculates the positive difference between a value and its lower-valued neighbor. The list of such

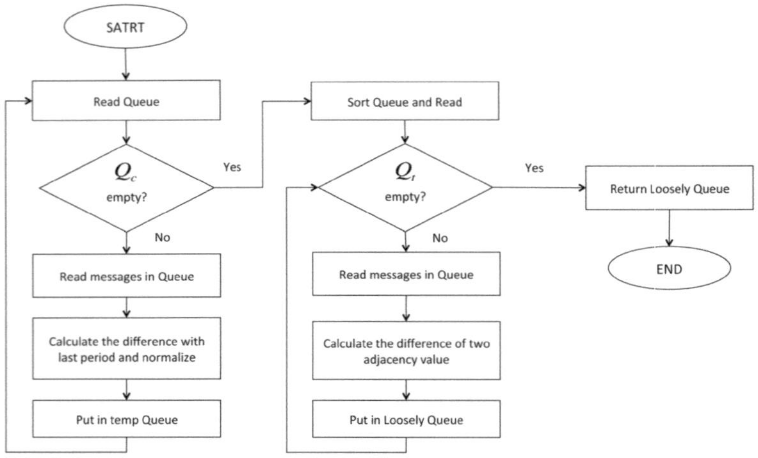

Figure 5.
LC flow.

differences is as follows: [46.112, 166.768, 110.889, 64.343, 222.650]. This list is input into Q_l and returned.

After compacting the data, the concentrators send the data to the MDMS. When the MDMS receives the data, it executes the LC recovery (LCR) algorithm to extract the raw data. **Figure 6** illustrates the LCR algorithm, which is the reverse of the LC algorithm. First, the LCR algorithm inputs the received data into Q_l; the algorithm then recovers those values prior to subtraction from a lower-valued neighbor. These recovered values are then input into Q_t. Second, the LCR algorithm denormalizes each value in Q_t and adds to it the value of the usage in the previous period. Finally, the MDMS outputs all values into Q_c and returns them.

Consider the values for the example presented in **Table 1**. First, the LCR algorithm recovers the compacted data; specifically, the MDMS inputs [46.112, 166.768, 110.889, 64.343, 222.650] into Q_l. Subsequently, the LCR algorithm recovers the

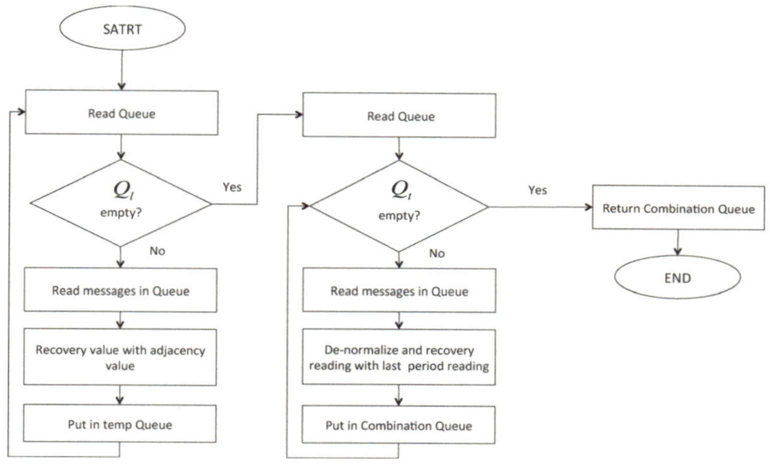

Figure 6.
LC recovery procedure.

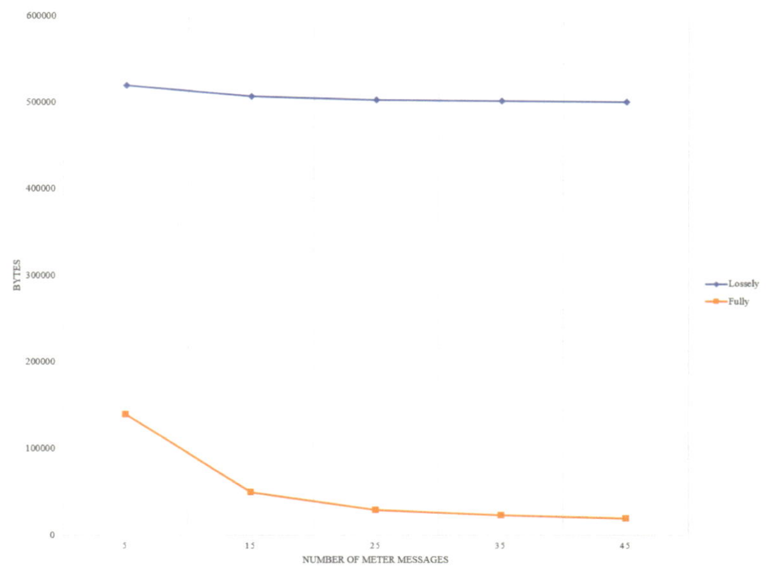

Figure 7.
FC and LC analysis.

values prior to subtraction from a lower-valued neighbor; specifically, the MDMS obtains [46.112, 212.880, 323.769, 388.112, 610.762] and inputs this list into Q_t. Finally, the LCR algorithm denormalizes the data by multiplying each value by 10^{-3}; specifically, the MDMS obtains the list of raw data [46.112, 212.880, 323.769, 388.112, 610.762].

3.2.3 Result analysis

First of all, all of messages sent from meters were encoded in ASCII. FC and LC compaction performance are provided in this section. The **Figure** 7 shows the compression ratio of the FC and LC method. The vertical scale label present the total bytes used before compaction and the horizontal scale label is the number of meter messages compacted when compaction. As the **Figure** 7 shows, FC has better performance than LC. However, FC does not support recovery, so it can achieve the highly compaction ratio. On the contrary, LC is the trade-off algorithm when recovery function is required and have demand of compression. If the system only need the sum of the energy usage, then FC would be the proper way for compressing the data.

4. Compression technologies

The choice of compression algorithm has become a critical issue because most IoT devices have hardware limitations, particularly in their low-power-consumption microprocessors. The best compression algorithm provides the best compression ratio given the hardware specifications of an IoT device [18, 19]. This section discusses two well-known approaches to data compression [20].

4.1 Lempel–Ziv–Markov chain algorithm

The Lempel–Ziv–Markov chain algorithm (LZMA) is a member of the LZ-family, and it is based on the LZ77 algorithm, which uses a dictionary-based scheme. The LZMA yields excellent compression ratios without being demanding on hardware, making it suited to IoT environments by virtue of its exclusive dictionary structure [19].

4.2 Prediction by partial matching

Prediction by partial matching (PPM) can predict a subsequent pattern using present or previous symbols. Moreover, PPM can be used with a Markov model to construct a compression algorithm. For example, the RAR algorithm, developed by Eugene Roshal and Alexander Roshal, uses PPM and the Lempel–Ziv–Storer–Szymanski algorithm to achieve impressive compression [20].

5. Conclusions

Smart grids will be key infrastructure considering the rapid developments in IoT technology. This chapter presents data compression techniques, such as combination and compaction, developed for reducing the communication data load in a smart grid. These techniques not only reduce the frequency between instances of transmission but also considerably reduce data volume. Moreover, FC and LC can be used in different situation. FC provides extremely compact ratio for the user who have less bandwidth for transport the meter data. LC can be used when the system requires the raw data of the energy usage and having sufficient network bandwidth. In addition, some well known data compression techniques are also introduce in the chapter. Proposed algorithm can be implement with the compression technique provided above to decrease the volume of the meter data. Also, this chapter explains recent advances in smart grid technology. Readers can build on the aforementioned algorithms to formulate novel contributions of their own.

Author details

Chih-Wei Hsu and Sun-Yuan Hsieh*
National Cheng Kung University, Taiwan

*Address all correspondence to: hsiehsy@mail.ncku.edu.tw

References

[1] Xi Fang, Satyajayant Misra, Guoliang Xue, and Dejun Yangand, "Smart grid-the new and improved power grid: a survey," IEEE Communications Surveys & Tutorials, vol. 14, no. 4, pp.944–980, October 2012.

[2] Nanpeng Yu, Sunil Shah, Raymond Johnson, Robert Sherick, Mingguo Hong and Kenneth Loparo, "Big data analytics in power distribution systems, " 2015 IEEE Power Energy Society Innovative Smart Grid Technologies Conference (ISGT), June 2015.

[3] Ruofei Ma, Hsiao-Hwa Chen, Yu-Ren Huang, and Weixiao Meng, "Smart grid communication: its challenges and opportunities," IEEE Transactions on Smart Grid, vol. 4, no. 1, pp. 36–46, March 2013.

[4] D. Alahakoon, X. Yu, "Smart Electricity Meter Data Intelligence for Future Energy Systems: A Survey, "IEEE Transactions on Industrial Informations, vol. 12, no. 1, February 2016.

[5] Melike Erol-Kantarci, and Hussein T. Mouftah, "Energy-efficient information and communication infrastructures in the smart grid: a survey on interactions and open issues, "IEEE Communications Surveys & Tutorials, vol. 17, no. 1, pp. 179–197, March 2015.

[6] Z. Fan, P. Kulkarni, S. Gormus, C. Efthymiou, G. Kalogridis, M. Sooriyabandara, Z. Zhu, S. Lambotharan, and W. H. Chin, "Smart grid communications: overview of research challenges, solutions, and standardization activities, "IEEE Communications Surveys & Tutorials, vol. 15, no. 1, pp. 21–38, February 2013.

[7] M. Ghofrani, M. Hassanzadeh, M. Etezadi-Amoli, and M. Fadali, "Smart meter based short-term load forecasting for residential customeers, "In proceedings of North American Power Symposium (NAPS), pp. 1–5, August 2011.

[8] S. Wang, L. Cui, J. Que, D. Choi, X. Jiang, S. Cheng, and L. Xie, "A randomized response model for privacy preserving smart metering," IEEE Transactions on Smart Grid, voll.3, no. 3, pp. 1317–1324, September 2012.

[9] V. Cagri Gungor, Dilan Sahin, Taskin Kocak, Salih Ergut, Concettina Buccella, Carlo Cecati, and Gerhand P. Hancke, "Smart grid technologies: communication technologies and standards," IEEE Transactions on Industrial Informatics, vol. 7, no. 4, pp. 529–542, November 2011.

[10] Ye Yan, Yi Qian, Hamid Sharif, and David Tipper, "A survey on smart grid communication infrastructures: motivations, requirements, and challenges," IEEE Communications Surveys & Tutorials, vol. 15, no. 1, pp. 5–20, February 2013.

[11] David Bernaudo, Larry Barto, Shannon Spizzirri, and Terry Penn, "SmartGrid/AEIC AMI interoperability standard guidelines for ANSI C12.19 / IEEE 1377 / MC 12.19 end device communications and supporting enterprise devices, networks and related accessories," The Association of Edison Illuminating Companies, Meter and Service Technical Committee report, ver. 2, May 2010.

[12] E. E. Queen, "Adiscussion of smart meters and RF exposure issues," Edison Electric Institute, Washington, C.C, A Joint Project of the EEI and AEIC MEter Committee, March 2011.

[13] B. Karimi, V. Namboodiri, and M. Jadliwala, "On the scalable collection of metering data in smart grid through message concatenation," In preceedings of IEEE International Conference on

Smart Grid Communications, pp. 318–323, October 2013.

[14] B. Karimi, V. Namboodiri, and M. Jadliwala, "Scalable meter data collection in smart grids through message concatenation," IEEE Transactions on Smart Grid Communications, vol. 6, no. 4, pp. 1697–1706, July 2015.

[15] PRIME Alliance, "PRIME: powerline intelligent metering evolution," Available from: http://www.prime-alliance.org [Accessed: 2016-5-2].

[16] H. Li, Z. Han, L. Lai, R. Qiu, and D. Yang, "Efficient and reliable multiple access for advanced metering in future smart grid," In proceedings of IEEE International Conference on Smart Grid Communications, pp. 440–444, October 2011.

[17] Sun-Yuan Hsieh, Dr. Hemant Kumar Gianey, Chih-Wei Hsu, Cheng-Han Yeh, and Geng-Hua Zhang, "Novel Scheme for Reducing Communication Data Traffic in Advanced Metering Infrastructure Networks," Submission Committing.

[18] K. Barr and K. Asanovic, "Energy Aware Lossless Data Compression," in Proceedings of 1st International Conference on Mobile systems, Applications, and Services (MobiSys), pp. 231–242, 2003.

[19] M. Ringwelski, C. Renner, A. Reinhardt, A. Weigel, V. Turau,"The Hitchhiker's Guide to Choosing the Compression Algorithm for Your Smart Meter Data," 2nd IEEE ENERGYCON conference & Exhibition, 2012.

[20] Sun-Yuan Hsieh, and Geng-Hua Zhang, "A Strategy for Real-Time Energy Data Compression to Reduce Data Traffic based on Smart Grid AMI Networks," Submission Committing.

Chapter 8

Energy Management and Optimal Power Scheduling in a Smart Building under Uncertainty

Dimitrios Thomas and Evangelos Kotsakis

Abstract

In this Chapter, we consider a microgrid with a certain number of distributed energy resources (DER) components connected to an office building (in a university campus) provided with electricity by a utility company. We develop the initial version of the energy management system which is responsible for the optimal energy scheduling of the microgrid's distributed energy resources. These resources include a photovoltaic (PV) installation, a Storage Energy System (ESS), a small Combined Heat and Power (CHP) unit, and a fleet of electric vehicles (EVs) used for work-related trips. The mobility behavior of the EVs fleet is modeled considering deterministic realizations of the probabilistic distributions used for the arrival/departure, and the time EVs remain parked. To investigate the impact of renewable generation and load unpredictability on the energy management system (EMS) operation, PV production and electric load are modeled under uncertainty using actual smart meters data for the scenarios formulation. We also assume that each DER component, through an EMS, can communicate and control the power exchange from and towards this component and that, two way communication with the utility company can be reached through aggregators using advanced metering equipment. We also consider a simplified thermal model that provides a specific level of thermal comfort to the building's occupants, by meeting the predicted heating load. The energy produced by the DERs can be sold back to the grid by the microgrid manager and/or it can be stored for future utilization.

Keywords: energy management system, smart grid, electric vehicles, distributed energy resources, optimization

1. Introduction

Buildings have become the major energy consumers over the world as they consume around 40% of total end-use energy [1]. In Europe, the Directive on Energy Performance of Buildings establishes a "nearly Net Zero Energy buildings" (NZEBs) as the aim for all new buildings from 2020 [2]. In recent literature, more and more studies consider nZEBs as part of a smart grid or a micro-grid (MG) and identify trends on energy management techniques and technological solutions for electric power system management. The main advantages of nZEBs have been identified to be the integration of renewable energy sources; the integration of energy storage mechanisms such as plug-in electric vehicles and the implementation of zero-energy concepts such as net zero source energy, net zero energy costs and net zero emissions.

The renewable energy exploitation is one of the most important aspects of NZEBs. Renewable Energy Sources (RES) are those sources of energy that can be derived from natural processes and thus can be replenished continuously such as solar energy, wind energy, biomass, hydropower etc. The wind and solar energies are mostly used in green buildings modeling and design [3] but they come with a number of issues that have to be taken into consideration. The wind energy systems may not be technically feasible at all sites due to the low wind speeds and/or to high unpredictability with respect to solar energy. In addition, the availability of a specific resource depends each time on the corresponding season and may also vary during the day [4]. NZEBs, either as standalone or as parts of a Net Zero Energy District, could help improving the energy performance of an electrical grid by shifting loads and reducing peak demands. Buildings, as one of the most important contributors involved in a smart grid, can deliver useful information such as energy behaviors, power demand and the corresponding load shifting potentials for grid control and optimization [5].

A microgrid is an electric system of limited extent, typically the suburban/district level, that includes distributed generation (i.e., solar, wind, cogeneration, electric vehicles, etc.), consumers and storage facilities, and operates by intelligently managing its own costs and production capacity to ensure a level of quality service. It is connected to the global grid but is designed to operate independently if necessary (islanded mode). Microgrid can be understood as a case of a more general concept called 'Smart grid', collecting a set of technological solutions for electric power system management. Its localized nature allows responding efficiently and accurately the energy needs and ensuring adequate levels of quality, safety, security, reliability, and availability. It is able of being disconnected from the global network for several hours without loss of service while ensuring voltage and frequency stability. In addition, the proximity of the sources of production to the consumption allows reducing energy transmission losses. Thus, the use of such a system (mainly decentralized) has as an aim to gain flexibility and adaptability with respect to the classical centralized power system model.

The development and the extensive utilization of building automation systems, Information and Communication Technologies (ICT) and grid energy management system facilitates the bidirectional communication between buildings and a grid which can be widely established and therefore be used for interacting and optimizing the power supply and the demand. This chapter attempts to address the major issues that are related to the design and optimization of grid-connected nearly and/or net zero energy buildings as parts of a smart grid and on which several scholars/researchers have been working the last years.

In this work, a microgrid with a certain number of DER components connected to an office building (in a university campus) provided with electricity by a utility company is considered. These components include a PV installation, a Storage Energy System (ESS), a small Combined Heat and Power (CHP) unit, and a fleet of electric vehicles (EVs) used for work-related trips. The mobility behavior of the EVs fleet is modeled considering deterministic realizations of the probabilistic distributions used for the arrival/departure and the time EVs remain parked. PV production and electric load are modeled under uncertainty. We use actual data from smart meters to formulate the scenarios. We also assume that each DER element can, through an EMS controller, to communicate and control the power exchange from and towards this component. We also consider that two-way communication with the utility company can be achieved via aggregators using advanced metering infrastructure. The energy generated by the DERs can be sold to the grid by the microgrid building-manager, and/or it can be stored for future utilization. The recommended EMS configuration is shown in **Figure 1**.

2. Methods and scenario construction

2.1 Photovoltaic and electric load scenarios

To classify PV and electric load production, yearly data-measurements from smart meters installed in Walloon region, Belgium, have been used. The smart meters communicate with the utility company server every 15-min providing the updated PV and load measurements. The 15-min datasets were merged to formulate 8760 hourly readings (365 24-hour PV generation and load profiles). The total PV capacity is 50 kVA. The original datasets are shown in **Figure 2**.

We use the scenario reduction technique introduced in [6] to construct the scenarios. A script developed in Matlab based on [6] is utilized to aggregate the two sources of uncertainty into one. That is, a discrete probability has been assigned to each one of the generated scenarios. Every scenario comprises two 24-hour vectors where each vector corresponds to a specific profile (one vector for PV production and one for load demand). Moreover, this scenario construction technique considers the potential correlation within the data. The latter is very important as, for example, a sunny day with increased PV production is expected to affect the load demand downwards and vice-versa. Moreover, one may notice that the PV profiles of **Figure 2a** look asymmetric and seem to have been shifted towards the left side of the time axis. This is due to the minimum cut-in voltage level required from the power electronics of the inverter to start being operational.

It is important that the final number of generated scenarios retain most of the relevant information on the stochastic process contained in the original scenario sets, while significantly reducing its cardinality. A very large number of scenarios

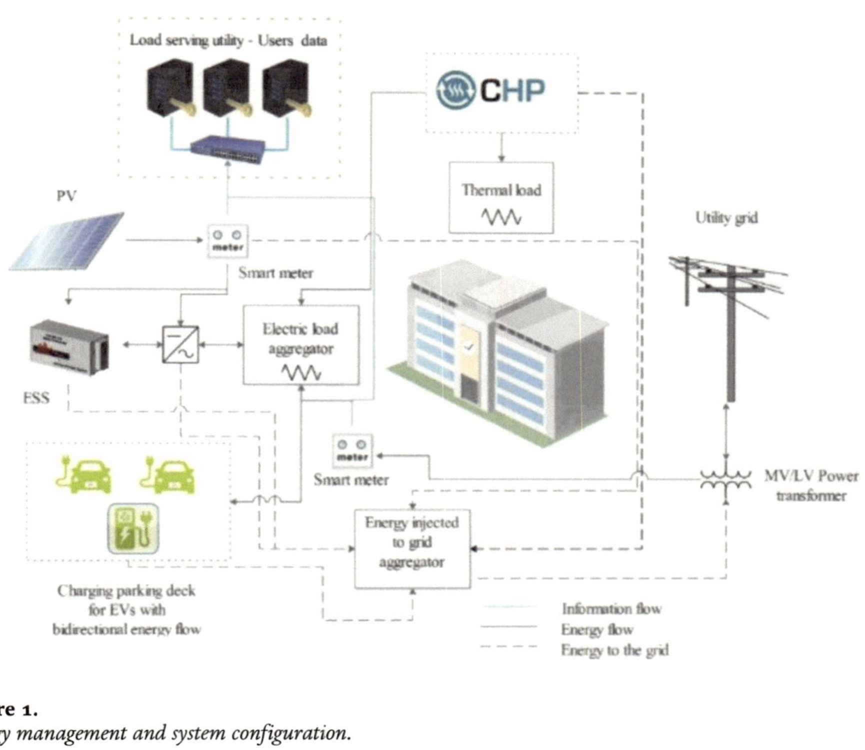

Figure 1.
Energy management and system configuration.

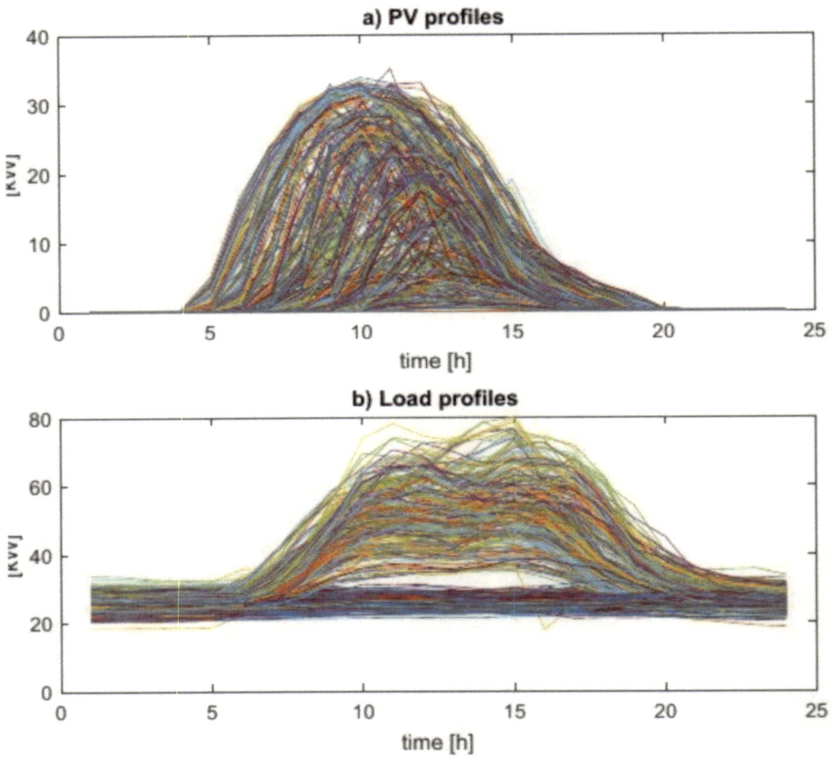

Figure 2.
The 365 original profiles for (a) PV production, and (b) electric load demand.

Numb of scenarios	6 scen.	12 scen.	24 scen.	48 scen.
TESC ($)	26.06	20.07	16.68	15.64
SD ($)	23.22	24.33	24.35	24.45
Elapsed time (s)	0.09	0.11	0.23	0.42

Table 1.
Parameters related with the number of scenarios.

may result in a computationally intractable associated stochastic programming problem which would require both increased time and computational resources to be solved. On the other hand, a small number of scenarios might not be representative of the original data sets. Thus, in order to decide the appropriate number of scenarios we take into consideration the total expected system cost (TESC), its standard deviation (SD), and the total computational time, as shown in **Table 1**. Simulations take place on an Intel Core i7-5500U CPU @ 2.4 GHz with 16 GB memory.

We can see in **Table 1** that the TESC decreases considerably from the 6 to 12 scenarios, and from 12 to 24. On the other hand, the cost reduction from the 24 to 48 scenarios is smaller. The standard deviation of the TESC increases somehow from the 6 to 12 scenarios, but it remains relatively constant in the rest scenario cases. Finally, one may notice that the computational time needed to obtain the optimal solution is increased around 100% in both cases, from the 12 to 24 and from 24 to 48 scenarios. Considering all the information above, the case of 24 scenarios provides a favorable trade-off between a satisfactory scenario representation and a computationally tractable problem. One should also note that the constructed scenarios are not equiprobable, but probability weighted. The 24 scenarios for PV generation and load demand are illustrated in **Figure 3**.

For the deterministic approach, we used the average yearly profiles (obtained from the original datasets in **Figure 2**) for both PV production and the electric load demand. These profiles are illustrated in **Figure 4**.

2.2 EVs driving schedule

When connected to the microgrid, the charging and discharging behaviors of the EVs make them considered as either power supplies (when discharging) or power loads (when charging). Here, the EVs selected for the fleet are used for work-related trips and it is also assumed that the mobility behavior with the EVs remains similar as with conventional vehicles.

In this work, the mobility behavior profiles for a fleet of 30 EVs are generated. In Belgium, 82% of the population has fixed working hours and shifts [7]. Usual working hours are considered from 8 am to 6 pm but they are not binding. The arrival time distribution is fitted in the form of chi-square distribution [8] with its probability density function given by: $f(t_{\text{arr},i}) = \frac{t_{\text{arr},i}^{(\upsilon-2)/2} \backslash e^{-t_{\text{arr},i}/2}}{2^{\upsilon/2}\Gamma(\upsilon/2)}$ where $\Gamma(\cdot)$ is defined as $\Gamma(a) = \int_0^\infty t^{a-1}e^{-1\text{dt}}, a > 0$ with $\upsilon = 4$ degrees of freedom [9], and $t_{\text{arr},i}$ is the arrival time for the i-th EV. The detention time of the EVs connected to the microgrid conforms to the normal distribution with a mean of 8 hours and a variance equal to 4 hours $N(8, 2^2)$ assuming that the working time of most people is 8 hours. The initial state-of-energy of the EVs is decided by applying the uniform distribution with values between 0.3 and 0.8. One should also note that for the numerical

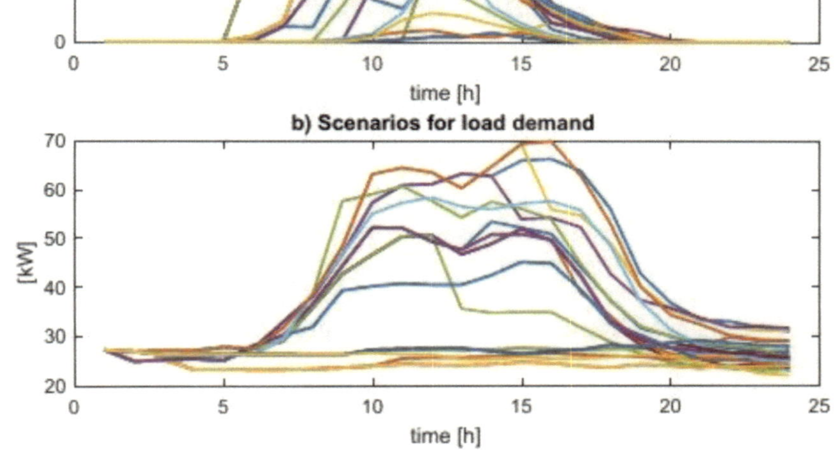

Figure 3.
The 24 representative scenarios for (a) PV production, and (b) electric load demand.

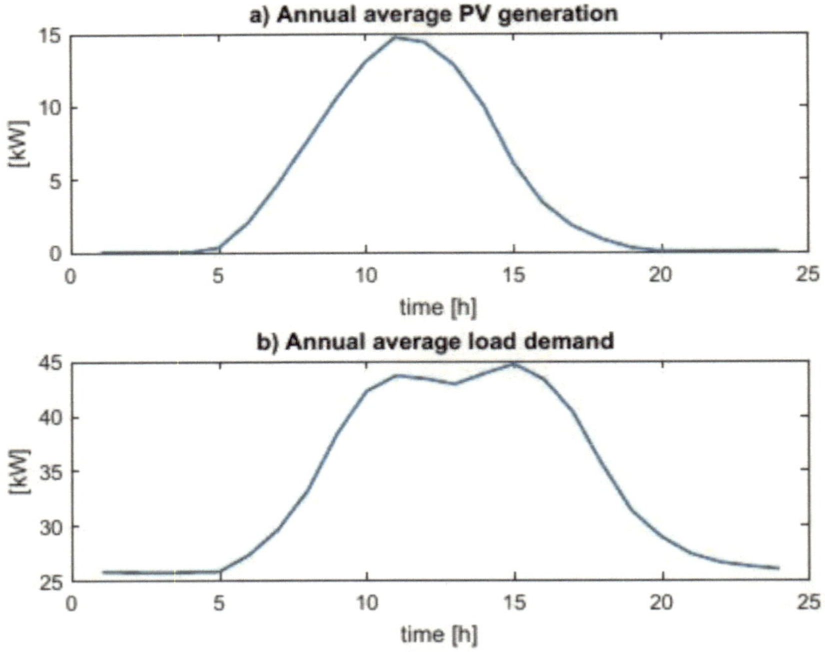

Figure 4.
The average yearly profiles for (a) PV production, and (b) electric load demand.

evaluation in the Results section, we used deterministic realizations of all the uncertain characteristics of the EVs (arrival, stay duration, initial state-of-energy).

2.3 Simplified thermal model

To simulate the thermal performance of a building, engineers developed, among other tools, the thermal network method. Thermal networks have been used to study the internal mass effects [10], appliances, indoor air temperature and heating load [11] for different buildings. In addition, they represent a comprehensible idea about the heat transfer phenomena in buildings with a simple systematic formulation of the problem. In the thermal network method, the whole mass of the system is accumulated in finite number of nodes, which are connected to thermal capacitances. The heat transfer between two nodes occurs through thermal resistances. It has been shown, that the functionality of control systems can be improved by the implementation of the thermal network method and the system identification approach [12].

System identification is an approach to construct mathematical models of dynamic systems by means of measurements of the system's input and output signals. The system identification needs the measured input and output signals from the system, a model structure, and an estimation method to estimate values for the adjustable parameters in the selected model structure. In a dynamic system, the output signal depends on both the instantaneous values of its input signals and on the initial conditions. In fact, a model is a mathematical relationship between a system's input and output variables. Differential or difference equations, transfer functions, and state-space equations are common methods to describe a dynamic system. The RC model method describes the system with ordinary differential equations that can be easily represented with the state space method.

Obtaining a good model of the system depends on how well the measured data reflects the behavior of the system. For this purpose, the measured data must capture the dynamics of the system. It is necessary to measure the right variables with enough accuracy and duration to capture the dynamics of interest. In general, to supply an appropriate dataset, the following inputs that excite the system dynamics are important: data duration to capture the important time constants, a detailed analysis of signal-to-noise ratio, and finally measuring the outputs at appropriate sampling intervals [13].

The use of the RC model method provides the structure of the model, but not the numerical values of its parameters. Afterwards, it is possible to represent the system with a state-space model and estimate the values of its parameters from the data. This approach is known as gray-box modeling. The system identification approach refers to methods and algorithms that estimate the model parameters by minimizing the error function (cost function – the mean square error), as shown below between the model output and the measured data.

$$V(\theta) = \frac{1}{L}\sum_{t=1}^{L} e^T(\theta,t)e(\theta,t), \tag{1}$$

where L is the number of data samples, $e(\theta, t)$is a given error vector at time t and parametrized with θ. Then, parameters are obtainable with minimizing $V(\theta)$ with respect to the parameter vector θ. After the model is estimated, quality metrics represent the quality of identified models.

$$\text{NRMSE} = 100 \times \left(1 - \frac{\sqrt{\left(y_{\text{measured}} - y_{\text{model}}\right)^2}}{\sqrt{\left(y_{\text{measured}} - \bar{y}_{\text{measured}}\right)^2}}\right). \tag{2}$$

The MATLAB® system identification toolbox is used in this work to minimize the cost function of Eq. 2 and to estimate the model parameters. MATLAB uses various minimization algorithms to perform the optimization. In our case, the 'auto' algorithm is used for the search method to minimize the cost function and to estimate model parameters, as it determines the optimized trajectory among different techniques at each iteration.

The simplified thermal model presented in [14] is used in this study to obtain the thermal load for the university building. The building is simulated using TRNSYS software utilizing weather data from the Uccle meteonorm file (Belgium). It has a heavy structured envelope and the buildings material properties are presented in [14]. Here a 4R2C model is proposed and used to simulate the thermal performance of the building. The corresponding proposed thermal network is represented in **Figure 5**.

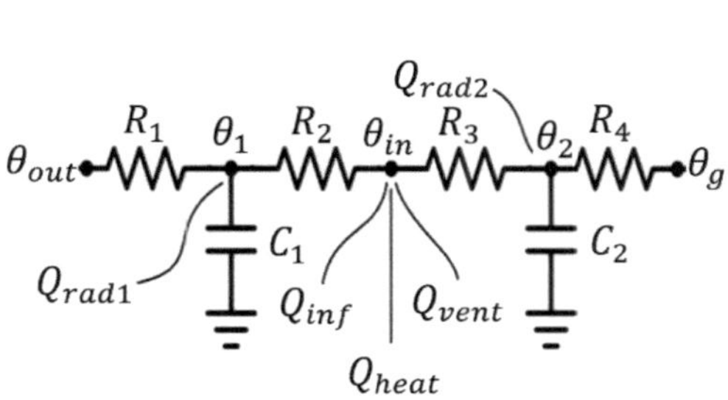

Figure 5.
The proposed thermal network.

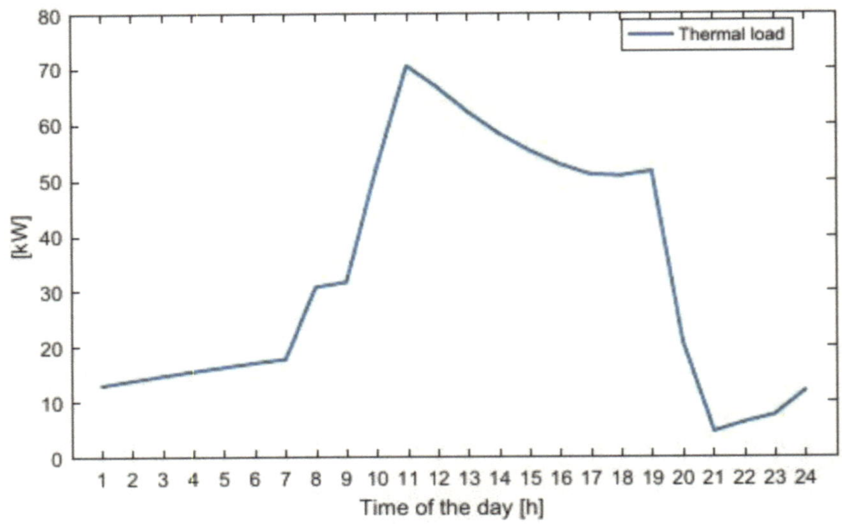

Figure 6.
Daily thermal load prediction.

To determine the parameters in the thermal network, the system identification approach has been used. Data from TRNSYS have been used as the information matrix for the model to be trained. To identify the model's parameters, the Matlab system identification toolbox is utilized. The information matrix contains one-month data. The model identification determines the values of each resistance and capacitance to achieve the highest fitness between the thermal network and the information matrix. Then, the identified model can predict the thermal performance of the building for a predetermined period of days.

To formulate a daily thermal load profile, so as it can be used by the EMS for its 24-hours scheduling horizon, the average heating load of the predicted working days is calculated. The calculated thermal load offers a temperature approximately around 22°C during working hours (from 9 am to 6 pm). The daily thermal load prediction is illustrated in **Figure 6**.

The thermal load is low during the night and the early morning hours and starts increasing around 8 am. This is necessary, so as the targeted thermal comfort level to be achieved in the office building during the working hours. The thermal load is covered by CHP's thermal production.

3. Mathematical formulation

The mathematical formulation of the EMS is presented in this Section. The objective function which minimizes the total expected system cost is given by Eq. (3) below:

$$\text{Minimize}_{\Xi} \sum_{t,\omega} \pi_\omega \left(p_{t,\omega}^{\text{grid,in}} \varepsilon_t^{\text{buy}} + p_{t,\omega}^{\text{CHP,tot}} \varepsilon^{\text{gas}} - p_{t,\omega}^{\text{grid,out}} \varepsilon_t^{\text{sell}} \right) + \sum_{i,t,\omega} \pi_\omega \left(p_{i,t,\omega}^{\text{EV,ch}} + p_{i,t,\omega}^{\text{EV,dis}} \right) C^{\text{EV,deg}}$$
$$+ \sum_{t,\omega} \pi_\omega \left(p_{t,\omega}^{\text{PV,grid}} \lambda_{\text{PV}} + p_{t,\omega}^{\text{ESS,grid}} \lambda_{\text{ESS}} + p_{t,\omega}^{\text{CHP,grid}} \lambda_{\text{CHP}} + \sum_i p_{i,t,\omega}^{\text{EV,grid}} \lambda_{\text{EV}} \right), \tag{3}$$

where Ξ is the set that contains all the decision variables of the problem. The time horizon is 24 hours. One should notice that since the considered time step is one hour, power and energy values coincide. This is a mixed integer linear optimization problem (MILP).

The expected cost function (3) is a probability-weighted mean of all the scenarios considered. It minimizes the power requested from the grid $p_{t,\omega}^{grid,in}$ at price ε_t^{buy} and the total power $p_{t,\omega}^{CHP,tot}$ for the CHP at a gas price equal to ε^{gas}. On the other hand, it maximizes the energy sold back to the grid $p_{t,\omega}^{grid,out}$ at price ε_t^{sell}, by those microgrid components that are able to generate energy (PV, ESS, EVs, and CHP). Parameter π_ω expresses the discrete probability assigned to each scenario ω. The second term and the third term of Eq. (3) are penalty factors. Specifically, the second term applies a small cost $C^{EV,deg}$, every time an EV charges $p_{i,t,\omega}^{EV,ch}$ or discharges $p_{i,t,\omega}^{EV,dis}$. This is to ensure that no unnecessary EV charges/discharges take place and to prevent EVs' battery degradation. The penalty is normalized for every kWh of battery charging/discharging and it does not depend on how frequently the EVs charge or discharge.

Finally, the third term of Eq. (3) introduces a prioritization mechanism in the form of a penalty factor. Parameters λ_{PV}, λ_{ESS}, λ_{CHP}, and λ_{EV} obtain excessively small positive values, so that the total cost function is not affected. The values of these parameters work as an artificial penalty and are determined by assumptions depending on which source is preferred by the EMS to give priority to selling the energy back to the grid. The smaller the relative value of parameter λ for a specific resource, the higher the priority in the EMS to sell the available energy from this resource first.

3.1 PV modeling

Eq. (4) enables the actual power generated by the PV to be utilized in three different directions. A portion can be sold directly to the grid $\left(p_{t,\omega}^{PV,grid}\right)$, another portion can be used to cover the building's load needs $\left(p_{t,\omega}^{PV,build}\right)$, and the third option allows PV energy to be stored into the ESS $\left(p_{t,\omega}^{PV,stored}\right)$ and used at a later time frame. The sum of all the PV power variables must be less than or equal to the PV generation $P_{t,\omega}^{PV,gen}$ at every time step t and scenario ω.

$$p_{t,\omega}^{PV,grid} + p_{t,\omega}^{PV,build} + p_{t,\omega}^{PV,stored} \leq P_{t,\omega}^{PV,gen} \quad \forall t, \omega \tag{4}$$

3.2 ESS modeling

The ESS operation is characterized by Eq. (5)–(10). The actual power provided by the ESS when discharges can be either sold back to the grid $\left(p_{t,\omega}^{ESS,grid}\right)$ or used to cover a portion of the building load demand $\left(p_{t,\omega}^{ESS,build}\right)$. as shown in Eq. (5). Eqs. (6) and (7) establish the limitations for the charging $\left(p_{t,\omega}^{ESS,ch}\right)$ and discharging $\left(p_{t,\omega}^{ESS,dis}\right)$ power of the ESS with the assistance of binary variable $\xi_{t,\omega}^{ESS}$. Constraint (8) ensures that the total power towards the ESS does not violate its maximum charging rate. The state-of-energy (soe) for the ESS is expressed by Eq. (9) and Eq. (10), while Eq. (11) sets the minimum and the maximum allowed limits for the ESS soe to prevent a deep discharge of the battery.

$$p_{t,\omega}^{ESS,grid} + p_{t,\omega}^{ESS,build} = p_{t,\omega}^{ESS,dis}\eta^{ESS,dis} \quad \forall t, \omega \tag{5}$$

$$0 \leq p_{t,\omega}^{\text{ESS,ch}} \leq \xi_{t,\omega}^{\text{ESS}} P^{\max,\text{ch}} \ \forall t,\omega \quad (6)$$

$$0 \leq p_{t,\omega}^{\text{ESS,dis}} \leq \left(1 - \xi_{t,\omega}^{\text{ESS}}\right) P^{\max,\text{dis}} \ \forall t,\omega \quad (7)$$

$$p_{t,\omega}^{\text{ESS,ch}} + p_{t,\omega}^{\text{PV,stored}} + p_{t,\omega}^{\text{CHP,stored}} \leq P^{\max,\text{ch}} \ \forall t,\omega \quad (8)$$

$$soe_{t,\omega}^{\text{ESS}} = SOE^{\text{ESS,ini}} + \eta^{\text{ESS,ch}} p_{t,\omega}^{\text{ESS,ch}} + p_{t,\omega}^{\text{PV,stored}} + p_{t,\omega}^{\text{CHP,stored}} - p_{t,\omega}^{\text{ESS,dis}} \ \forall \omega, t = 1 \quad (9)$$

$$soe_{t,\omega}^{\text{ESS}} = soe_{t-1,\omega}^{\text{ESS}} + \eta^{\text{ESS,ch}} \backslash p_{t,\omega}^{\text{ESS,ch}} + p_{t,\omega}^{\text{PV,stored}} + p_{t,\omega}^{\text{CHP,stored}} - p_{t,\omega}^{\text{ESS,dis}} \ \forall \omega, t > 1 \quad (10)$$

$$SOE^{\min} \leq soe_{t,\omega}^{\text{ES}} \leq SOE^{\max} \ \forall t,\omega \quad (11)$$

3.3 EVs modeling

The EVs operation is described in Eq. (12)–(18). Eq. (12) ensures that the discharge power of the EVs is either injected back to the grid $\left(p_{i,t,\omega}^{\text{EV,grid}}\right)$ and/or used to cover a part of the building load demand $\left(p_{i,t,\omega}^{\text{EV,build}}\right)$. Constraints (13) and (14) set a limit on the charging $\left(p_{i,t,\omega}^{\text{EV,ch}}\right)$ and discharging power $\left(p_{i,t,\omega}^{\text{EV,dis}}\right)$ of the EVs with the assistance of the binary variable $\xi_{i,t,\omega}^{\text{EV}}$. For each EV i in every scenario ω, the available state can be charging, discharging, or remaining in idle state. Eqs. (15)-(18) refer to the state-of-energy of the EVs. More specifically, Eq. (15) defines the state-of-energy of the EVs for initial conditions, while Eq. (16) describes the state-of-energy of each EV for the rest time steps. In Eq. (16), the state-of-energy of the current time interval for an EV is equal to the previous state plus the energy deriving from charging the EV battery (if charging) minus the energy that is subtracted if the EV battery is discharging. The nominal capacity of an EV's battery is 24 kWh (Nissan Leaf) and the rated charging power of an individual charger deployed in the parking lots is 7.68 kW (SAE-J1772, level 2, 208–240 VAC) with a charging and discharging efficiency of 90% [15]. To avoid EVs' batteries over-charge and over-discharge, Eq. (17) limits the batteries' lowest state-of-energy at 20% of the EVs nominal capacity. Finally, constraint Eq. (18) sets the minimum state-of-energy for each EV upon its departure time.

$$p_{i,t,\omega}^{\text{EV,grid}} + p_{i,t,\omega}^{\text{EV,build}} = p_{i,t,\omega}^{\text{EV,dis}} \eta^{\text{EV,dis}} \ \forall i, t \in \left[T_i^{\text{arr}}, T_i^{\text{dep}}\right], \omega \quad (12)$$

$$0 \leq p_{i,t,\omega}^{\text{EV,ch}} \leq \xi_{i,t,\omega}^{\text{EV}} P^{\text{EV,max,ch}} \ \forall i, t \in \left[T_i^{\text{arr}}, T_i^{\text{dep}}\right], \omega \quad (13)$$

$$0 \leq p_{i,t,\omega}^{\text{EV,dis}} \leq \left(1 - \xi_{i,t,\omega}^{\text{EV}}\right) P^{\text{EV,max,dis}} \ \forall i, t \in \left[T_i^{\text{arr}}, T_i^{\text{dep}}\right], \omega \quad (14)$$

$$soe_{i,t,\omega}^{\text{EV}} = SOE_i^{\text{EV,arr}} + \eta^{\text{EV,ch}} p_{i,t,\omega}^{\text{EV,ch}} - p_{i,t,\omega}^{\text{EV,dis}} \ \forall i, t = T_i^{\text{arr}}, \omega \quad (15)$$

$$soe_{i,t,\omega}^{\text{EV}} = soe_{i,t-1,\omega}^{\text{EV}} + \eta^{\text{EV,ch}} p_{i,t,\omega}^{\text{EV,ch}} - p_{i,t,\omega}^{\text{EV,dis}} \ \forall i, t \in \left(T_i^{\text{arr}}, T_i^{\text{dep}}\right], \omega \quad (16)$$

$$SOE^{\text{EV,min}} \leq soe_{i,t,\omega}^{\text{EV}} \leq SOE^{\text{EV,max}} \ \forall i, t \in \left[T_i^{\text{arr}}, T_i^{\text{dep}}\right], \omega \quad (17)$$

$$soe_{i,t,\omega}^{\text{EV}} \geq SOE_i^{\text{EV,dep}} \ \forall i, t = T_i^{\text{dep}}, \omega \quad (18)$$

3.4 CHP modeling

The utilization of small-sized CHP turbines is typical for covering thermal load demand and has been often proposed in literature as a distributed energy resource [16].

The equations that describe the operation of the CHP microturbine are presented in Eqs. (19)-(24) below.

$$p_{t,\omega}^{\text{CHP,el}} + p_{t,\omega}^{\text{CHP,th}} = p_{t,\omega}^{\text{CHP,tot}} \eta^{\text{CHP,ovrl}} \ \forall t, \omega \tag{19}$$

$$p_{t,\omega}^{\text{CHP,el}} = p_{t,\omega}^{\text{CHP,tot}} \eta^{\text{CHP,el}} \ \forall t, \omega \tag{20}$$

$$p_{t,\omega}^{\text{CHP,th}} = p_{t,\omega}^{\text{CHP,tot}} \eta^{\text{CHP,th}} \ \forall t, \omega \tag{21}$$

$$P^{\text{CHP, min}} \leq p_{t,\omega}^{\text{CHP,tot}} \leq P^{\text{CHP, max}} \ \forall t, \omega \tag{22}$$

$$p_{t,\omega}^{\text{CHP,th}} \geq P_{t}^{\text{build,th}} \ \forall t, \omega \tag{23}$$

$$p_{t,\omega}^{\text{CHP,el}} = p_{t,\omega}^{\text{CHP,grid}} + p_{t,\omega}^{\text{CHP,build}} + p_{t,\omega}^{\text{CHP,ESS}} \ \forall t, \omega \tag{24}$$

Constraint (19) states that the total power $p_{t,\omega}^{\text{CHP,tot}}$ generated by the CHP consists of its electrical $\left(p_{t,\omega}^{\text{CHP,el}}\right)$, and its thermal $\left(p_{t,\omega}^{\text{CHP,th}}\right)$ production. Eqs. (20) and (21) relate the electrical and thermal production of the CHP with their corresponding efficiencies. Constraint (22) imposes the limits to the CHP's minimum and maximum operation state. Eq. (23) ensures that the thermal load demand is met at any time by the CHP operation, while Eq. (24) describes the possible directions towards the electrical production of the CHP can be directed. More specifically, Eq. (24) states that a portion of the electric power produced is used to cover the building's electrical load demand, another portion can be stored to the ESS for later exploitation, while an amount of CHP energy can be directly sold back to the grid. The main criterion based on which the CHP size has been selected, is its ability to fully cover the thermal load of the building during the whole day, including the peak time periods.

3.5 Power constraints

The total power injected to the grid is described in Eq. (25). The total power injected to the grid at time t and for each scenario ω consists of the power provided by the PV, the ESS, the CHP, and the sum of the power derived from EVs discharging and intended for the grid. In this work it is assumed that all the available energy to be injected into the grid can be acquired by the utility serving company at the time it is produced.

$$p_{t,\omega}^{\text{grid,inj}} = p_{t,\omega}^{\text{PV,grid}} + p_{t,\omega}^{\text{ESS,grid}} + p_{t,\omega}^{\text{CHP,grid}} + \sum_{i} p_{i,t,\omega}^{\text{EV,grid}} \ \forall t, \omega \tag{25}$$

The power balance equation is defined in Eq. (26) below.

$$\begin{aligned} & p_{t,\omega}^{\text{grid,in}} + p_{t,\omega}^{\text{PV,build}} + p_{t,\omega}^{\text{ESS,build}} + \sum_{i} p_{i,t,\omega}^{\text{EV,build}} + p_{t,\omega}^{\text{CHP,build}} \\ & = P_{t,\omega}^{build} + p_{t,\omega}^{\text{ESS,ch}} + \sum_{i} p_{i,t,\omega}^{\text{EV,ch}} \ \forall t, \omega \end{aligned} \tag{26}$$

Constraint (26) forces the balance between the input and the output electric power of the EMS in each time interval. More specifically, it is stated in Eq. (26) that the total load consisting of the office-building electric load demand, the charging needs of the ESS and the sum of the charging needs for the EVs is covered by the power requested from the grid and/or by the combined procurement of power provided by the PV, the ESS, the sum of discharging power of the EVs, and the CHP.

Finally, Eq. (27) and Eq. (28) realize the logic of power exchange.

$$p_{t,\omega}^{\text{grid,in}} \leq L\,\xi_{t,\omega}^{\text{grid}} \;\; \forall t, \omega \tag{27}$$

$$p_{t,\omega}^{\text{grid,out}} \leq L\left(1 - \xi_{t,\omega}^{\text{grid}}\right) \;\; \forall t, \omega \tag{28}$$

When the EMS needs to draw power from the grid, power is not allowed to be injected into the grid at the same time, and vice versa. The limitations in power exchange are imposed by parameter L which corresponds to the local line capacity. To avoid the installation of extra power facility infrastructure for the EMS, the potential limits of the university-building dedicated medium-voltage to low-voltage (MV/LV) transformer are used. The apparent power of the transformer is 160 kVA with MV input 10.5 kV and LV output 400 V. Assuming a whole building's power factor of 0.9, the actual (useful) power that can be drawn from the grid at any time is 144 kW. This constraint can be also time-dependent and be imposed to lower values, for example, by an aggregator responsible for coordinating multiple microgrids owning EMS or by the utility company itself responsible for the smooth operation of electrification in the area.

4. Numerical results and discussion

To examine the effectiveness of the proposed EMS algorithm, the impact of different case studies on total system cost is evaluated. The proposed EMS framework is a mixed integer linear problem modeled in GAMS v.24.7.1 and solved by the IBM CPLEX Optimizer v.12.6. The time required to find the optimal solution varies from a few seconds to several minutes, depending on the model. The optimality gas has been set at 1.0E-04.

The electric load demand and PV scenarios are given in **Section 2.1** along with the deterministic day-ahead (DA) forecasts. The thermal load demand prediction is shown in **Section 2.3**. It should be noted that as the PV generation data came from actual smart metering measurements, no study regarding the positioning and the installation of the PV panels was performed.

The bidirectional energy flows between the utility company and the end-user (the building-microgrid manager in this case) assume the utilization of smart-metering approach. The day-ahead time-varying price signal which represents the electricity cost at each time interval t is depicted in **Figure 7**. A time-varying rate has also been applied for the energy sold back to the grid. This rate is 20% lower

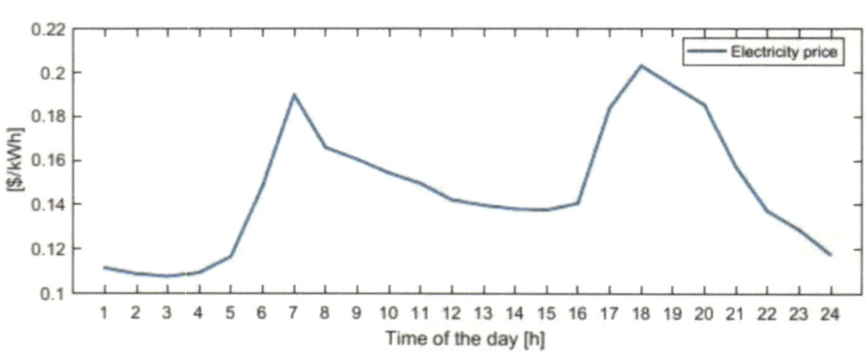

Figure 7.
Day-ahead electricity price forecast.

than the aforementioned time-varying price signal, based on the assumption that the utility company would not buy energy at a more expensive rate than it would sell it. In this study, no other incentive-based scheme (e.g., selling green certificates for renewables) apart from the utility company price signal is applied.

The ESS consists of a battery group with a total capacity of 80 kWh. The maximum charging/discharging rate is 40 kW with corresponding power electronics efficiency of 0.88. The minimum allowed state-of-energy of the ESS has been set to 10 kWh (12.5% of max ESS capacity) to prevent deep battery discharging. The initial state-of-energy of the ESS is 40kWh.

The thermal efficiency $\left(\eta^{CHP,th}\right)$ of the CHP microturbine is 0.51 and the electric efficiency $\left(\eta^{CHP,el}\right)$ 0.36 resulting in an overall efficiency $\left(\eta^{CHP,ovrl}\right)$ of 0.87. The overall efficiency of the CHP is kept constant regardless its load for sake of simplicity. The rated power of the CHP is 150 kW, and to avoid start-up costs, a minimum state of 10 kW has been set for the CHP operation.

As mentioned earlier, a bidirectional energy flow concept for EVs and their potential V2B and V2G capabilities could significantly reshape the current perception of power systems. The first step is their integration into the smart grid (or microgrid). The EVs are equipped with constantly bigger battery capacities increasing thus their potential contribution as DERs. The EVs could either be granted to (University's or a company's) personnel for commuting purposes under the form of a third-party contract and/or they could be privately owned. In both cases, it would make sense to assume that the EV users would be willing to allow the building-microgrid operators to use their batteries' capacity but they would not prefer to have a lower state-of-energy upon departure compared to their arrival. In addition, in the case of self-owned EVs, possible monetary benefits for the EV owners may be needed for motivating them to opt-in the EMS scheme.

In our base case study, the first business model is considered, namely the EVs are provided to the personnel and, in exchange, the EVs' users have to participate in the EMS framework. It is considered here that the final state-of-energy of the EVs should be at least equal to their initial one. We have also considered $\lambda_{PV} < \lambda_{ESS} < \lambda_{CHP} < \lambda_{EV}$ assigning a higher priority to the energy coming from PV to be sold to the grid, afterwards the energy from ESS, then the energy from CHP, and finally the energy from the EVs. The reason that the lowest priority has been assigned to EVs is to have as few charging/discharging cycles for the EVs as possible to prevent battery degradation.

First, we consider the total system cost (TSC), as shown in **Table 2**.

The first case corresponds to an operation of the microgrid without the presence of an EMS and thus, no optimization takes place. That is, the loads cannot be shifted and are always met. In addition, as the EVs should depart at least having the same battery state of energy as the one they had when arrived, charging/discharging of the EVs are not activated. The ESS operation is also omitted, as its charging / discharging cannot be coordinated due to the absence of an EMS. Finally, when

Case	Description	Total system cost
1	No EMS in operation (average of all historical data)	59.47
2	With EMS in operation (average of all historical data)	13.51
3	Expected mean of all 24 scenarios	16.68
4	Most probable scenario of the 24 (prob. 9.3%)	58.11

Table 2.
Total system cost across all case [$].

there is a net energy consumption at time t, electricity is bought at price ε_t^{buy}, while when there is a net supply of energy to the grid, it is sold at price ε_t^{sell}. The second case corresponds to the minimization of the total system cost for one day of microgrid operation. The annual average values of the PV and electricity load historical data have been considered, among others, as input parameters. The third case minimizes the expected mean cost considering the 24 PV and electricity load probability-weighted scenarios, as these are determined in **Section 2.1**. A cost distribution for all scenarios considered in case 3 is shown in **Figure 8**. Finally, the fourth case minimizes the total system cost for the microgrid considering the most probable scenario (scenario 23 out of the 24). It should be noted that all cases apart from case 1 assume the presence of an EMS in microgrid's operation.

The importance of considering an EMS in microgrid's operation is depicted in the TSC results across all cases, as shown in **Table 2**. First, the total system cost of case 1, where no EMS is assumed, is 340% higher compared to case 2, where an EMS is present coordinating the microgrid operation (from \$13.51 to \$59.47). The expected TSC for case 3 is 23% higher compared to case 2 due to the impact of some extreme scenarios on the final result. Moreover, the total cost distribution across all the different scenarios (**Figure 8**) implies that the final total system payoff for the majority of the scenarios is positive in terms of cost (a positive value declares a cost, while a negative one declares a profit). Finally, one may notice that the TSC for the most probable scenario, as seen in case 4, is much higher compared to the other two cases (case 2 and 3) in which an EMS is also present on microgrid's operation. The reason is that for this particular scenario, the PV generation and the building load demand are very different compared to the corresponding annual average values, as these are considered for case 2 (**Figure 4**). More specifically, the projected PV generation in the most probable scenario is much lower than the yearly average, as presented in case 2. On the contrary, the building load demand is higher than the average. Therefore, the results presented in this Section should be interpreted taking this context into account.

To analyze a few more aspects of the optimization results and examine the individual scheduling of each DER, as it is decided by the EMS, we compare the microgrid's operation under two different case studies: case 2, which from now on

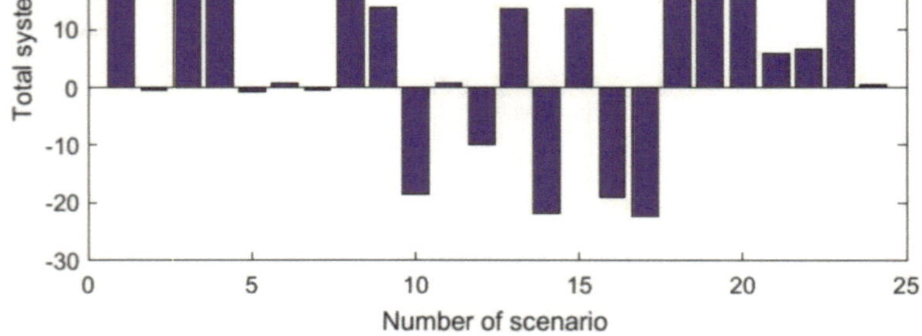

Figure 8.
Total system cost distribution for the 24 scenarios.

will be referred to as simply the **deterministic** case, and case 4, which will be referred to as the **most probable scenario**. (**Figure 9**).

Figure 10 presents the total power requested by the EMS from the grid and injected back to it for the deterministic approach and the most probable scenario.

There are many observations one might make regarding **Figure 10**. First, notice that the power requested from the grid is zero during the whole 24-hour time horizon for the deterministic approach. This implies that the microgrid can fully cover its electric load demand using its own distributed energy resources. In addition, it is able to inject a great portion of its produced energy back to the grid. From 1 pm to 4 pm though, the microgrid neither requests nor injects power back to the grid. This means that the produced energy is entirely used to cover the local microgrid load demand.

On the other hand, we can see that during the most probable scenario, the microgrid draws power from the grid from around 9 am to 5 pm which indicates that the microgrid's distributed energy resources cannot fully cover the load demand during that period. This is mostly due to the limited daily PV production assumed in this scenario in combination with a higher than average electric load demand. In addition, one may notice that the total power injected back to the grid is much lower in the most probable scenario.

To better understand how EMS coordinates the operation of the microgrid's components, **Figure 11** presents the decomposition of the total power injected to the grid for the involved DERs (PV, ESS, CHP, and EVs).

In both the deterministic and the most probable scenario, CHP is the DER that injects most of the power back to the grid. We can see that in the deterministic case PV also contributes, especially during the noon hours. The ESS is more active in the case of the most probable scenario, while one might notice that the EVs are not used at all as a potential source for energy to be injected to the grid. This happens mainly due to the lowest prioritization factor EVs have for selling energy back to the grid as

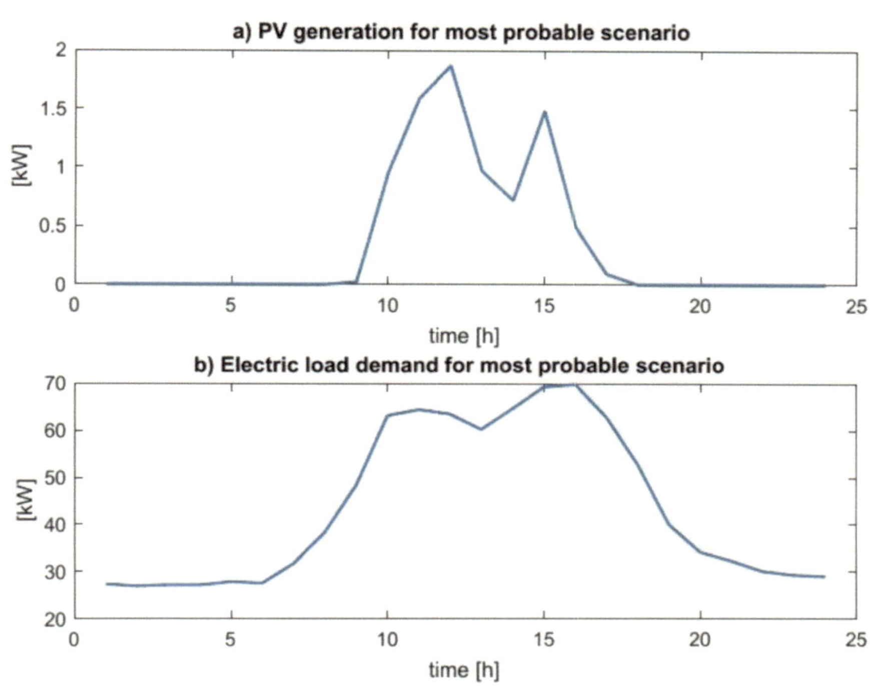

Figure 9.
Most probable scenario for (a) PV production, and (b) electric load demand.

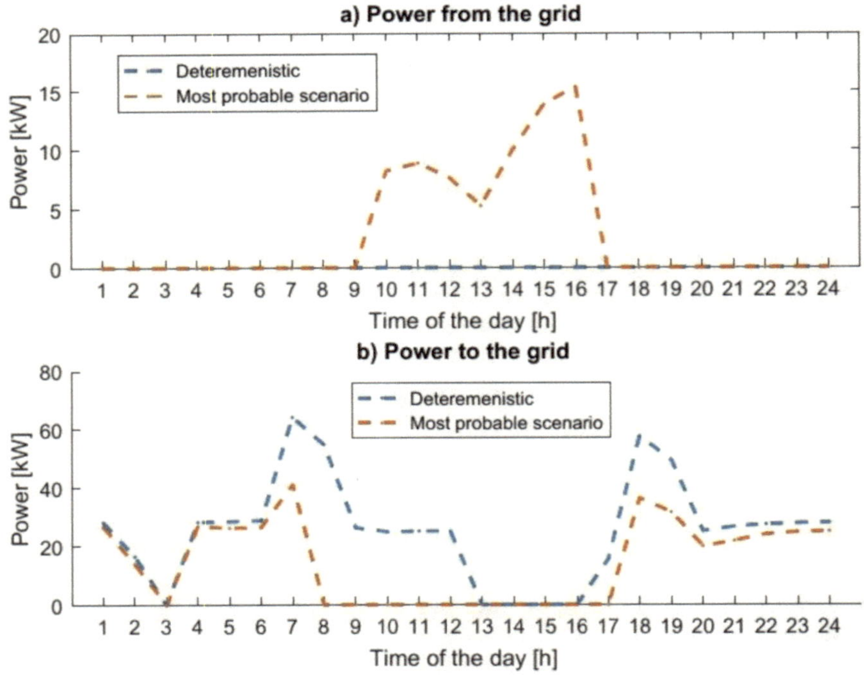

Figure 10.
Power requested from and injected to the grid for the most probable scenario and the deterministic approach.

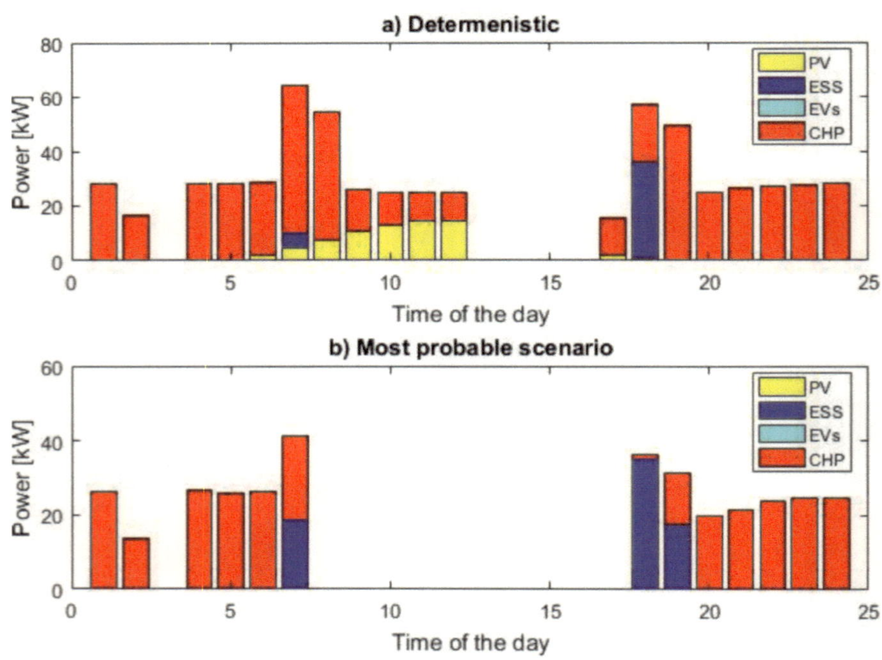

Figure 11.
Decomposition of power injected to the grid for the (a) deterministic approach, and (b) the most probable scenario.

described earlier, but also due to the penalty that has been set to prevent EVs battery degradation. Finally, we can observe that in both the deterministic and the most probable scenario, the EMS tries to inject most of the power back to the grid during the peaks of electricity price (around 7 am and 6 pm as shown in **Figure 7**) to maximize the reward.

Figure 12 shows the decomposition of the projected PV generation for the deterministic approach and the most probable scenario.

In the deterministic case study, PV production is mostly sold to the grid (early and noon hours) or stored in the ESS for future exploitation (afternoon hours). Only a small portion at 1 pm is used to cover the building's load demand. On the contrary, in the most probable scenario, all the produced PV energy is used to meet the building's load demand.

Figure 13 shows how the electric power produced by the CHP is divided among the grid, the ESS, and the local building load. Like the PV, most of the CHP electric production in the most probable scenario is used to cover the building's load. Moreover, we can see that the EMS tries to inject most of the CHP's produced energy back to the grid, during the electricity price peak hours. Finally, in both cases a smaller amount of the CHP's produced energy is stored in the ESS for future implementation. The thermal load demand parameter $P_t^{\text{build,th}}$ is not an uncertainty-related parameter and thus, remains the same in each scenario. The thermal load demand is met by variable $p_{t,\omega}^{\text{CHP,th}}$, as stated in Eq. (23).

Storage is an important distributed energy resource for the system. As stated in Eqs. (9)-(10), the ESS can either be charged from the grid, the PV, and the CHP. When discharging, its energy can be either injected into the grid and/or cover a portion in building's load demand. **Figure 14** shows the decomposition of the ESS available energy for the deterministic approach, as well as for the most probable scenario.

Figure 14 can be better analyzed taking into account **Figure 15**, which demonstrates the evolution of the ESS state of energy for the two aforementioned case studies.

We can see that in both deterministic and most probable scenario cases, the ESS is mainly active during two distinct period of times, in the morning (between 7 am and 8 am), and in the afternoon (between 5 pm and 7 pm). There are two main

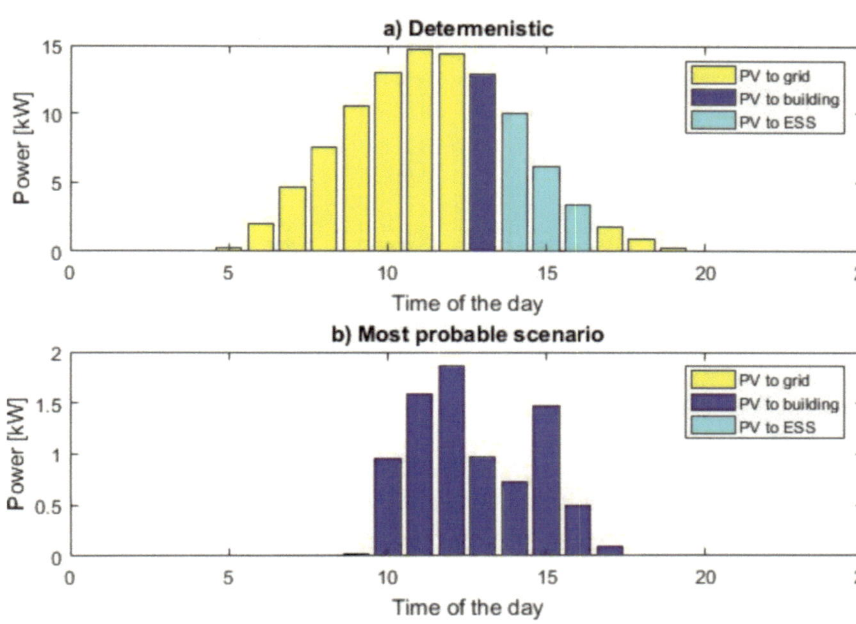

Figure 12.
Decomposition of PV production for the (a) deterministic approach, and (b) the most probable scenario.

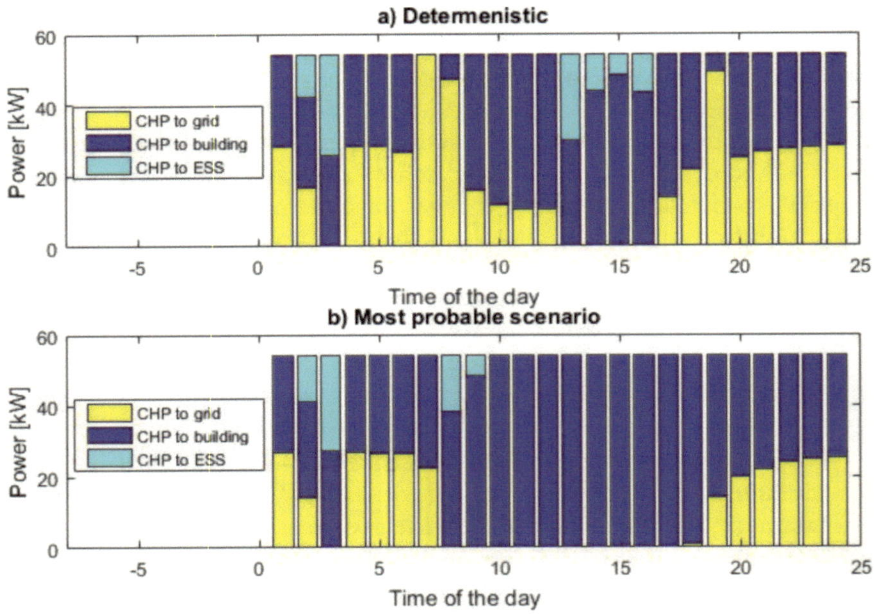

Figure 13.
Decomposition of CHP electric produced power for the (a) deterministic approach, and (b) the most probable scenario.

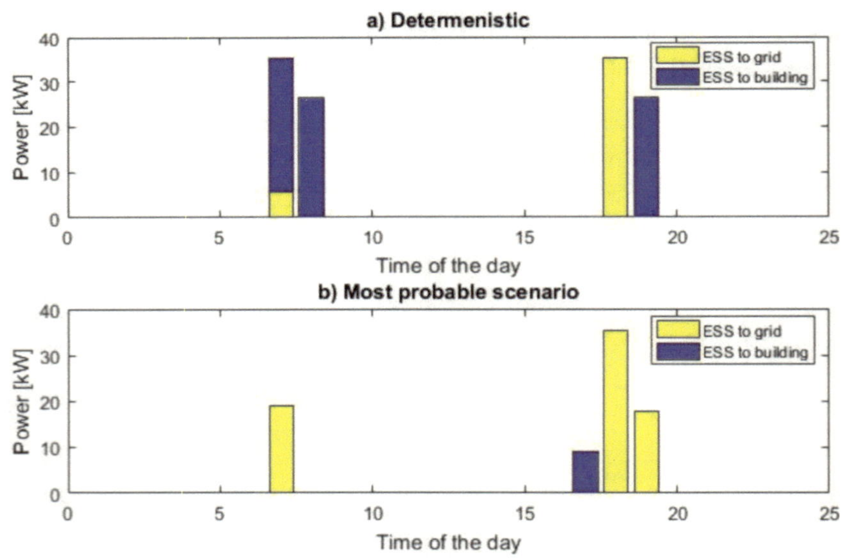

Figure 14.
Decomposition of ESS provided power for the (a) deterministic approach, and (b) the most probable scenario.

observations one may make regarding the ESS operation. First, the ESS uses two discharge cycles in the deterministic approach, while it only discharges once in the most probable scenario. The relatively high PV generation considered in the deterministic scenario is responsible for this second cycle of charge/discharge. Looking at **Figure 12**, we notice that PV production during the afternoon hours is mostly directed to the ESS. Second, the ESS covers mainly the building's load demand in the deterministic case, while in the most probable scenario the ESS injects most of its energy back to the grid.

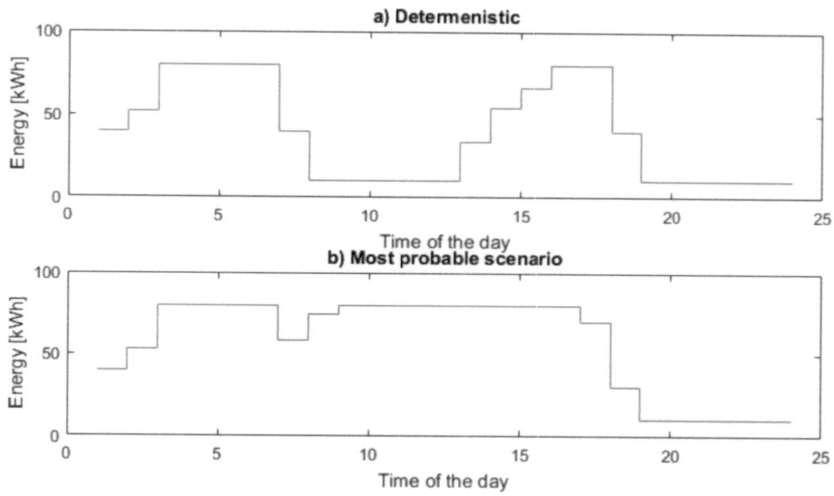

Figure 15.
State of energy for the ESS for the (a) deterministic approach, and (b) the most probable scenario.

EVs constitute the third available DER in the microgrid but contrary to the rest DERs (PV, ESS, and CHP), they are not actively involved in microgrid's energy exchange. The EVs battery degradation cost on the one hand, and the lowest energy prioritization factor that has been assigned to them on the other hand, do not make the an attractive alternative power source for the EMS (in terms of cost). Nevertheless, the EVs can always be used as a back-up ancillary power source in case of an emergency situation.

5. Sensitivity analysis

Sensitivity analysis is used to study the robustness of the solution to a linear programming model. If there is cause for concern regarding the accuracy of the data used, sensitivity analysis is undertaken to determine the way the solution might change if the data were different. When the solution does not change (or when the nature of the solution does not change, as when the basis remains optimal), one may assume that the proposed solution is appropriate.

Dual variables, also known as *shadow prices*, are of great interest in the solution of a linear optimization problem. A dual variable is reported for each constraint. The dual variable is only positive when a constraint is binding. The dual price can be defined as "*the improvement in the objective function value if the constraint is relaxed by one unit*". In the case of a less-than-or-equal constraint, such as a resource constraint, the dual variable gives the value of having one more unit of the resource represented by that constraint. In the case of a greater-than-or-equal constraint, such as a minimum production level constraint, the dual variable gives the cost of meeting the last unit of the minimum production target. The units of the dual prices are the units of the objective function divided by the units of the constraint. To obtain the values of the dual variables, we first solve the MILP to find the optimal allocation. Next, we remove the integrality constraints and insert equality constraints that force the integer variables to assume their optimal values in the resulting linear program [17].

The following example presents how the dual variable of a constraint can be used for the sensitivity analysis. **Figure 16** shows the dual prices of constraint Eq. (4) for the 24 hours of the daily time horizon. One should recall that this is a

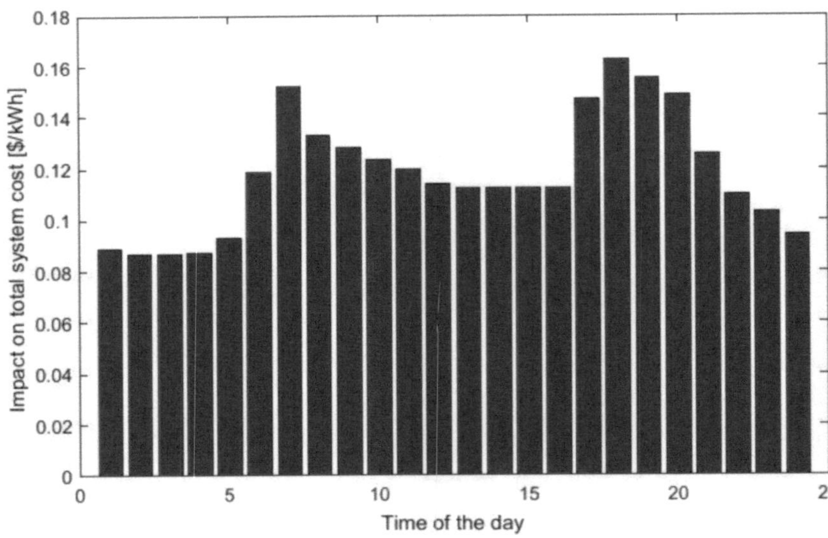

Figure 16.
Value of sensitivity factor: Dual variable corresponding to the upper bound of constraint (4) [$/kWh].

resource constraint, and specifically it bounds the actual power generated by the PV to be less-than-or-equal-to the maximum PV generation, as this is defined by parameter $P_t^{PV,gen}$. Also note that this example refers to the EMS operation of case 2, as presented in **Table 2**.

This value implies the sensitivity of the system cost with respect to the actual PV power utilized by the system. Note that the positive value for this dual variable means that the total system cost decreases with the additional availability of PV power. More specifically, it indicates the decrease in the total system cost that corresponds to the increase of the available PV generation by 1 kWh. The fact that the value of the dual variable is positive during the whole day implies that additional PV potential has always positive impact on the total system cost, regardless the time of the day. However, one might also notice that there some time periods (7 am, from 5 pm to 9 pm), where the extra PV power would be more beneficial for the system compared to the rest time periods. In a similar way, one could evaluate the impact of the relaxation of the rest important resources to the total system cost.

6. Conclusion and future work

The transition to the new "smart" era requires the utilization of smart technology through comprehensive and efficient energy management functions. We propose in this study, a two-way communication energy management framework for a microgrid in a university campus including local renewable energy sources, a storage system, a combined heat and power small turbine, and a fleet of EVs used for work-related trips. Two-way energy exchange is allowed using net metering technology. The developed MILP framework incorporates an optimizer which decides the power exchange among the DER components of the microgrid and the grid, exploiting the V2B and V2G capabilities of the distributed energy resources. It also provides a specific level of thermal comfort to the building's occupants by meeting the predicted heating load. The formulation of an EMS model which takes into account the PV and load variability is very important if we want to consider the impact of planning for one scenario, and having another scenario occurs.

To overcome this challenge, actual smart metering data for a period of one year have been used to construct a number of potential scenarios. The PV and load demand data are classified using a scenario construction technique, leading to the formulation of 24 different PV and electric load scenarios, each one represented by a designated probability. The importance of considering an EMS in microgrid's operation is depicted in the total system cost across all cases. Results confirm that the EMS substantially decreases the total system cost by optimally coordinating and scheduling the microgrid operation. An additional significant remark is that the majority of the total daily system's cost is due to the natural gas expenses required for the operation of the CHP microturbine. Finally, we compare the optimal scheduling of the microgrid's DERs under the deterministic case and the most probable scenario. The most probable scenario assumes a lower PV production and a higher building electric load demand than the average values considered in the deterministic case, resulting in a substantially different energy scheduling for the DERs. It is worth noting that under the deterministic approach and the current design, the microgrid seems to be self-sufficient in terms of covering its energy demand. However, this is not the case under the most probable scenario approach, where the microgrid relies also on grid energy to meet its load demand, on top of the energy production of its own DERs. Suggestions for future work include the introduction of additional stochasticity parameters (e.g., electricity price) and the integration of power flow constraints into the optimization problem.

Author details

Dimitrios Thomas[1,2*] and Evangelos Kotsakis[2]

1 Department of Electrical Engineering, University of Mons, Mons, Belgium

2 European Commission, Joint Research Centre (JRC), Ispra, Italy

*Address all correspondence to: thomasdimitrios@gmail.com

References

[1] D. Kolokotsa, D. Rovas, E. Kosmatopoulos, and K. Kalaitzakis, "A roadmap towards intelligent net zero- and positive-energy buildings," *Solar Energy*, vol. 85, no. 12, pp. 3067–3084, 2011, doi: 10.1016/j.solener.2010.09.001.

[2] Y. Lu, S. Wang, Y. Zhao, and C. Yan, "Renewable energy system optimization of low/zero energy buildings using single-objective and multi-objective optimization methods," *Energy and Buildings*, vol. 89, pp. 61–75, 2015, doi: 10.1016/j.enbuild.2014.12.032.

[3] C. Xia, Y. Zhu, and B. Lin, "Renewable energy utilization evaluation method in green buildings," *Renewable Energy*, vol. 33, no. 5, pp. 883–886, 2008, doi: 10.1016/j.renene.2007.10.005.

[4] P. Nema, R. K. Nema, and S. Rangnekar, "A current and future state of art development of hybrid energy system using wind and PV-solar: A review," *Renewable and Sustainable Energy Reviews*, vol. 13, no. 8, pp. 2096–2103, 2009, doi: 10.1016/j.rser.2008.10.006.

[5] X. Xue, S. Wang, Y. Sun, and F. Xiao, "An interactive building power demand management strategy for facilitating smart grid optimization," *Applied Energy*, vol. 116, pp. 297–310, 2014, doi: 10.1016/j.apenergy.2013.11.064.

[6] N. Growe-Kuska, H. Heitsch, and W. Romisch, "Scenario reduction and scenario tree construction for power management problems," in *2003 IEEE Bologna Power Tech*, Bologna, Italy, pp. 152–158.

[7] J. van Roy, N. Leemput, F. Geth, J. Buscher, R. Salenbien, and J. Driesen, "Electric Vehicle Charging in an Office Building Microgrid With Distributed Energy Resources," *IEEE Trans. Sustain. Energy*, vol. 5, no. 4, pp. 1389–1396, 2014, doi: 10.1109/TSTE.2014.2314754.

[8] D. Wang, X. Guan, J. Wu, P. Li, P. Zan, and H. Xu, "Integrated Energy Exchange Scheduling for Multimicrogrid System With Electric Vehicles," *IEEE Trans. Smart Grid*, vol. 7, no. 4, pp. 1762–1774, 2016, doi: 10.1109/TSG.2015.2438852.

[9] W. Guan, Y. Tan, H. Zhang, and J. Song, "Distribution system feeder reconfiguration considering different model of DG sources," *International Journal of Electrical Power & Energy Systems*, vol. 68, pp. 210–221, 2015, doi: 10.1016/j.ijepes.2014.12.023.

[10] S. Wang and X. Xu, "Parameter estimation of internal thermal mass of building dynamic models using genetic algorithm," *Energy Conversion and Management*, vol. 47, 13–14, pp. 1927–1941, 2006, doi: 10.1016/j.enconman.2005.09.011.

[11] H. Park, M. Ruellan, N. Martaj, R. Bennacer, and E. Monmasson, "Generic thermal model of electrical appliances in thermal building: Application to the case of a refrigerator," *Energy and Buildings*, vol. 62, pp. 335–342, 2013, doi: 10.1016/j.enbuild.2013.02.061.

[12] T. Dewson, B. Day, and A. D. Irving, "Least squares parameter estimation of a reduced order thermal model of an experimental building," *Building and Environment*, vol. 28, no. 2, pp. 127–137, 1993, doi: 10.1016/0360-1323(93)90046-6.

[13] L. Ljung, *System identification*. New Jersey: Prentice Hall, 1987.

[14] A. Bagheri, V. Feldheim, D. Thomas, and C. S. Ioakimidis, "Coupling building thermal network and control system, the first step to smart buildings," in *2016 IEEE*

International Smart Cities Conference (ISC2), Trento, Italy, 2016, pp. 1–6.

[15] P. Richardson, D. Flynn, and A. Keane, "Optimal Charging of Electric Vehicles in Low-Voltage Distribution Systems," *IEEE Trans. Power Syst.*, vol. 27, no. 1, pp. 268–279, 2012, doi: 10.1109/TPWRS.2011.2158247.

[16] D. Thomas, A. Bagheri, V. Feldheim, O. Deblecker, and C. S. Ioakimidis, "Energy and thermal comfort management in a smart building facilitating a microgrid optimization," in *IECON 2017 - 43rd Annual Conference of the IEEE Industrial Electronics Society*, Beijing, pp. 3621–3626.

[17] R. P. O'Neill, P. M. Sotkiewicz, B. F. Hobbs, M. H. Rothkopf, and W. R. Stewart, "Efficient market-clearing prices in markets with nonconvexities," *European Journal of Operational Research*, vol. 164, no. 1, pp. 269–285, 2005, doi: 10.1016/j.ejor.2003.12.011.

Chapter 9

Cognitive Dynamic System for AC State Estimation and Cyber-Attack Detection in Smart Grid

Mohammad Irshaad Oozeer and Simon Haykin

Abstract

The work presented in this chapter is an extension of our previous research of bringing together the Cognitive Dynamic System (CDS) and the Smart Grid (SG) by focusing on AC state estimation and Cyber-Attack detection. Under the AC power flow model, state estimation is complex and computationally expensive as it relies on iterative procedures. On the other hand, the False Data Injection (FDI) attacks are a new category of cyber-attacks targeting the SG that can bypass the current bad data detection techniques in the SG. Due to the complexity of the nonlinear system involved, the amount of published works on AC based FDI attacks have been fewer compared to their DC counterpart. Here, we will demonstrate how the entropic state, which is the objective function of the CDS, can be used as a metric to monitor the grid's health and detect FDI attacks. The CDS, acting as the supervisor of the system, improves the entropic state on a cycle to cycle basis by dynamically optimizing the state estimation process through the reconfiguration of the weights of the sensors in the network. In order to showcase performance of this new structure, computer simulations are carried out on the IEEE 14-bus system for optimal state estimation and FDI attack detection.

Keywords: false data injection, cognitive dynamic systems, cognitive control, AC state estimation, smart grid

1. Introduction

The Cognitive Dynamic System (CDS) is an organized physical model and research tool that is based on certain features of the brain. Following its first introduction in [1], it was later expanded in [2] leading to its first applications in cognitive radio [3] and cognitive radar [4]. Since then, CDS has progressed enormously to give rise to Cognitive Control (CC) [5] and Cognitive Risk Control (CRC) [6] as two of its particular functions. Using those principles, the CDS was first merged in [7] with the Smart Grid (SG) to form a new structure, based on the DC state estimation model, that shows tremendous potential for handling the possible problems that the SG will be facing in the near future. Furthermore, in [8], the construct presented in [7] was expanded to include a more complex CRC that is closer to the brain. In that paper, it was proven how this new approach can to be used to mitigate the problem of cyber-attack in the SG. From a neuroscience perspective, the CDS is founded on Fuster's paradigm of cognition comprising of the following five principles: perception-action cycle (PAC), memory, attention, intelligence and language [9]. In its simplest form, the CDS is built on two main components: the perceptor, on one side, and the executive on the other with the feedback channel uniting them together. In [7], it was shown that the integration of the over-arching function of CDS, CC, with the SG, is well adapted for slowly progressing cyber-physical systems. In this chapter, the construct presented in [7], where the DC-estimation model was involved, will be re-engineered to be able to carry out AC state estimation optimally and also be able to detect cyber-attacks. In order to do so, the perceptor of the CDS will incorporate a generative model that will allow it to sense and control the environment indirectly. Moreover, in order to bring forward the cognitive ability of the CDS and make it compatible with the current nonlinear state estimation in SG, the steps involved in the state estimation process will be re-engineered in a novel way. It will also be shown how the entropic state, which is the objective function of the CDS, will be instrumental in implementing a control-sensing mechanism that is capable of identifying and handling bad measurements. We will also show how this entropic state serves as the basis for detecting False Data Injection attacks (FDI) in SG.

1.1 Smart grid

The next generation of engineering systems consisting of the Internet of Things (IoT) and Cyber-physical systems (CPSs) are currently paving the way towards the fourth industrial revolution [10]. As those systems are gradually occupying a more prominent role in our daily lives, through applications in critical infrastructures such as electrical power grids or transportation systems, the cyber-security aspects of those systems will also grow in importance [11]. In the context of this chapter, emphasis will be laid upon the SG and its most dangerous threat known as False Data Injection (FDI) attacks. More specifically, compared to our previous research where the DC model for state estimation was investigated [7], focus will be laid upon on the AC model, which is more a realistic representation of the smart grid, and the introduction the CDS for a new way of control and FDI attack detection.

Making use of all the new generation of sensing, monitoring and control strategies, the SG is forecasted to be a more powerful entity than the traditional power grid in many facets such as reliability and efficiency [12, 13]. In the SG, the Supervisory Control and Data Acquisition systems (SCADA) is responsible for monitoring and processing the main control actions by collecting meter measurements from remote terminal units (RTUs) consisting of different field devices or sensors. Through a process known as state estimation, those measurements are then processed and analyzed for errors and inconsistencies after being transmitted to a control center [14, 15]. The state variables that are calculated by this process usually consist of the voltage magnitudes and angles of the different busses in the system [16]. The measurements used for state estimation are the currents, real and reactive power flows, power injections and voltage magnitudes and angles. In the DC model, the state variables are the bus angles only while in the more complex AC model, the voltage magnitudes and angles of the different busses in the network are estimated. Weighted Least Squares (WLS), introduced by Schweppe [14], is the technique used for the power system state estimation using those measurements. In order to enhance the accuracy of the estimated states, another process, known as Bad Data Identification, is carried out to remove bad measurements. Bad measurements are erroneous measurement readings that will impact state estimation negatively. The most commonly applied bad data identification techniques are the Chi-Squared Tests and Largest Normalized Residual Test [15, 17]. Those statistical tests rely on the residuals between the estimated states and the measurement residuals to identify the bad data. In the case of an FDI attack, bad data, which can bypass the previously mentioned tests, is introduced into the system such that the estimated states can be modified stealthily. Those bad data are maliciously crafted offsets to measurements that are injected to the sensor readings so as the state estimation process is influenced in a particular way. Consequently, with the incorrect calculated states, bad control decisions will be applied.

Although FDI attacks have been a popular topic of research over the past years [18], most of the works, e.g., in [10–13, 19], investigated the FDI attacks on the DC model. Few works have been published on the AC model and those attacks [18, 20, 21]. Nevertheless, the DC model is just a simplified representation of the nonlinear AC state estimation model. There are major differences between the two models that could explain why the AC model has been unpopular. Firstly, in the nonlinear state estimation model, the estimated states are obtained after undergoing iterations, while in the DC model, those states are obtained in closed-form. Moreover, the linear state estimation relies on active power flow analysis [16, 22, 23]. On the other hand, the AC model uses both active and reactive power flow analysis. Furthermore, the state variables in the DC model consist of the voltage angles only while the states in the AC model consist of both the voltage angles and magnitudes. Consequently, these differences raise the complexity and computational expense of nonlinear state estimation as a topic of research when it comes to FDI attacks [24]. In fact, DC based FDI attacks can be detected by AC-based data detection techniques [20]. Hence, since the AC model is commonly applied in power systems, finding a way to detect these attacks and mitigating them under that environment is going to be very important for the coming years.

1.2 Contribution and organization

The main contributions of this chapter can be summarized as follows:

i. The architectural architecture of the CDS, tailored for AC state estimation and FDI attack detection in the SG, is presented. Compared to our earlier work in [7], which was based on the DC model, we will show how that construct can re-engineered with the goal of nonlinear state estimation and computational efficiency in mind. Consequently, it will be shown how the CDS allows for optimal state estimation with relatively less computations, using the principles of cognition rooted in the brain.

ii. To expand on our previous research, the entropic state will be re-introduced for two purposes namely; (1) it serves as a metric of the grid's health on a cycle to cycle basis and (2) it is used in the detection of FDI attacks. The optimization of the entropic state is the goal of the cognitive controller residing in the executive of the CDS. The latter does this by selecting the most optimal actions that will maximize the available information from one PAC to the next. Simulations are performed on the IEEE 14-bus network to show the efficiency of this new approach using the CDS. By learning which measurements to prioritize and which ones to neglect, the CDS showcases a new way of control for bad data correction and FDI attack detection with the SG being the topic of application.

The rest of this chapter is organized as follows: In Section 2, the basic concepts of state estimation and data detection for the AC model will be presented and contrasted. The mathematics of FDI attacks for this model will also be demonstrated. Section 3 expands on the structure of the CDS for the SG. Since this research is an extension of [7], the material presented in that paper will be re-engineered for this new application. In the context of the CDS, the SG is considered as the environment with which it interacts. Section 4 gives a discussion on the application and simulation results of this approach on the IEEE 14-bus network. It will be shown how this new structure is able to handle the two problems of bad data detection and FDI attack detection simultaneously. Finally, Section 5 concludes this paper by highlighting the key results and presenting new avenues of research for this novel construct.

2. Preliminaries

2.1 Weighted least squares state estimation

In order for the Energy Management System (EMS) to operate properly, it is important for the SCADA to provide the latter with the required measurement data so that correct control decisions can be applied in real-time. However, as those signals are often contaminated with noise, filtering is carried out by both the state estimator and the bad data detector to obtain the most accurate states. However, since power systems comprise of an overdetermined system whereby redundant measurements are taken, the filtering process allows the discarding of those erroneous measurements that will be detrimental for state estimation.

2.2 AC model

The states of a power system refer to the bus voltages angle θ and bus voltage magnitudes V. In the case of the DC model, the states are restricted to the bus angles only and the measurements consist of the real power flows and injections. Additionally, it is assumed that prior knowledge relating to the bus magnitudes is available and those are taken to be close to unity. After choosing a reference bus and setting it to zero radians, state estimation in the linear system is simplified to only estimating the n bus voltage angles $[\theta_1, \theta_2, \dots, \theta_n]^T$. The DC power flow model has been a popular research tool for power engineers and smart grid cyber-security researchers as it serves as a linearization and approximation of the AC power flow model [14, 25–27]. In fact, this substitution to the AC model has been widely accepted for reasons such as guaranteed faster convergence and reduced algorithmic complexities [28].

In the AC model, the nonlinear power flow equations are fundamental for state estimation since they indicate the link between the measurements and the estimated states. In this model, the active and reactive power for the transmission line between busses k and m are given by

$$P_{km} = V_k^2 g_{km} - V_k V_m g_{km} \cos(\theta_{km}) - V_k V_m b_{km} \sin(\theta_{km}) \tag{1}$$

$$Q_{km} = -V_k^2 b_{km} + V_k V_m b_{km} \cos(\theta_{km}) - V_k V_m g_{km} \sin(\theta_{km}) \tag{2}$$

Additionally, for each bus k, it is calculated using the following equations:

$$P_k = V_k \sum_{m \in S_k} V_m \left(-g_{km} \cos(\theta_{km}) - b_{km} \sin(\theta_{km})\right) + V_k^2 \sum_{m \in S_k} g_{km} \tag{3}$$

$$Q_k = V_k \sum_{m \in S_k} V_m \left(-g_{km} \sin(\theta_{km}) - b_{km} \cos(\theta_{km})\right) - V_k^2 \sum_{m \in S_k} b_{km} \tag{4}$$

where $S_k \subset S$ is the set of all busses that have lines connected to bus k and g_{km} and b_{km} are the conductance and susceptance of the line between busses k and m respectively. θ_{km} denotes and the phase angle difference between bus k and bus m. In AC power flow estimation, the nonlinear relationship between the state variables and the measurements is described as follows:

$$\mathbf{z} = \mathbf{h}(\mathbf{x}) + \mathbf{e} \tag{5}$$

where

- $\mathbf{x}$ is the n vector of the true states (voltage magnitudes and angles)
- $\mathbf{z}$ is the m vector of measurements (active and reactive power flows, active and reactive power injections, voltage magnitudes and angles)
- $\mathbf{h}$ is the m x n Jacobian matrix (relates measurements to states)
- $\mathbf{h(x)}$ is the m vector of nonlinear function linking measurements to states
- $\mathbf{e}$ is the m vector of measurement errors
- m is the number of measurements
- n is the number of variables

H in (5), also known as the Jacobian matrix, is a matrix that defines the theoretical calculations that relates the states to the measurement vector **z** and therefore serves as a mathematical description of the power system. These equations are also referred to as the power flow equations and are described as vectors inside **H**. While in the DC model, those entries consists of a set of linear functions of the state variables, those functions are nonlinear as far as the AC model is concerned. The determination of the state variables is done according to the following criteria:

$$\min J(\mathbf{x}) = (\mathbf{z} - \mathbf{h}(\mathbf{x}))'\mathbf{W}(\mathbf{z} - \mathbf{h}(\mathbf{x}))' \tag{6}$$

W in (6), is a diagonal matrix that contains the measurement weights. These are based on the reciprocals of the measurement error variance σ:

$$\mathbf{W} = \mathbf{R}_{\mathbf{z}}^{-1} = \begin{bmatrix} \sigma_1^{-2} & \cdots & \cdots & \cdots \\ \cdots & \sigma_2^{-2} & \cdots & \cdots \\ \vdots & \vdots & \ddots & \vdots \\ \cdots & \cdots & \cdots & \sigma_m^{-2} \end{bmatrix} \tag{7}$$

where **Rz** is the covariance matrix of the measurement. The performance index J (**X**) is then differentiated to obtain the first order optimal conditions which can be solved using iterative methods, such as Honest Gauss Newton method, Dishonest Gauss Newton method and Fast Decoupled State Estimator [23]. The first order optimality condition of (6) to be solved is then expressed as:

$$\frac{\partial J(\mathbf{x})}{\partial \mathbf{x}}|_{\mathbf{x}=\hat{\mathbf{x}}} = -2\mathbf{F}_h^T(\hat{\mathbf{x}})\mathbf{W}(\mathbf{z} - \mathbf{h}(\hat{\mathbf{x}}))' = 0 \tag{8}$$

where $\mathbf{F}_h$ is the Jacobian matrix derived from $\mathbf{h}(\mathbf{x})$ and the $\hat{\mathbf{x}}$ is the estimated state vector. In the case of the CDS, the state estimation process is modified slightly in order to remain compatible with the planning stages in the executive, which will be discussed later. Therefore, for the first t_s cycles, state estimation proceeds similar to the iterative procedures mentioned previously. As from t_s, the preceding calculated state of the AC state estimator, $\mathbf{x}_{k-1}$, is used as the initial guess for the current cycle with any of those iterative techniques. Moreover, the number of iterations is also limited to N_s iterations to save on computational resources.

2.3 Bad data detection

During the state estimation process, faulty measurements have to be detected and identified to be removed as they lead to erroneous calculated states. However, the statistical properties of these errors simplify their detection and identification. In order to determine those errors, the estimated measurements, $\hat{\mathbf{z}}$, are first calculated from (5) using the following equation for the AC case:

$$\hat{\mathbf{z}} = \mathbf{h}(\hat{\mathbf{x}}) \tag{9}$$

The individual estimated measurement error is then obtained using:

$$\hat{e}_j = (z_j - \hat{z}_j) \tag{10}$$

As these errors follow a zero mean Gaussian distribution [16], techniques such as the Chi-Squares test and normalized residual have been the most common ones applied for their detection [27]. When Chi-squares test is applied, it is assumed that the state variables are mutually independent from each other and the errors follow a normal distribution. The test involves a number of iterative steps that depend on the number of degrees of freedom of the system, sum of squares $\hat{f}$ and a critical value corresponding to α satisfying the inequality:

$$\hat{f} < \chi^2_{(k,\alpha)} \tag{11}$$

where k is the appropriate number of degrees of freedom and α is a specified probability. Thus, $\hat{f}$ will be large when a large number of bad measurements are present. However, since k is large in power systems, this method allows for the removal of those measurements that are responsible for the largest standardized residuals.

2.4 False data injection attacks

FDI attacks (also known as Bad Injection attacks) is a special category of attacks targeting the SG, whereby bad measurements are injected such that they are able to bypass the bad data detection methods discussed previously. While FDI attacks can also target other cyber-physical systems, various forms of these attacks and consequences have been investigated in [11, 12, 15, 16, 28–38]. In this paper, FDI attacks will be simulated using assumptions from [26], whereby it is assumed that the system parameters and topology (system Jacobian) is known to the attackers, and [18], where a mathematical formulation for simulating the FDI attack in the AC model is provided. Additionally, FDI attacks satisfying the first assumption regarding prior knowledge of the system have been proven to result in more disastrous consequences. Moreover, in [17], the authors demonstrate how an attacker, using that knowledge of the system matrix $\mathbf{H}_{m\times n}$, can inject an attack vector $\mathbf{a}_{m\times 1}$ to the measurement vector $\mathbf{z}_{m\times 1}$ that remains undetected from the detection techniques mentioned previously. Consequently, with the insertion of $\mathbf{a}_{m\times 1}$, the new corrupted measurement signals $\mathbf{z}'_{m\times 1}$ takes the following form:

$$\mathbf{z}'_{m\times 1} = \mathbf{z}_{m\times 1} + \mathbf{a}_{m\times 1} \tag{12}$$

Hence, this will result in the calculation of an incorrect system state vector $\mathbf{x}'_{m\times 1}$ instead of the original state $\mathbf{x}_{m\times 1}$. The difference between those states is denoted as $\mathbf{c}$ and is calculated as follows:

$$\mathbf{x}' = \mathbf{x} + \mathbf{c} \tag{13}$$

For the AC model, it is shown in [18] that the attack vector will remain undetected when it satisfies the condition:

$$\mathbf{a} = \mathbf{h}(\mathbf{x}_a) - \mathbf{h}(\mathbf{x}) \tag{14}$$

It is then proven as follows:

$$\begin{aligned}
\mathbf{r}_{attack} &= \mathbf{z}' - \mathbf{h}(\mathbf{x}') \\
&= \mathbf{z} - \mathbf{h}(\mathbf{x}') + \mathbf{h}(\mathbf{x}) - \mathbf{h}(\mathbf{x}) \\
&= \mathbf{z} + \mathbf{a} - \mathbf{h}(\mathbf{x}') + \mathbf{h}(\mathbf{x}) - \mathbf{h}(\mathbf{x}) \\
&= \mathbf{r} + \mathbf{a} - \mathbf{h}(\mathbf{x}') + \mathbf{h}(\mathbf{x}) \\
\mathbf{r}_{attack} &= \mathbf{r}_{normal} \;\; (\text{since,} \;\; \mathbf{a} = \mathbf{h}(\mathbf{x}_a) - \mathbf{h}(\mathbf{x}))
\end{aligned} \tag{15}$$

Consequently, in the case of nonlinear state estimation, it is more complicated to implement the FDI attack. Compared to the attack in the DC case [17], where the attacker only required knowledge of the Jacobian matrix, in the AC model, the latter is now additionally required to have some prior knowledge of the current states of the system. While it is more complicated to meet those conditions, it is still shown in [18] that such an attack is possible and the consequences can be disastrous. In both the DC and AC model, the calculation of wrong state variables, caused by this attack, can start a domino effect of incorrect control decisions leading to dire consequences. As this type of attack targets state estimation in the SG predominantly, the vector **a** can be inserted physically by tampering with the meters or wirelessly by injecting the offsets when the readings are transmitted to the SCADA. Hence, the substation state estimator (SSE), which is also an important component of the SG, will also be the target of such attacks as it plays an essential role in state estimation at the substations.

3. Architectural structure of CDS for smart grid

From a neuroscience perspective, the CDS is the entity that matches Fuster's paradigm [9] the closest as far as cognition is concerned. Basically, the CDS is made up of four components namely; environment, perceptor, executive and feedback channel. Moreover they are arranged in a very particular way. The feedback channel links the perceptor and executive, which are situated on two opposite sides. The environment finally closes the global feedback channel whereby the entire CDS is contained within it. Since the focus of this chapter is the nonlinear state estimation and FDI attack in the SG, the AC state estimator will be considered as the environment with which the CDS interacts since it is the recipient of the measurements in the network. By acting as the supervisor of the network, the CDS empowers the state estimator, through CC, with the cognitive ability to learn during every PAC which measurements to prioritize for optimal state estimation and which to ones to discard. **Figure 1** shows the complex diagram whereby the CDS and AC state estimator are brought together for meeting the goals mentioned previously. In the next subsections, it will be elaborated how the arrangement and the role of each constituent plays a major role for goal-oriented action on the SG.

3.1 Perception-action cycle

When the environment is free of uncertainty, the PAC is responsible for updating the CDS with new information from the environment for every cycle. Thus, with the continuous acquisition of new information from this global feedback loop, the information extraction ability of the perceptor is constantly being improved with each successive cycles. Consequently, this sets up an uninterrupted cyclic directed flow of information from the perceptor to the executive to lead the PAC with the most optimal actions to be performed on the environment. As a result, this hypothesis for a goal-focused scenario is then modified with new information gained from the PAC to allow the executive to improve its current ability to achieve the primary goal that it was designed for.

3.2 Perceptor

Similar to the concept of Percept in the agent of AI [39], both in the brain and CDS, a perception process is performed on incoming measurements. The perceptor of the CDS extracts useful information from the noisy measurements, which

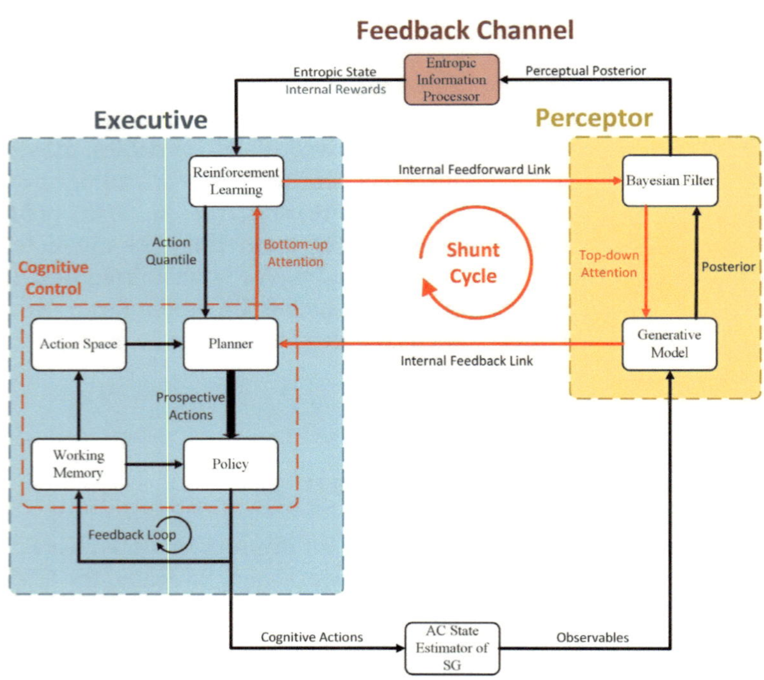

Figure 1.
Architectural structure of CDS for the nonlinear SG.

subsequently the executive uses to optimize its actions and improve the information gain for the next cycles. Those actions, performed by the executive under CC, are called cognitive actions. However unlike the role of the percept in AI, the perceptor perceives the environment directly and extracts relevant information from it which in turn the cognitive controller, residing in the executive, uses to sense the environment indirectly. In order to perform its function, the perceptor is made up of the generative model and the Bayesian filter, which are reciprocally coupled to each other.

3.2.1 Generative model

As defined in [6], the first component of the perceptor for the CDS is conceptually the *Bayesian generative model* [6], which acts as classifier for the observables received from the environment. However, in [7], it was argued that due to the dynamic nature of the SG, the Bayesian generative model would not be suitable for this specific application. Due to the complexity of the SG and its adoption for almost all applications, it is of upmost importance to detect anomalies or cyber-attacks as soon as possible before they can infect the network further, thereby starting a domino effect of cascaded problems throughout the entire network and end users. Therefore, inspired from quickest detection theory, the generative model proposed for the perceptor was based on cumulative sum (CUSUM) and is written as follows:

$$\mathbf{B}_k = \sum_{i=k-L}^{k} \mathbf{x}_i \tag{16}$$

Where k refers to the current cycle number, L is the window over which the past states is being accumulated, $\mathbf{B}_k$ is the vector retaining the cumulative sum for each cycle and $\mathbf{x}_i$ is the vector of the states' output from the AC state estimator for the cycle i. While CUSUM-based detection methods has been very effective in detecting FDI attacks in [40, 41], they fall short when the attacker has prior knowledge of the threshold applied. Indeed, the latter can then craft an attack that remains undetected. However, the CDS allows us to bypass this problem through the use of the dynamic *entropic state*, as will be elaborated later. The entropic state is the foundation of control and attack detection in this CDS structure adapted for the SG. Lastly, the CUSUM based generative model also possesses some other desirable traits such as the smoothing out of noise operating under the slow dynamics of the SG.

3.2.2 Bayesian filter

The second component of the perceptor is the Bayesian filter, which is coupled to the generative model. Although the equations describing the SG for state estimation are nonlinear in nature, we can linearize the state estimates using the *Kalman* filter and assuming that it is operating under additive white Gaussian noise [42]. Since we are assuming that the power system is quasi-static in nature in this paper [43–45], we can use the well-known Kalman filter as the Bayesian filter in the perceptor. The Kalman filter is based on the state-space model which operates on a pair of equations known as the Process equation and the Measurement equation respectively. Moreover, under quasi-static assumptions, we can assume that the state variables $\mathbf{x}$ at the time $k+1$ will only deviate by a small amount from its previous values at its previous cycle k. Consequently we can simplify this relationship to the following equation:

$$\mathbf{x}_{k+1} = \mathbf{x}_k + \omega_k \tag{17}$$

where ω_k is independent Gaussian noise vector with zero mean. Based on (17), we can propose the measurement equation as follows:

$$\mathbf{Y}_k = \mathbf{L}_k\mathbf{B}_k + \omega_k \tag{18}$$

and the covariance of matrix ω_k as:

$$\mathbf{R} = \text{diag}\left[\sigma_\omega^2\right], \sigma_\omega^2 = \text{var}[\omega_i] \tag{19}$$

As we are assuming that the system is operating under quasi-static conditions, a random walk model can be employed as the process equation as follows:

$$\mathbf{B}_{k+1} = \mathbf{F}_k\mathbf{B}_k + \mathbf{v}_k \tag{20}$$

where $\mathbf{v}_k$ is the process noise vector which is assumed to be statistically independent and zero mean. The covariance matrix of $\mathbf{v}_k$ is:

$$\mathbf{Q} = \text{diag}\left[\sigma_\text{v}^2\right], \sigma_\text{v}^2 = \text{var}[\text{v}_i] \tag{21}$$

Referring to (18) and (20), the system matrix $\mathbf{L}_k$ and the predictive transition matrix $\mathbf{F}_k$ are assumed to be identity respectively. In regards to the measurement and process equations mentioned previously, the computational steps of the Kalman filter starts with some predefined initial estimates of the states $\hat{\mathbf{B}}_{k|k}$, and predicted error covariance, $\mathbf{P}_{k|k}$, which are used for the time update steps as follows:

The predicted estimated states of the generative model and predicted error covariance, $\hat{\mathbf{B}}_{k+1|k}$ and $\mathbf{P}_{k+1|k}$ respectively, are calculated using the following equations:

$$\hat{\mathbf{B}}_{k+1|k} = \mathbf{F}_{k+1,k}\hat{\mathbf{B}}_{k|k} + \mathbf{v}_k \tag{22}$$

$$\mathbf{P}_{k+1|k} = \mathbf{F}_{k+1,k}\mathbf{P}_{k|k}\mathbf{F}_{k+1,k}^T + \mathbf{Q} \tag{23}$$

When the next cycle starts, those two estimates are then used for the measurement update stages to calculate the Kalman gain, $\mathbf{K}_k$, filtered accumulated estimate, $\hat{\mathbf{B}}_{k|k}$, and to update the process covariance matrix, $\mathbf{P}_{k|k}$, according to the equations below:

$$\mathbf{K}_k = \mathbf{P}_{k|k-1}\mathbf{L}_k^T\left(\mathbf{L}_k\mathbf{P}_{k|k-1}\mathbf{L}_k^T + \mathbf{R}\right)^{-1} \tag{24}$$

$$\hat{\mathbf{B}}_{k|k} = \hat{\mathbf{B}}_{k|k-1} + \mathbf{K}_k\left(\mathbf{Y}_k - \mathbf{L}_k\hat{\mathbf{B}}_{k|k-1}\right) \tag{25}$$

$$\mathbf{P}_{k|k} = \mathbf{P}_{k|k-1} - \mathbf{K}_k\mathbf{L}_k\mathbf{P}_{k|k-1} \tag{26}$$

As a result, through the iteration of the time update and measurement update steps, the preceding *a posteriori* estimates are used to predict new *a priori* estimates.

3.3 Feedback channel

The feedback channel has very distinctive roles in the CDS as it completes the PAC by bringing together the perceptor and the executive. It is mainly related to control and cyber-attack detection in the SG. In order for the CDS to supervise the SG, the feedback channel holds the entropic-information processor, which is tasked with calculating the *entropic state* and internal rewards during reinforcement learning in the executive. This will be elaborated in sub-Section 3.4 (Executive) where it is more relevant to the role of the executive during planning.

3.3.1 Entropic-information processor

The directed cyclic flow of information from the perceptor to the executive is known as the *entropic state of the perceptor*. The entropic state is built on the principles of the perceptual posterior, which can be viewed as the incoming filtered posterior embodying the essence of the generative model, Kalman filter and entropy, which is derived from *Shannon's information theory* [46]. The entropic state at time k, in this architecture is calculated using:

$$\frac{h_{k|k} = \mathrm{Tr}\left\{\mathbf{P}_{k|k-1} - \left(\mathrm{diag}\left\{\hat{\mathbf{B}}_{k|k-1} - \mathbf{Y}_k\right\}^2\right)\right\}}{\mathrm{Tr}\left\{\mathbf{P}_{k|k-1}\right\}} \tag{27}$$

where Tr represents trace operator, diag{.} is the diagonal operator and $h_{k|k}$ is the entropic state. In [7], the efficiency of (27) for control and cyber-attack detection was proven and illustrated. For this reason, it will be retained for the CDS architecture being elaborated. Mathematically, (27) simplifies the information between the filtering-error covariance $\mathbf{P}_{k|k-1}$ and the error between the state estimate $\hat{\mathbf{B}}_{k|k-1}$ and current states calculated at cycle k into a single metric. The denominator of (27) normalizes the equation such that $h_{k|k}$ can only take values ranging from 0 to 1 when the environment is operating in the absence of uncertainty. The degree of disturbance affecting the SG can then be characterized through the entropic state; the lower $h_{k|k}$ is, the greater the amount of disturbance or uncertainty in the system. Since the SG will be facing different situations during its operation such as the normal day to day routine and cyber-attacks, we can further dissociate the entropic state with the two following important properties:

i. When the environment is operating in the absence of uncertainty, $h_{k|k}$ will always be positive because of the probabilistic representation of the uncertainties.

ii. When uncertainties are present, $h_{k|k}$ will fluctuate around values which are less than 1. Thus, to distinguish between normal uncertainties, such as process disturbance, due to the probabilistic nature of the environment, and abnormal uncertainties, such as cyber-attack, a suitable threshold γ can be chosen such that if $h_{k|k}$ is below γ, then this would indicate presence of attack and to switch on CRC.

3.4 Executive

From a design perspective, the Executive is the most important entity of the CDS as it is responsible for control of the SG in the absence of uncertainty. With this goal in mind, it consists of Reinforcement Learning (RL) and Cognitive Control (CC), which can be further subdivided into the action space, planner, working memory and policy.

3.4.1 Reinforcement learning: Bayes-UCB

Asides from its role in the calculation of the entropic state during each PAC, the feedback channel is also involved in the calculation of internal rewards during the planning stages of the RL [39] algorithm in the executive. RL in the CDS is based on the current entropic state at each cycle which is subsequently used to optimize an objective function for optimal control in the network. Before we elaborate on the pivotal role of RL with the other components of the executive, Bayes-UCB [47] RL algorithm will be covered briefly in order to give an overview on how it operates. Bayes-UCB represents the current state of the art from a class of multi-armed bandit algorithms called UCB algorithms [48], which are based on the principle of optimism in the face of uncertainty. In this approach to the multi-armed bandit problem, the algorithm updates the estimate of the reward distribution for each action using a Bayesian method. The action that will be applied is then chosen according to the one that will yield the highest reward. Consequently, Bayes-UCB algorithm is an index policy that uses the prior distribution to pick a dynamic quantile of the posterior estimates for the index for each action. Hence, at each discrete time t, the algorithm will select the action A_t that satisfies the following condition:

$$A_t = \underset{a}{\text{argmax}} q_a(t) = Q\left(1 - \frac{1}{t(\log(t))^c}, \lambda_a^{t-1}\right) \tag{28}$$

where $Q(\alpha, \pi)$ refers to the quantile of order α of the distribution π. Moreover, by assuming that the rewards follow a Bernoulli distribution, and when the prior distribution of each action is Beta(1,1), [49] shows that (28) can be further simplified. To maintain consistency of the used notations in this paper, (28) can be reduced to:

$$A_k = \underset{a}{\text{argmax}} q_a(k) = Q(1 - \frac{1}{k(\log(k))^c}; \text{Beta}(S_a(k) + 1, N_a(k) - S_a(k) + 1) \tag{29}$$

where k is the PAC cycle number, S_a is the cumulative reward for action a, N_a is the number of times action a has been chosen and **c** is real parameter. As the CDS is a construct that draws its origin from the neuroscience of the brain, it is to be emphasized that Bayes-UCB shares many common traits to the Bayesian approach of decision making in human brains [50]. Following this brief coverage of Bayes-UCB, it will be shown in the next section, pertaining to Cognitive Control, how the RL algorithm integrates the system configuration **H** of the power grid, the generative model of the perceptor and the process model in the Kalman filter together for optimal state estimation.

3.4.2 Cognitive control

CC can be considered in many ways as the heart of the CDS as it brings together all the components, described so far, for goal oriented action on the SG. CC is made up of two important modules namely the *planner* and the *policy*. The planner is involved in the extraction of a set of prospective actions from the action-space A and their evaluation during the planning cycles (i.e., shunt cycles [6] in CDS terminology) during each PAC. Consequently, under the influence of attention from one PAC to the next, the policy learns the most appropriate actions yielding the maximum rewards to be applied. In the context of the SG, the action space consists of discrete weight values that can be attributed to the different meters. Thus, under the influence of attention, the CDS will learn the optimal weight values for the different meters for optimal state estimation. Those meters, which are detrimental for the state estimation, will be assigned lower weight values while those, which are crucial, will be given larger weight values as the CDS keeps learning about its environment to better perform its set goal. Planning in CC brings together all the other modules previously discussed. The process starts with the selection of a randomly chosen prospective action $a_k^{i,j}$ which represents weight value a^i for meter j during cycle k. This hypothesized weight value is then applied virtually to the weight matrix $\mathbf{W}$ in (5) and (6) to form $\mathbf{W}_k^{i,j}$. $\mathbf{W}_k^{i,j}$ is then used to calculate a new planned state estimate, $\hat{\mathbf{x}}_k^p$, using the same procedures mentioned in the last paragraph of Section 2.2. Thus, the same preceding calculated state of the AC state estimator, $\mathbf{x}_{k-1}$, is used as the initial guess for the current cycle using any of those iterative techniques cited. However, the number of iterations is limited to N_p iterations this time around. Due to the different weight matrices being examined, each iteration of using a $\mathbf{W}_k^{i,j}$ will also involve a different hypothesized gain, $\mathbf{G}_k^p$, during planning. Since state estimation is computationally costly, by doing this process with a restricted number of iterations in the methodology explained, this allows the CDS to learn during the planning stages at a lower resource cost. With $\hat{\mathbf{x}}_k^p$ denoting the planned state estimate using the modified weight matrix with the hypothesized weight, the planned cumulative sum involving $\hat{\mathbf{x}}_k^p$ is then calculated:

$$\mathbf{B}_k^p = \sum_{i=k-L}^{k-1} \mathbf{x}_i + \hat{\mathbf{x}}_k^p \tag{30}$$

where $\mathbf{B}_k^p$ is the planned cumulative sum involving $\hat{\mathbf{x}}_k^p$ instead of $\hat{\mathbf{x}}_k$. Using this new cumulative sum, a planned entropic state, $h_{k|k}^p$, is subsequently calculated as follows:

$$h_{k|k}^p = \frac{\det\left\{\mathbf{P}_{k|k-1} - \left(\operatorname{diag}\left\{\hat{\mathbf{B}}_{k|k-1} - \mathbf{B}_k^p\right\}^2\right)\right\}}{\det\left\{\mathbf{P}_{k|k-1}\right\}} \tag{31}$$

The presence of uncertainties in the environment, whether stochastic or probabilistic, will cause a deviation in the output of the generative model of the perceptor from the estimated hidden state of the Kalman filter. Hence, the goal of (31) is to reduce this divergence by finding the best configuration weights for the respective meters. This condition is satisfied whenever the $\mathbf{W}_k^{i,j}$ generates $h_{k|k}^p$ closer to the optimal value of 1, which implies that the planned estimated state of the AC state estimator reduces the propagated variation in the generative model.

3.4.3 Internal rewards

Moving forward with equations that describe the planning steps, the stage is now set to define the relationship between the previous steps and the calculation of the internal rewards during RL. The hypothesized internal rewards, $r_k^{i,j}$, associated with each prospective action $a_k^{i,j}$, for cycle k can be written as:

$$r_k^{i,j} = h_{k|k}^p - h_{k|k} \tag{32}$$

As it can be seen from (32), the objective of RL, when operating under CC, is to minimize the amount of uncertainty in the SG by searching for an improved weight configuration during every PAC that will result in a better entropic state than the previous cycle. In other words, RL attempts to restrict the amount of uncertainty or disturbance during the state estimation process to the range computed by the Kalman filter in the perceptor. Referring back to the steps described so far that led to (32), we can see that the CDS, as defined in this specific architecture, learns from the past and present actions to pick the best actions for the future. To assist in this task, after undergoing the shunt cycles during every PAC, the working memory holds temporarily the actions that have achieved the highest quantile from Bayes-UCB in (29) and applies them to the system before starting the next PAC. Thus, when the next PAC starts and a new set of prospective actions are evaluated according to their quantile values, if any of those actions achieves a higher quantile than the quantile of its respective meter in the working memory, then the higher achieving action will replace that previously considered best action. This way of performing control in the SG can also be viewed from a Bandit perspective, whereby it can be considered as a Contextual Bandit problem where every cycle presents new situations to be faced. According to those conditions, the actions performed on the SG will modify the system configuration to a new set point, from which the RL algorithm will have to adapt. This then continues on until the CDS is brought to rest. The complete algorithm of the methodology presented in this chapter can be found in [51] where it is integrated with a cyber-attack mitigation strategy known as Cognitive Risk Control, which was not discussed in this chapter. In [51], a greater discussion on the parameters and its selection is provided and contrasted with other popular cyber-attack detection methods.

4. Computational experiments

In this section, two different experiments are carried out to show the capability of CC in this new CDS architecture adapted for the smart grid. The first experiment shows CC's potential for optimal state estimation by using the optimization of the entropic state as objective function. In the second experiment, the capability of the entropic state as an attack detector will be demonstrated in four different scenarios based on the amount of information an attacker has and his access to the sensors. As IEEE bus networks have generally been used as benchmarks for evaluation in the other papers previously referenced and relating to this topic, the IEEE 14-bus network will be used for assessing the architecture proposed in this chapter. Since this particular network comprises of a large number of measurements and states, the results for the two different experiments will focus on certain aspects of the network that are relevant to the actual simulation. For both experiments, the data used to simulate the network configuration comes from the 14-bus case file in *MATPOWER* [53] which is an Electric Power System Simulation and Optimization Tools for MATLAB and Octave. Moreover, in order to bring about the modification for the AC state estimation algorithm, the doSE function of *MATPOWER* was modified for the requirements of the architecture. Originally, the algorithm uses Honest Gaussian Newton method with a maximum number of iterations of 100 and error tolerance of 10^{-5}. It also uses a *Flat Start* initialization each time the function is called. During the *Flat Start*, all the values of the different states for the initial guess is set to 1 unit.

4.1 Cognitive control for BDD

In the first experiment, the measurement signals relating to the state values were available from the case data in *MATPOWER* [52]. For this simulation, a noisy version of those signals was then generated with a signal-to-noise ratio (SNR) of 20 dB to create $\mathbf{z}$. From the case data, 39 measurement signals are used to calculate the 29 state values of the IEEE 14-bus network, half of which are the voltage magnitudes and the other are the voltage angles for the different busses involved. The total duration of this experiment is 2000s. The parameter L, which is the window over which the past states is being accumulated, of the generative model of the perceptor was set to 20. In regards to the initialization of the Kalman filter, the initial estimates of the values to be received from the generative model are assigned a value of 0 and the diagonal elements of $\mathbf{Q}$ were set to 0.0324. Those of $\mathbf{R}$ were assigned a value of 0.01. On the executive side of the CDS for CC, the action space is made up of 156 actions, whereby each meter can be assigned a weight value from the following: 25, 50, 100, and 200. The goal of this experiment is to highlight this architecture's properties in terms of adaptability and robustness towards optimal state estimation to changing conditions. Consequently, in order to create a perturbation in the system, the SNR of the following meters is changed to 5 dB at the mentioned times: $\mathbf{t} = \mathbf{1000s}$ for meter 2 and $\mathbf{t} = \mathbf{1200s}$ for meter 15. This simulated context can be viewed as meter malfunction or a random attack, where the attacker only has limited access to meters to perform his task. In this simulation, CC is started at $\mathbf{t} = \mathbf{300s}$. As mentioned in the earlier sections, CC is not started at $\mathbf{t} = \mathbf{0s}$ as some time (cycles) have to be allowed so that the Kalman filter can settle on the track in order for the algorithm to be operated effectively.

Referring to **Figure 2**, it can be seen that CC makes the whole network dynamic, whereby the executive of the CDS is assigning the best weight values for the meters for optimal state estimation on a cycle to cycle basis. Consequently, the cognitive controller shows it ability to learn from the current and past cycles to choose the best actions for future. Moreover, the constant modification of the weight values adds another level of nonlinearity on top of the already very complex and nonlinear AC state estimator. While this may appear to be over-complicated at first, the results show that this is not only feasible but it also makes the SG more powerful. As it can be seen in **Figure 2**, at the first instance of meter malfunction for meter 2 at $\mathbf{t} = \mathbf{1000s}$, this has virtually no effect on this system at all as the CDS has assigned a lower weight value to that meter compared from the rest. While **Figure 2** shows the graphs of weight values for the some of the meters pertinent to this simulation, it is left to reader to realize that all the meters are undergoing weight reconfigurations every cycle. Thus, the different respective weight values for the meters are not all the same since the cognitive controller is adapting to the probabilistic nature of the noisy signals continuously. It is also shown that the algorithm is able to apply more than one action during each PAC under a stable manner. At $\mathbf{t} = \mathbf{1200s}$, when meter 15 starts malfunctioning, we can now really see the capability of the architecture. As shown in **Figure 2**, it takes only a couple of cycles for the cognitive controller to learn and adapt to the new situation by lowering the weight assigned to meter 15

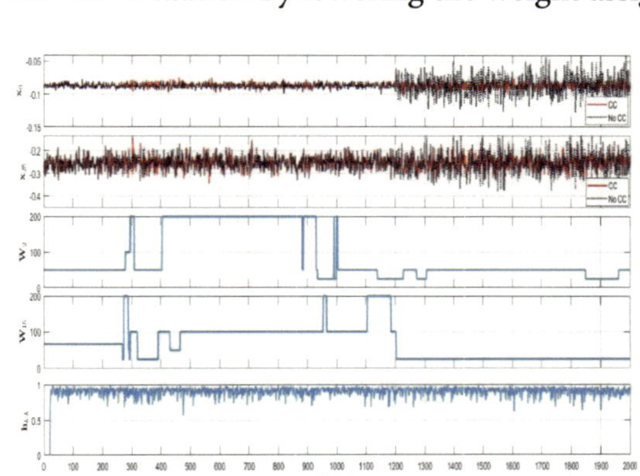

Figure 2.
Graphs of some affected states, weights and entropic state.

and compensating for it by boosting the other meters. Thus, we can see that state 6 is the most affected and state 25 is also afflicted to a lower extent. Compared to the traditional AC state estimator, the CC algorithm is able to keep this perturbation under control as demonstrated in the referenced figure. Consequently, this shows the robustness of the algorithm to adapt and act according to the evolving situations. Although some of those weight values are changed at a later point in time, this is due to the frequentist approach of the Bayes UCB coupled with the probabilistic origin of the noise. As a result of those reconfigurations in earlier situations, this highlights the cognitive ability of the controller to trust certain meters more than the others. This simulation demonstrated CC's ability to pick the best set of meters for state estimation on the go. Referring back to **Figure 2**, it can be seen the CC has performed better than the traditional algorithm. Lastly, the Chi-squares test was not implemented in this experimented as it is based on statistical properties of the signals while the approach proposed is rooted on the principle of cognition of the brain.

4.2 Cyber-attack detection

In this section, the dual property of the entropic state for FDI cyber-attack detection will be demonstrated. Previously, it was shown how the latter is an objective function for the normal running of CC under the absence of uncertainty whereby it is always positive. However, when the presence of uncertainties are no longer probabilistic, such as when an attack takes place, the entropic state will also enable early detection of such attacks. In all the cases, it is assumed that the attacker has knowledge of current states of the system. Although many specialized attacks such as replay attack or Distributed Denial of Service (DDoS) attack exist, four broad categories of FDI attacks will be considered as follows:

i. Case 1: Here we assume that the intruder has perfect knowledge of the network configuration **H** and full access to meters to commit the perfect FDI attack as described earlier. The remaining cases consider more realistic scenarios whereby the hacker faces some restrictions.

ii. Case 2: In this scenario, the intruder still has full knowledge of **H** but has limited access to meters in the grid. To simulate this attack, half of the rows of the attack vector **a** are zeroed to represent the inability to access those sensors.

iii. Case 3: Here the circumstances of case 2 are flipped around; the intruder has access to all the meters but incomplete knowledge of **H**. To carry out the attack, the entries of **H**, used to craft the attack vector, are altered in some way as an indication of the lack of information. Depending on the amount of incomplete information, this attack can have different effects. In order to simulate the attack here, some noise are added to most of the non-zero entries of the attacker's **H**. However, it is important to mention that if the attack vector was generated using an **H** matrix where the zero entries have also been altered, as a representation of more lack of information from the attacker's side, then consequences will vary. In the lower extreme, the attack will still be feasible and detected by the entropic state. In the most extreme cases, the state estimation process will not coverage and fail.

iv. Case 4: Finally, a rogue attack combining case 2 and 3 is considered. The attacker has both imperfect knowledge of **H** and constrained access to the sensors in the grid. In order to simulate this attack, the conditions used in those two cases were combined to create the attack vector.

The mentioned attacks in those different situations were simulated on the IEEE 14 bus network as shown in **Figures 3–6**. In all of the mentioned cases, the hacker's goal is to deflect the value of two of the voltage magnitudes by −0.3 and 0.4 units respectively and one voltage angle by 0.3 radians. Since attack data is not publicly available, the parameters in the *MATPOWER* package will be used to simulate the IEEE 14 bus network.

In all four attack cases, the attack is started at $\mathbf{t = 500s}$. The same parameters were used as in the previous simulation. Additionally, the property of h_k will be demonstrated as a stand-alone utility in the absence of CC. While CC is originally defined for tackling control when the uncertainties are probabilistic and h_k is

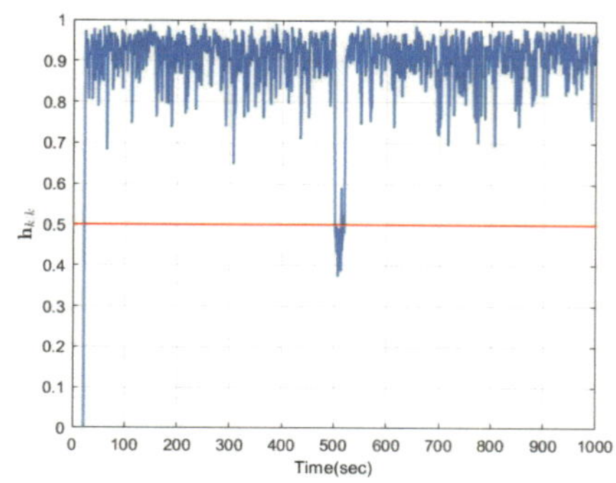

Figure 3.
Case 1.

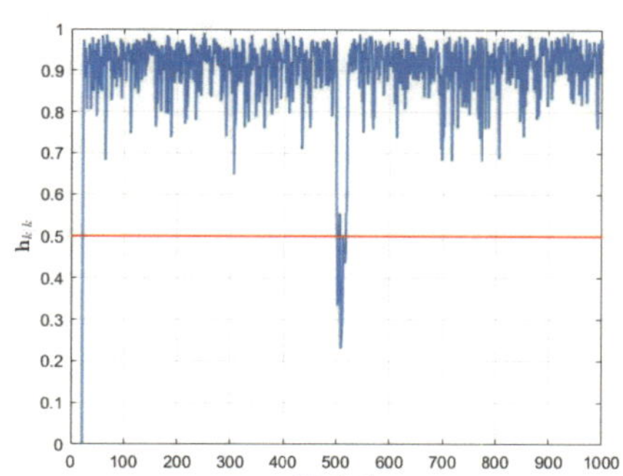

Figure 4.
Case 2.

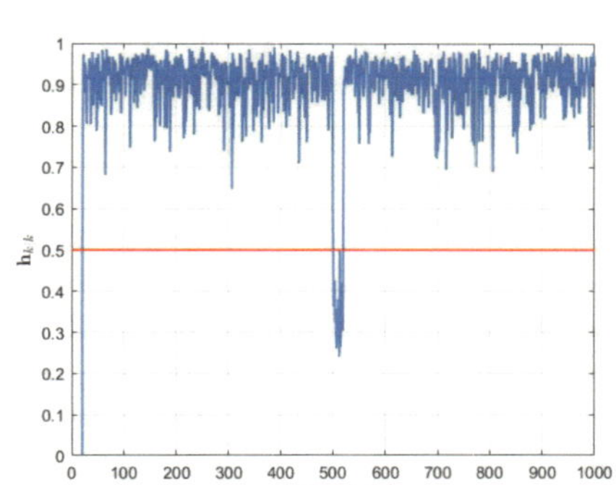

Figure 5.
Case 3.

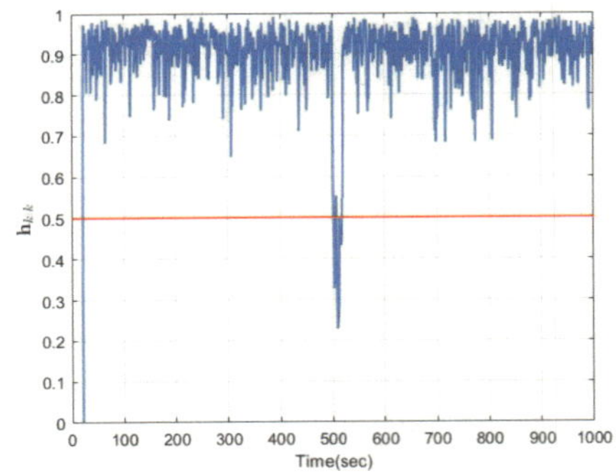

Figure 6.
Case 4.

positive, the CDS has to expand its structure its to include CRC to be able to bring risk under the control in the presence of the cyber-attacks. The implementation of CRC to this architecture can be found in chapter 4 of [51]. The results pertaining to the simulation of the attacks presented earlier are shown in **Figures 3–6**. In all four cases, by assigning a suitable γ, the attack was detected. Furthermore, it can also be seen that as the hacker has less and less information on the current grid, it becomes easier to detect the deflection as the entropic state becomes more negative. The results also displays the efficiency of the generative model, whereby the attack propagates throughout the cumulative sum up to a certain point before the Kalman filter gets back on the current track. This propagation causes h_k to become increasingly negative which consequently lends the property of detection. All the computational experiments were carried out on a system running Windows 10 with an Intel i7-8750H processor. The computational running time of the first experiment was around **40s** and the second experiment took ranged from the shortest time of **3s** for case 1 up to the longest time of **17s** for case 3 and 4. This increase in time for these two specific cases has mostly to do with the increased number of iterations required from the AC state estimator when the sensor data has lost some coherence due to the random attack vector generated as a result of lack of information.

If the CDS architecture proposed in this paper is applied in a medium or large-scale power system, the computational complexity will be lesser compared to the other current detection methods, such as the ones mentioned earlier. A greater elaboration of this technique compared to the other detection methods can be found in [7]. Moreover, the application of the CDS for an application such as the SG is revolutionary as it is a dual system catering to both the control and attack detection aspects of the SG. The main parameter of interest that needs to be scaled up for a more complex grid will be the number of shunt cycles since more meters will have to be evaluated. Nevertheless, it is recommended to keep the action space small so as to make planned rewards, during planning, distinguishable from each other. Another important hyper-parameter in the system, especially for FDI attack detection, are the values in the **Q** matrix. Unlike many tracking applications such as the simulation carried out in [5], which was supported by a mathematical formulation [53], this is not the case in our system. Thus, the contents of **Q** has be defined by the designer depending on the required sensitivity of the system towards disturbances. In order to find proper values for **Q**, prior simulations can be carried out using past historical data. Usually, it is recommended to start with very small values, like the ones used in the simulations carried out in this paper, and then tuning until the desired performance is obtained. Lastly, as the SG is scaled up, that hyper-parameter will have to be increased to reflect the circumstances of a bigger power system.

As voltage fluctuations are common occurrence disturbances in power systems, the second simulation was designed to provide the reader a greater intuition on how the algorithm is able to distinguish between what constitutes a perturbation and the normal condition. When the states of the AC state estimator is experiencing important fluctuations, this is propagated to the generative model and therefore affects the entropic state as a result. Since $h_{k|k}$ serves as an embodiment of the grid's performance, it was illustrated in the earlier simulation how those perturbation would cause a decline in the entropic state. Since the objective function of CC is to always bring $h_{k|k}$ as close as possible to 1, the optimization of $h_{k|k}$ allows CC to reduce fluctuations in the system and keep state estimation under control. Additionally, it was shown in **Figure 2** that when the attack occurred, this caused the estimated states to experience greater deviation. This was then propagated to the generative model and the Kalman filter as result, thereby causing a large drop in $h_{k|k}$ for a number of cycles. This was then successfully detected through the use of the threshold γ. Consequently, those experiments showcases the importance of each of the individual roles of the different components of the CDS and how they work together for goal oriented action on the SG.

5. Conclusion

This chapter covered the following points:

i. This is the first time that a CDS structure has been proposed for handling the nonlinear version of the SG. While previous research in this field, which were focused on bringing the CDS and the SG together, were based on the DC model, the AC model is a more realistic approach to the SG. Consequently, the new construct, which was described in the chapter, shows a lot of potential at tackling the future problems that the grid will face in the coming years as it becomes increasingly interconnected with the other aspects IT such as IoT.

ii. While there are some tradeoffs to be made due to the already inherent computational complexity of the AC state estimation algorithms, it was shown that the CC is revolutionary in the sense that it allows the application of multiple actions during every PAC while still maintaining the stability of state estimation.

iii. The CDS tailored for the AC model of the SG, proposed in this chapter, is a unique architecture that is able to make the SG more powerful by providing a new kind of control and cyber-attack detection, that are both based on cognition from the brain's perspective.

In this chapter, a new CDS based architecture was united with the SG in order to tackle the issues of nonlinear state estimation and cyber-attack detection through CC. Computational experiments were carried out to show the individual benefits of CC for optimal state estimation and FDI attack detection respectively. Moreover, it was also discussed how the algorithm and the parameters can be adjusted so that it can be scaled up to work with bigger networks. In those bigger networks comprising of a large number of meters, a function approximator such as a Neural Network [54] can employed to simplify some of the computations involved. Although this chapter focused on the problems of control on state estimation and cyber-attack in the SG, the architecture covered in this paper, can also be formulated to work for other similar applications where state estimation is critical such as Vehicular Radar Systems. In order, to adapt the CDS for other applications, the mathematics involving the perceptor and the executive will have to be adjusted accordingly depending on the final goal of the different intended systems.

Author details

Mohammad Irshaad Oozeer*† and Simon Haykin†
McMaster University, Hamilton, Canada

*Address all correspondence to: oozeeri@mcmaster.ca

† These authors contributed equally.

References

[1] S. Haykin, "Cognitive dynamic systems [Point of view]," Proc. IEEE, vol. 94, no. 11, pp. 1910–1911, Nov. 2006.

[2] S. Haykin, "Cognitive Dynamic Systems: Radar, Control, and Radio", Proc. IEEE, Point of View Article, vol. 100, no. 7, pp. 2095–2103, July 2012.

[3] S. Haykin, "Cognitive radio: Brain-empowered wireless communications," IEEE J. Sel. Areas Commun., vol. 23, no. 2, pp. 201–220, Feb. 2005.

[4] S. Haykin, "Cognitive radar: A way of the future," IEEE Signal Process.

[5] M. Fatemi and S. Haykin, "Cognitive control: Theory and application," IEEE Access, vol. 2, pp. 698–710, Jun. 2014.

[6] S. Haykin, J. M. Fuster, D. Findlay, and S. Feng, "Cognitive risk control for physical systems," IEEE Access, vol. 5, pp. 14 664–14 679, Jul. 2017.

[7] M. I. Oozeer and S. Haykin, "Cognitive Dynamic System for Control and Cyber-Attack Detection in Smart Grid," in IEEE Access, vol. 7, pp. 78320–78335, 2019. doi: 10.1109/ACCESS.2019.2922410

[8] M. I. Oozeer and S. Haykin, "Cognitive Risk Control for Mitigating Cyber-Attack in Smart Grid," in IEEE Access, vol. 7, pp. 125806–125826, 2019. doi: 10.1109/ACCESS.2019.2939089

[9] J. M. Fuster, "Cortex and Mind: Unifying Cognition", Oxford University Press, 2003.

[10] Y. Wang, M. Amin, J. Fu, H. Moussa, "A novel data analytical approach for false data injection cyber-physical attack mitigation in smart grids", IEEE Access 2017.

[11] A. Humayed, J. Lin, F. Li, and B. Luo, "Cyber-Physical Systems Security – A Survey", IEEE Internet of Things Journal, vol. PP, no. 99, pp. 1–1, 2017.

[12] J. Hao, R.J Piechocki, D. Kaleshi, et al: 'Sparse malicious false data injection attacks and defense mechanisms in smart grids', IEEE Trans. Ind. Inf.s, 2015, 11, (5), pp. 1–12 (doi: 10.1109/TII.2015.2475695).

[13] X. Fang, S. Misra, G. Xue, D. Yang, "Smart grid - the new and improved power grid: A survey", IEEE Commun. Surveys Tutorials 2012.

[14] F. C. Scheweppe and J. Wildes, "Power system static-state estimation, Part I: Exact model," IEEE Trans. Power App. Syst., vol. PAS-89, no. 1, pp. 120–125, Jan. 1970.

[15] K. P. V. Priya, J. Bapat, "Bad Data Detection in Smart Grid for AC model", IEEE Indicon 2014.

[16] J J. Grainger and W D. Stevenson JR., "Power System Analysis 1st Edition", McGraw-Hill Series in Electrical and Computer Engineering, 1994.

[17] Y. Liu, P. Ning, M. Reiter, "False data injection attacks against state estimation in electric power grids", ACM CCS pp. 21–32 2009.

[18] Md A. Rahman, and H. Mohsenian-Rad, "False data injection attacks against nonlinear state estimation in smart power grids." In 2013 IEEE Power and Energy Society General Meeting, pp. 1–5. IEEE, 2013.

[19] S. Sridhar, A. Hahn, M. Govindarasu, "Cyber-physical system security for the electric power grid", Proc. IEEE vol. 99 no. 1 pp. 1–15 Jan. 2012.

[20] H. Zhu and G. B. Giannakis, "Robust power system state estimation

for the nonlinear AC flow model," in Proc. IEEE North Amer. Power Symp., 2012, pp. 1–6.

[21] G. Hug and J. A. Giampapa, "Vulnerability assessment of AC state estimation with respect to false data injection cyber-attacks," IEEE Trans. Smart Grid, vol. 3, no. 3, pp. 1362–1370, Sep. 2012.

[22] A. Abur and A. Gómez-Expósito,"Power System State Estimation Theory and Implementation", 2004.

[23] A. Monticelli, "State Estimation in Electric Power System A Generalized Approach", Springer Science+Business Media New York, 1999.

[24] M. Jin, J. Lavaei, and K. H. Johansson, "Power grid ac-based state estimation: Vulnerability analysis against cyber attacks." IEEE Transactions on Automatic Control 64.5 (2018): 1784–1799.

[25] Z. Yu, W. Chin, "Blind false data injection attack using PCA approximation method in smart grid", IEEE Trans. Smart Grid vol. 6 no. 3 pp. 1219–1226 May 2015.

[26] J. Kim, L. Tong, and R. Thomas, "Subspace methods for data attack on state estimation: A data driven approach," IEEE Transactions on Signal Processing, vol. 63, no. 5, pp. 1102–1114, March 2015.

[27] L. Liu, M. Esmalifalak, Q. Ding, V. Emesih, and Z. Han, "Detecting false data injection attacks on power grid by sparse optimization," IEEE Transactions on Smart Grid, vol. 5, no. 2, pp. 612–621, March 2014.

[28] A. Anwar, A. N. Mahmood, M. Pickering, "Modeling and performance evaluation of stealthy false data injection attacks on smart grid in the presence of corrupted measurements", J. Comput. Syst. Sci. vol. 83 no. 1 pp. 58–72 2016.

[29] J. Jiang, Y. Qian, "Defense mechanisms against data injection attacks in smart grid networks", IEEE Commun. Mag. vol. 55 no. 10 pp. 76–82 Oct. 2017.

[30] P. McDaniel, S. McLaughlin, "Security and privacy challenges in the smart grid", IEEE Security Privacy vol. 7 no. 3 pp. 75–77 May/Jun. 2009.

[31] R. Deng, G. Xiao, R. Lu, H. Liang, A. V. Vasilakos, "False data injection on state estimation in power systems—Attacks impacts and defense: A survey", IEEE Trans. Ind. Informat. vol. 13 no. 2 pp. 411–423 Apr. 2017.

[32] D. Wang, X. Guan, T. Liu, Y. Gu, Y. Sun, Y. Liu, "A survey on bad data injection attack in smart grid", Proc. IEEE PES Asia-Pac. Power Energy Eng. Conf. pp. 1–6 2013.

[33] K. Manandhar, X. J. Cao, F. Hu, Y. Liu, "Combating false data injection attacks in smart grid using kalman filter", Proceedings of International Conference on Computing Networking and Communications Communications and Information Security Symposium pp. 16–20 2014.

[34] K. Manandhar, X. Cao, F. Hu, Y. Liu, "Detection of faults and attacks including false data injection attack in smart grid using kalman filter", IEEE Trans. Control Netw. Syst. vol. 1 no. 4 pp. 370–379 Dec. 2014.

[35] P.Y. Chen, S. Yang, J. A. McCann, J. Lin, X. Yang, "Detection of false data injection attacks in smart-grid systems", IEEE Commun. Mag. vol. 53 no. 2 pp. 206–213 Feb. 2015.

[36] Y. Liu, L. Yan, J. Ren, D. Su, "Research on efficient detection methods for false data injection in smart grid", International Conference on

Wireless Communication and Sensor Network (WCSN) pp. 188–192 December 2014.

[37] D. B. Rawat, C. Bajracharya, "Detection of false data injection attacks in smart grid communication systems", IEEE Signal Process. Lett. vol. 22 no. 10 pp. 1652–1656 Oct. 2015.

[38] Y. Gu, T. Liu, D. Wang, X. Guan, Z. Xu, "Bad data detection method for smart grids based on distributed state estimation", Proc. IEEE Int. Conf. Commun. pp. 4483–4487 2013.

[39] R. S. Sutton and A. G. Barto, "Reinforcement Learning", Cambridge, MA, USA: MIT Press, 1998.

[40] Y. Huang et al., "Real-time detection of false data injection in smart grid networks: An adaptive CUSUM method and analysis", IEEE Syst. J. vol. 10 no. 2 pp. 532–543 Jun. 2016.

[41] S. Li, Y. Yilmaz, X. Wang, "Quickest detection of false data injection attack in wide-area smart grids", IEEE Trans. Smart Grid vol. 6 no. 6 pp. 2715–2735 Nov. 2015.

[42] R. E. Kalman, "A New Approach to Linear Filtering and Prediction Problems,". Journal of Basic Engineering, 82: 34–45, 1960

[43] A. S. Debs, R. E. Larson "A dynamic estimator for tracking the state of a power system", IEEE Trans. PAS vol. PAS-89 pp. 1670–1673 September/ October 1970.

[44] E. A. Blood, M. D. Ilic, J. Ilic, B. H. Krogh, "A Kalman filter approach to quasi-static state estimation in electric power systems", 38th North American Power Symposium pp. 417–422 2006 2006.

[45] A Saikia, RK Mehta, "Power system static state estimation using Kalman filter algorithm", EDP Sciences. 2016; 7: 1–7.

[46] C. E. Shannon, "A mathematical theory of communication", Bell Syst. Tech. J., vol. 27, no. 3, pp. 379–423, Jul./ Oct. 1948.

[47] E. Kaufmann, O. Cappé and A. Garivier, "On Bayesian upper confidence bounds for bandit problems" Proc. Int. Conf. Artif. Intell. Stat. pp. 592–600 2012.

[48] G. Burtini, J. Loeppky, and R. Lawrence, "A survey of online experiment design with the stochastic multi-armed bandit", CoRR, abs/ 1510.00757, 2015.

[49] E. Kaufmann, "Analysis of bayesian and frequentist strategies for sequential resource allocation", Machine Learning [cs.LG]. Télécom ParisTech, 2014. English. ¡NNT : 2014ENST0056¿. ¡tel-01413183¿

[50] P. Reverdy, V. Srivastava, N. E. Leonard, "Modeling human decision-making in generalized Gaussian multi-armed bandits", Proc. IEEE vol. 102 no. 4 pp. 544–571 Apr. 2014.

[51] Oozeer, M.I., 2020. Cognitive Dynamic System for Control and Cyber Security in Smart Grid (Doctoral dissertation). URL: https://macsphere.mcmaster.ca/handle/11375/25551

[52] R. D. Zimmerman, C. E. Murillo-SÃ¡nchez, and R. J. Thomas, "MATPOWER: Steady-State Operations, Planning and Analysis Tools for Power Systems Research and Education," Power Systems, IEEE Transactions on, vol. 26, no. 1, pp. 12–19, Feb. 2011.(Digital Object Identifier: 10.1109/TPWRS.2010.2051168)

[53] D. J. Kershaw and R. J. Evans, "Optimal waveform selection for tracking systems," IEEE Trans. Inf. Theory, vol. 40, no. 5, pp. 1536–1550, Sep. 1994.

[54] S. Haykin, "Neural Networks and Learning Machines", 3rd ed. Prentice-Hall, 2009.

Section 4

Smart Mobility and Transportation

Chapter 10

Architecture of a Telemonitoring System for the Mobility of the Elderly in Wheelchairs Supported by Internet of Things Technologies as a Component of a Smart City

Nancy Edith Ochoa Guevara, Juan Sanchez Arteta, David Almesiga Riaño, Diego Sarmiento Vargas, Bryan Tunarosa Naranjo, Sandra Patricia Ochoa Guevara and Hector Edmundo Davila

Abstract

Digital transformation and the entry of new technologies in industry are changing the way of mobility and production in large cities. The industrial revolution 4.0 comes hand in hand with technology and advances in this field. The Internet of Things (IoT) is one of the innovations that offers the most versatility to industrial companies in the main cities as a relevant axis for supporting rural areas of the country. This Technology enables Cities to allow mobility and movement for all, regardless of their physical or mobility conditions. This chapter presents the proposal of the project "Safe mobility in conventional wheelchairs in public spaces from smart cities", in which the creation of a prototype of coupling to wheelchairs in people with disabilities in their lower extremities is expected. Through the caterpillar traction system, to facilitate the ascending and descending of stairs safely and reliably called Wheelchair Adapter. It is intended to review two relevant elements for this type of people, such as health and mobility. The design and implementation of an Architecture of a Telemonitoring System for Older Adults in Health and Mobility in their wheelchair supported by Internet of Things Technologies (IoT) generally called RobotUp_IoT. Its purpose is to efficiently monitor both the health and wheelchair movement of older adults with disabilities in their lower extremities. Therefore, an analytical and predictive methodology is proposed with the support of the Build Information Modeling (BIM) process and the 4.0 industry in the IoT technique, in order to build a conceptual 3D model and its generation of tests for its respective implementation and implementation of this architecture. Heelchair Adapter and it is expected to incorporate the health part through Telemonitoring for seniors between 2020 and 2022 contributing to other solutions and research in this regard.

Keywords: disability, obstacles, sensors, motors, quality of life, stairs, cloud

IntechOpen

1. Introduction

The Internet of Things, or IoT in English, is a tool created by the advances in Information and Communication Technologies. This technology can be adapted to a myriad of industrial processes. It also allows integrating a processing, storage and communication system between physical objects in a city that are connected to each other. Using sensors, the Internet of Things makes it possible to collect data in a simple way and for example send alerts if something does not work correctly. The potential of the Internet of Things is in the use of another tool of digital transformation: Big Data. An essential technology to facilitate decision-making based on the data collected.

Every time we live in a more interconnected world. The number of connected devices has been growing exorbitantly globally as technological improvements and cost reductions in wireless communications have enabled businesses, manufacturers, industries and smart cities to connect their products to the Internet of Things (IoT).

In this sense, citizens have at our supply a wide range of services that can make everyday life easier. There are numerous architectures to deploy IoT platforms in different contexts and levels, such as within the home, healthcare, business, municipal, national or even global level. The most direct impact on the lives of citizens involves our cities and interconnects people and services, in short, it transforms a city into a Smart City [1].

The "Smart City" or "SmartCity" can be seen as an instrumented, interconnected and intelligent urban ecosystem. Instrumented refers to the ability to capture and integrate real-time data on city life through the use of sensors and mobile devices (IoT technology). Interconnected means the integration of these data in an urban computing platform that allows the communication of said information between the various services of the city. Smart refers to the integration of complex analytics, modeling, optimization and visualization of services to make better operational decisions for the city.

Today there is an analysis of assistive technology services, according to the class and subclass of assistive technology products formulated. Therefore, there are assistive technology devices or products that are classified according to Smith [2], who presents high technology (high-tech) versus low technology. High technology refers to devices that are made up of electronic devices or that are not for everyday use today. This type of technology, having an exotic appearance, represents the idea of what people understand by technology; Examples are cell phones, electronic agendas, magnetic resonance imaging devices, positron emission tomography, bio-feed-back and neuropros theses [3, 4]. On the other hand, there is low technology, which includes less complex, more common devices with mechanical or electrical drive; here we can mention home appliances, calculators, manual wheelchairs, canes and splints (Ríos et al., 2007).

IoT technology is essential for the operation of the Smart City, through sensors in charge of collecting data on the state of the city and subsequently disseminating them among citizens. IoT consists of allowing things to connect to the Internet, this achieves the generation of information, and the interaction of the physical world with the virtual world [5]. Regarding the number of things connected, the evolution of this field is observed from 2015 to 2020. In 2020 the figure of 31 billion is reached worldwide and a value of more than one billion US dollars per year is projected from 2017 [6].

Defining assistive technology products according to their level or degree of technology is important, since the costs generated by their formulation; adaptation and training vary significantly between one group and another. For this reason, in the classification below, devices that are considered high-tech have been marked with an asterisk. Obviously, as technological development is so accelerated, many devices that are considered high-tech at the moment, in a few years may be considered low-tech, such as:

- Bimanual wheelchair: two evaluations by a specialist in physical medicine and rehabilitation and five sessions of physical therapy.

- Single-sided manual wheelchair: two evaluations by a specialist in physical medicine and rehabilitation and ten sessions of physical therapy.

- Manual-operated electric motor wheelchair: evaluation, formulation and approval by an interdisciplinary team is required, consisting of a minimum of one physician specializing in physical medicine and rehabilitation, physical therapist, occupational therapist, social worker, psychologist and a representative of the EPS. Approval will be given in an interdisciplinary meeting, where the number of sessions to learn how to manage it will be defined.

- Powered-drive electric motor wheelchair: evaluation, formulation and approval by an interdisciplinary team is required, consisting of a minimum of a specialist in physical medicine and rehabilitation, physical therapist, occupational therapist, social worker, psychologist and a representative of the EPS. Approval will be given in an interdisciplinary meeting, where the number of sessions to learn how to manage it will be defined.

- According to the Colombian population profile of DANE 2017, the elderly population has difficulties in appropriately affected lower limbs of 34,000 (9.7%) in men, 28,007 (8.5%) in women for a total of 62,007 (9.1%)

The IoT is of great importance being the first real evolution of the Internet; considered as a great leap that will lead to revolutionary applications that have the potential to significantly improve the way of living, learning, working and entertaining people [7].

The problem faced by older adults in wheelchairs is presented in two fundamental aspects such as health: presence of underlying tissue called pressure ulcers, as they remain in the same part of the body for a long time; Dermatitis that also responds to a lesion on the skin as a sequel to fecal and/or urinary incontinence; the lack of control of the supply of drugs in the time allocated; as well as the inadequate programming of the doctor's assistance for the control check-up and mobility in aspects such as: dependence to go from one place to another, obstacles impossible to overcome, going up or down steps with the wheelchair, among others.

These lead to family members or friends being concerned about the quality of life of their family member and on many occasions they go to the care of nurses or hospitalizations that what they do is affect the mood and health of the person at high costs for some of their relatives [8].

In **Figure 1**, the previous situation is observed, in terms of internal or external mobility of adults who are in wheelchairs. There it is visualized that at a given moment they do not know what to do or who to call, they only have to wait for their relatives or friends to help them in this regard.

Due to this problem, the following research question arises: How to contribute to improving the quality of life conditions and reduce the risk situations that the

Figure 1.
Situation of the elderly to achieve the ascent and descent of the stairs. Source: Authors.

elderly who are bedridden in a wheelchair may present, enabling timely attention, their care of permanent form and your independent mobility? Risk situations can be prevented with viable technological solutions such as an Architecture of a Telemonitoring System for Older Adults, which can improve the quality of life of older adults, since it will not only have benefits for them, it will also allow them to your family members feel calm and safe, because they will have a reliable way of verifying the state in which their loved ones are.

The objective of the study is to manage a monitoring service for the health and mobility of older adults who are bedridden in a wheelchair through Internet of Things (IoT) technology with the use of sensors for the detection of data, internet for the transmission of these and a mobile or web application for the visualization of the information, in order to support the prevention of new diseases and control the diseases that the elderly already suffer and at the same time achieve their independent mobility inside and outside your room.

2. Theoretical framework

According to an analysis of the experience of the Australian System of Care for Older Adults, which for Pérez [9] is considered one of the most advanced among experts and public policy analysts and in which he exposes three fundamental pillars, among which is the development of efficient operational skills, highlighting as one of these efficient skills the importance of technology and innovation as key factors to improve the effectiveness of the system. Successful experiences of technological innovation that range from the implementation of computerized systems to solutions such as tableware at home, which defines it as a type of permanent video-conference with the patient at home, reminding her to take her medicines, offering her a health channel.

An e-Health project at the national level is the Mobile Telehealth System, designed in the College of Engineering of Antioquia by Garcia and Torres [10], it consists of a system composed of three main blocks that are: Input subsystem which by means of Sensor collects the patient's vital signs and sends them to another block, the Local subsystem, which is made up of different devices and elements that receive this

DOI: http://dx.doi.org/10.5772/intechopen.94074

information and send it to the third block, the Remote Subsystem, which is responsible for storing the information in a database of data and sample it through a website.

In Colombia, this type of project related to telemedicine and e-Health has had very little demand, it is a recent technology and they are projects that are carried out as a commitment to innovation in these fields of health and mobility. At the regional level, no reports of projects aimed at monitoring older adults were found, this serves as motivation for the implementation of projects such as the Architecture for the Telemonitoring System for Older Adults who are in wheelchairs mentioned in this research and to be able to carry out a significant contribution for the elderly in the city of Bogotá.

3. Methodology

The approach of this applied research of an analytical and predictive type with the use and use of IoT technology, with the support of a descriptive study that allows monitoring the elderly who is in a wheelchair, through detection of your data and in this way send a notification message to alert your family, friends or doctor to avoid complications that endanger the health and mobility of older adults.

An evaluation instrument called a survey will be established to a sample of Twenty-five (25) people who are elderly in wheelchairs in their homes in the city of Bogotá in the town of Tunjuelito, during the period of 2017–2018, at as well as an unstructured interview with a focus group of relatives and friends with elderly people in permanent wheelchairs. A data analysis is established and the disposition of the technology to the need to implement the architecture referenced in this research is determined by means of the Rational Unified Process Architecture (RUP). This process consists of the realization of a software from UML, which will be built in four phases: start, elaboration, construction and transition [11].

A partial investigation will be implemented on the population directly affected by mobility in this case of people disabled in wheelchairs as older adults [12].

3.1 Design of the methodological proposal

It will be made up of four fundamental stages such as:

- **Stage 1**. Technique of IoT and Telemonitoring for Older Adults in wheelchairs under the aspect of health and mobility.

- **Stage 2**. Identification of the hardware and software requirements necessary for the implementation of the Telemonitoring System for Older Adults located in the wheelchair as a Hub platform.

- **Stage 3.** Design of an architecture of a Telemonitoring system with the use of the IoT technique for older adults who are in wheelchairs, in order to provide security and a low-cost mechanism due to the use of available technologies that allow the user to stay informed at all times of the health status of his family member or friend.

- **Stage 4.** Implementation of the functional prototype of the Telemonitoring architecture, through testing and adjustment in order to correct the weak points of said architecture

4. Expected results

It is expected that by applying IoT technology in this research, it is possible to improve the quality of life of older adults who are in a wheelchair [13]. For this reason, a domain model will be created in which the concepts and associations considered important in a Telemonitoring Architecture for older adults who are in wheelchairs from health and mobility will be identified [14].

This monitoring is done by means of the wheelchair with the use of sensors generating medical information, which in turn presents early alerts, which can be received by the attendants or the medical person through a mobile device, where will make the reminders of medical appointments and medications to later receive notification when the due date of these is close. This chair will also allow the independent mobility of the elderly with disabilities, since it will have a wireless connection and will use the IoT for monitoring according to comfort, performance, maintenance needs and location.

In **Figure 2**, the Domain Model of the architecture of a Telemonitoring system for older adults is observed. in which it can be seen that everything begins with the doctor's assessment of the elderly, this allows determining that this patient requires constant monitoring, the monitoring is carried out by generating medical information, which in turn generates early alerts, said alerts can be received by caregivers or medical staff through a mobile device, in which it is also possible to carry out the management of reminders of medical appointments and medications to later receive notifications when the date of compliance with these is close.

In **Figure 3**, the three levels with which the Architecture proposed in this investigation will be worked are observed: Sensing. In this layer the necessary elements are implemented to carry out the obtaining of the data through the elements of sensors, wearables, among others; Networking and Data Communications: The data sensed in the previously described layer are sent through a medium for subsequent analysis, in this layer are the devices and elements necessary to send this data through the internet, in this layer you can find servers, databases, protocol adapters, among others, this layer acts as a link between the Sensing layer and the Services or Applications layer; and Services / Applications: This layer,

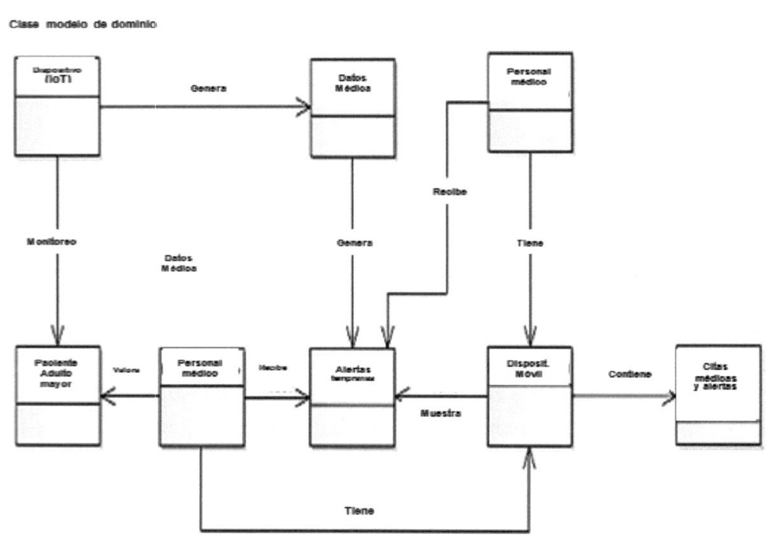

Figure 2.
Domain model of the architecture of a Telemonitoring system for older adults. Source: IEEE P2413 project, http://standards.ieee.org/innovate/iot/.

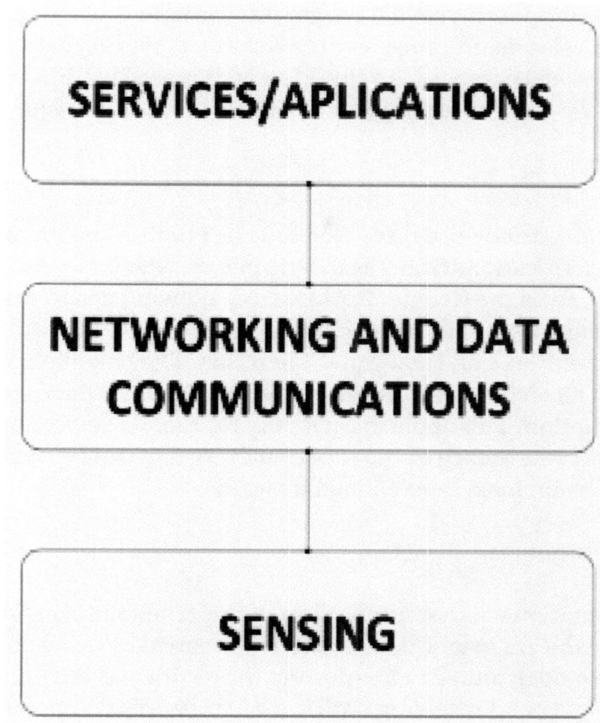

Figure 3.
Levels of an IoT architecture. Source: Authors.

Based on the elements of **Figure 3**, the prototype of the computational system is built based on the IoT business model, allowing the good performance of the proposed Architecture.

This IoT business model is made up of three fundamental layers:

- Detection Layer: It is achieved by means of the components in charge of the sensing information, as well as the information to be sent to the next layer of the model.

- Network and Data Communication Layer: It consists of a data knowledge base; a server on which the web services are stored; and connectivity that enables communication between the detection layer and the service / application layer.

- User layer: It is where it allows the hosting of the services to be offered by the system, it is known as services / applications, such as notifications, elderly information, movement indicators, parameterization of actions and administration of the platform.

For the mobility aspect, the wheelchair uses IoT technology that includes a Global SIM card, a communication center can be established in the cloud.

4.1 Examine the architecture

The different parts of the architecture must be examined and evaluated separately and allow ease in visualizing the complexity of the system, responding to interested parties, clients, programmers, engineers, among others. Therefore, the

4 + 1 views model proposed by Philippe Kruchten is used, such as the view: logic, deployment or development, processes, physics and scenario views. Through the IoT architecture model composed of three levels proposed by IEEE P2413, it will be possible to meet the objectives proposed in this research.

4.1.1 Mobile device

It is the technological device that will allow the family member or friend to receive alerts or notifications from the elderly person who is in the wheelchair. These alerts come from the Google GCM service, allowing you to perform an action and thus prevent an eventuality affecting the health or mobility of the Elderly. According to the authors Liu, Huang and Chen [14]. They highlight the importance of managing and receiving notifications for reminders of medical appointments and drug consumption. This application is only for mobile devices with Android operating systems, where each of the functionalities that can be carried out through the mobile application must be explained in detail.

4.1.2 Architecture evaluation: Case study

The implementation of a test mechanism of the Telemonitoring Architecture for Older Adults who are in wheelchairs should be carried out, which follows the guidelines of the model, aimed at monitoring the health and mobility status of the elderly, managing and receiving notifications of medical appointments and medicines, view statistics of alerts and information on the sensing of the elderly, obtaining the benefit of constant care, keeping a correct control of the assigned treatments, intervention in case of situations that put health at risk and mobility of the elderly. The purpose is to find the most outstanding benefits of the implementation of this service model such as: economy, security, reliability and risk reduction among others.

Regarding mobility, physical aspects of the chair should be evaluated, such as: seat position, cushion height, battery level and maintenance requirements and GPS location.

4.1.3 Product already obtained in the first phase of the project

The Wheelchair Adapter is a mechanism that adjusts smoothly and safely to conventional wheelchairs, helping people with disabilities in their lower extremities to navigate different terrain, mainly the ascent and descent of stairs.

4.1.3.1 Procedure

- **Base:** Resistant metal base which will keep the other components that make up the Wheelchair Adapter.

- **Gears:** Four gears which allow to adapt and generate the movement of the track system.

- **Caterpillar motors:** Two motors located on the internal sides of the metal base connected by chain systems to the shafts that carry the gears, which in turn are attached to the tracks.

- **Caterpillars:** Element in the form of an elongated chain, which allows the firm adherence of the Wheelchair Adapter to different terrains.

- **Ultrasound sensors:** Two sensors located on the front of the Wheelchair Adapter that allows it to be located in the environment it is in.

- **Support:** Metal plate which allows the conventional wheelchair to recline and generate stability in movement

- **Ramp:** Metal plate whose function is to provide quick and easy access when adapting the conventional wheelchair with the Wheelchair Adapter.

- **Adjustment system and Angle:** It allows locating the wheelchair and the person at an adequate and safe angle to start the ascent and descent of stairs, this is made up of a structure with a system of rails powered by two motors located behind the backrest.

- **Cortex microprocessor:** Microcontroller which saves and executes the programmed instructions for the sensors and motors that perform the movements of the Wheelchair Adapter.

- **Battery:** 7.2 volt battery which allows to power the motors and sensors of the Wheelchair Adapter.

- **Triangle rims:** Triangle-shaped structure with tires at each end that provides additional grip on difficult terrain, ascending and descending stairs

Movements

- 2-wire motor 393
- Shaft collar
- Shaft coupler
- Shaft, 3 ″long
- Plain bearing
- Spur gear, 12 teeth
- Spur gear, 60 teeth
- Spur gear, 84 teeth
- Wheel 4″

Structure

- Bar, 20 holes
- Chassis bumper (20 holes)
- Chassis rail (16 holes)
- Channel C, hole 1x2x1x15
- Channel C, hole 1x2x1x20

Electric

- 7.2V NiMH 3000mAh robotic battery
- Smart charger and power cord
- NiMH AAA rechargeable battery (6 pack)
- 8-Bay AA / AAA Smart Battery Charger
- Motor controller 29
- Battery strap (2-pack)

Electronics

- VEX ARM® Cortex®-based Microcontroller
- USB AA tether cable

Control

- VEXnet Joystick
- VEXnet 2.0 key
- VEXnet Battery Backup Holder

Hardware

- Screw, 8-32 x 1/4 ″long
- Screw, 8-32 x 1/2 ″long
- Screw, 8-32 x 1 1/2 ″Long
- Locking screw, 6-32 x 1/4 ″long
- Locking screw, 6-32 x 1/2 ″long
- Nut, 8-32 Keps
- Shaft Spacer, Slim (4.6mm)
- 4″Zip Ties
- Bearing accessory rivet

Equipment

- Tool, hex wrench (5/64 ″)
- Tool, hex wrench (3/32 ″)
- Tool, VEX open end wrench

Sensors

- Bumper switch (2-pack)
- Limit switch (2-pack)
- Potentiometer (2-pack)
- Line Tracker
- Ultrasonic range finder
- Optical Axis Encoder (2 Pack)

4.1.3.2 Final product explanation

In **Figure 4**, the Wheelchair Adapter, with the help of the motors located in the upper rear part, allow through a slow and safe movement that the rails located on the sides raise the platform where the conventional wheelchair is located, so that at the time of the ascent and descent is left with a 90 degree angle for safety and comfort.

In **Figure 5**, you can see the Wheelchair Adapter, with the help of the complex system of equilateral triangle-shaped rims, it allows you to have a better grip on each obstacle in your environment, generating more security when moving.

In **Figure 6**, the Wheelchair Adapter is observed, thanks to its caterpillar traction system and its 45 degree angle allows generating the ascent and descent of stairs and with its sophisticated caterpillar design it allows to face any terrain.

In **Figure 7**, the Wheelchair Adapter is observed, it presents the mobility operation of the chair during the ascent and descent, as mentioned above, the prototype generates a lot of stability thanks to its caterpillar system and its triangular wheels,

Figure 4.
Prototype of the chair of wheels.

Figure 5.
Chair of wheels with rims in shape of angle.

Figure 6.
Chair of wheels with design of caterpillars.

as well as generates security with the 90 degree position of the person by its side rails and total dependence thanks to the ultrasound sensors that identify the possible obstacles to overcome.

5. Conclusions

When investigating a little more about the difficulties that people with disabilities present in their lower extremities, there are some areas in which conventional wheelchairs do not have the necessary equipment, it is proposed to start with a prototype

Figure 7.
Up and down movement of the wheelchair. Source: Author.

which can be adjusted accordingly. Fast and safe way, with the least physical effort to conventional wheelchairs, which solves this problem and facilitates the movement of these people for the moment when ascending and descending stairs.

When starting with the adapter prototype designs, many ideas were postulated that were conclusive as the variables were established to generate a safe, comfortable and fast tool, resulting in an adapter prototype that not only meets its objective, but also that provides an expertise to the user of comfort, security and confidence.

It is expected to work with the cloud system for data processing and permanent connection with the wheelchair and computer-assisted algorithms for the mobility and health of the patient.

Author details

Nancy Edith Ochoa Guevara[1*], Juan Sanchez Arteta[2], David Almesiga Riaño[2], Diego Sarmiento Vargas[2], Bryan Tunarosa Naranjo[2], Sandra Patricia Ochoa Guevara[3] and Hector Edmundo Davila[4]

1 Unitec University Corporation/Uniremington University Corporation, Colombia

2 University Foundation Unipanamericana, Colombia

3 Logistics and Information Technology Center – SENA, Colombia

4 Remington University Corporation, Colombia

*Address all correspondence to: nancy.ochoa@unitec.edu.co, nancyochoa@uniremington.edu.co and ochoaguevara@gmail.com

References

[1] Haller, Stephen. Internet of Things: An Integral Part of the Future Internet. SAP presentation, 2009 http://services.futurelinternet.eu/images/1/16/A4_Things_Haller.pdf.

[2] Sebestyen, G., Hangan, A., Oniga, S., & Gal, Z. EHealth solutions in the context of Internet of Things. In 2014 IEEE International Conference on Automation, Quality and Testing, Robotics (AQTR) (pp. 1-6). IEEE, 2014.

[3] Jara, A. J., Zamora, M. A., & Skarmeta, A. F. An architecture based on internet of things to support mobility and security in medical environments. In Consumer Communications and Networking Conference (CCNC), 2010 7th IEEE (pp. 1-5). IEEE, 2010

[4] Rabanales Sotos, J., Párraga Martínez, I., López-Torres Hidalgo, J., Andrés Pretel, F., & Navarro Bravo, B. Tecnologías de la Información y las Telecomunicaciones: Telemedicina. Revista Clínica de Medicina de Familia, 4(1), 42-48, 2011.

[5] M. A. Simarro, P. Guzmán, M. A. Rodríguez, P. Arce, G. Piñero, A. González, and J. C. Guerri, "Open Standards for the Internet of Things," Waves, vol. 10, pp. 5-14, 2018.

[6] C. Harrison, B. Eckman, R. Hamilton, P. Hartswick, J. Kalagnanam, J. Paraszczak, and P. Williams. "Foundations for Smarter Cities," IBM Journal of Research and Development, vol. 54, no. 4, 2010.

[7] Evans, D. Internet de las cosas. Cómo la próxima evolución de Internet lo cambia todo. Naciones Unidas, 2011.

[8] ZAFAR, A., & WON, S. Abnormal human activity recognition system based on R-transform and kernel discriminant technique for elderly home care. Consumer Electronics, 58(4), 1843-1850, 2011.

[9] Pérez, S. El Sistema de Cuidados del Adulto Mayor en Australia, ¿un modelo para Chile?. Instituto de Políticas Públicas de la UDP, 2012.

[10] Garcia, J. E., & Torres, R. A. Telehealth mobile system. In Health Care Exchanges (PAHCE), 2013 Pan American (pp. 1-5). IEEE, 2013.

[11] Sommerville, I. Ingenieria del Software. En I. Sommerville, Ingenieria del Software (págs. 76,78). Pearson Educación, 2005.

[12] Departamento Administrativo Nacional de Estadística, (s.f). En Wikipedia. Recuperado el 27 de Abril de 2015 de http://es.wikipedia.org/wiki/Departamento_Administrativo_Nacional_de_Estad%C3%ADstica.

[13] A. González, R. Morales, A. Nieto, J. M. Chicharro, P. Pintado, V. Feliu (2007). Un Nuevo mecanismo para subir escaleras. Ciudad Real - España: Paper-1, 2007. https://www2.uned.es/ribim/volumenes/Vol11N1Enero_2007/V11N1A05%20Gonzalez.pdf.

[14] Chen S., Xu H., Liu D., Hu B., & Wang, H. A vision of IoT: Applications, challenges, and opportunities with china perspective. Internet of Things Journal, IEEE, 1(4), 349-359, 2014.

Chapter 11

Smart Growth and Transit Oriented Development: Financing and Execution Challenges in India

Alok Kumar Mishra and Shibani Mishra

Abstract

Cities today face burgeoning personalized vehicles as a consequence of neglected public transport and a spatial planning model isolated from transport planning. Transportation planning has been accorded a residual rank post spatial planning. This has prompted dispersed and automobile-centric growth of cities. The pursuit of more sustainable, liveable, congestion and pollution free cities resulted in the paradigm of New Urbanism and Smart Growth. Transit-oriented Development (TOD), an integral part of Smart Growth, has emerged as a paradigm in urban design. It aims at the concentration of development in or around a transit station or along a transit corridor. TOD could be a befitting reply to sprawl, congestion, pollution and provide an effective way to restructure existing cities. By integrating public transport and land use planning TOD provides ways to intensify agglomeration economies and weaken congestion diseconomies. TOD has several socio-economic and environmental benefits to its credit. The chapter looks at the various advantages of TOD and the challenges faced in its execution and financing. Further, several successful TOD practices from around the globe have been discussed to draw lessons for replication in India.

Keywords: new urbanism, smart growth, transit oriented development, agglomeration economies, congestion diseconomies

1. Introduction

In the urban context, the importance of transport stems from the fact that it contributes to the productivity of workers and competitiveness of firms. It widens labour markets and makes them inclusive. It saves travel time and costs to reach valued destinations – for work, education, shopping and leisure. Urban transport investments augment agglomeration economies by enhancing access to the economic mass, reducing congestion and channelizing residential and non-residential development in desired directions. They balance the location of jobs, housing and common facilities. Urban transport plays an important role in the working of cities, enhancing their efficiency, facilitating economic growth, generating value enhancements to finance planned urban development, and creating livable, competitive and sustainable cities.

In the last four decades, the issues of urban transport have come into sharp focus in many developed countries around the world due to the problems of their automobile-dependent model of urban development. They have been subject to high levels of traffic congestion, air pollution, accidents, damages to ecosystems and neighborhoods, segregation and adverse impacts on the quality of life in cities.

Rapid motorization has worsened traffic conditions, aggravated congestion and pollution levels in several cities around the globe. Apart from environmental concerns, traffic congestion is also detrimental to the economic health of cities by adding to the wastage of time and fuel and increasing the levels of emissions. It hampers productivity by delaying and hindering the movement of goods, raw materials as well as people.

The proliferation of personalized vehicles, lack of investment in public transport and implementation of a spatial planning model that promoted dispersed, automobile-centric development have been the primary factors behind the urban transport problems in countries. The search for ways of making urban communities provide a better quality of life and promoting sustainable cities led to the emergence of an urban design paradigm called 'New Urbanism' in North America and Western Europe in the 1970s and 1980s. This was followed by a theory of urban planning and transportation called 'Smart Growth'. This theory is founded on the following principles of urban design: (i) mixed land use; (ii) compact design; (iii) increased densities; (iv) housing opportunities and choices; (v) walkable and accessible neighborhoods; (vi) multiple transportation mode choices; (vii) neighborhood centres to foster social interaction; (viii) preservation of open space, farm land, natural beauty and critical environmental areas; (ix) strengthening of and directing development towards existing communities; (x) making development decisions predictable, fair, and cost-effective; and (xi) community and stakeholders' consultation in development decisions [1]. Transit-oriented Development (TOD), a key element of Smart Growth, aims at concentrating development around one or more transit stations or within a transit corridor.

TOD aims at compact, high density and mixed use development within easy walking or biking distance from a transit station, typically about 1 kilometer. Focused around a transit node, TOD facilitates access to public transit, thereby inducing people to walk, cycle and use public transport rather than personal vehicles. The selective concentration of development acts against sprawl, promotes agglomeration economies and mitigates congestion diseconomies. It also leads to increase in property values, reflecting the benefits to residents and businesses of diverse transportation options, and resultant automobile and parking cost savings [2]. Thus, TOD assists in the mobilization of value capture finance by harnessing the windfall gains accruing to land and property-owners. The key factors that support TOD include: land use and development policies promoting dense and compact development around transit nodes and discouraging such development in the areas without good access to public transport; development of public transit and provision of quality transit services; integration of transportation and land use; and application of other mobility management strategies. These factors jointly increase the cost-effectiveness and utility of TOD for consumers as well as businesses. TOD has the potential of becoming a powerful tool for planned development of cities and rural areas in developing countries. It not only improves connectivity between regions, but also saves a lot of time and costs of workers. It augments productivity and efficiency of economic agents. The case for transit-oriented development is well-argued in research [3].

The motivation behind this research is the need for India to move from an automobile-dependent to a public transportation-led and transit-oriented model of planned urban development. The current practice of master planning in India, rooted in the 1947 Town and Country Planning Act in the United Kingdom, has neglected urban transport. The model has not facilitated transportation-land use integration, transit-oriented development and value capture financing. Land use planning and transportation planning have been pursued as independent exercises, a prime example being Delhi. Cities have thus not been able to benefit from the interaction of transport and land use for sustainable urban development and adopt a robust mechanism of financing public transit. In this context, this paper explores the theory and international practice of New Urbanism, Smart Growth and TOD. It also examines the potential of TOD to raise revenues towards financing public transportation. The objective is to draw lessons from successful practices to strategize TOD policy for cities in India. Finally, the paper analyses the existing practices in Indian cities, identifies its inadequacies and suggests corrective measures.

The study is organized as follows. Section 2 deals with the challenges of urban mobility in India. Referring to the trends of urbanization, metropiltanization and motorization, it highlights the imperative for a public transportation-based strategy of urban development in the country. Section 3 discusses the paradigmTransit Oriented Development (TOD), adopted by several developed countries to address their problems of sprawl, inefficient urban form, excessive energy consumption, greenhouse gas emission, and environmental degradation. It makes a strong case for TOD as a dominant paradigm of urban planning and development in India. Section 4 presents some examples of successful TOD strategies practiced world-wide and draws lessons for urban transport development and land use planning, in general and TOD, in particular. Section 5 focuses on financing and execution practices of TOD internationally in the overall context of urban transport development to present a range of financing instruments that Indian cities could consider to promote TOD. Section 6 presents the emerging approaches towards TOD in India, referring to case studies, including projects, policies and plans in the offing. We specially focus on financing issues. Section 7 brings out the challenges of implementing TOD in India and indicates some directions for the design of a public transportation-led, transit-oriented and value increment financing-based strategy to address India's urbanization challenges. It also calls for an effective institutional structure for the implementation of TOD and suggests reforms in the regional and urban planning model being followed. Section 8 concludes.

2. India's urban mobility challenges

Bourgeoning travel demand, rapid motorization, rise in personalized vehicles, dwindling share of public transport, congestion, degradation environmental quality, rising number of road accidents and fatalities, fragmented institutional arrangements and chronic under-investment in transport infrastructure pose major hindrances to urban mobility. These are linked to the trends and patterns of urbanization, concentration of productive economic activity, income distribution structure in cities and motorization.

2.1 Urbanization trends and patterns

Urbanization in India is characterized by rising urban population and increased density in large cities. This has led to a rapid growth in travel demand. **Tables 1–4** present the trends and patterns of urbanization in India.

While the number of cities/towns in India increased by 3 times, urban population rose by 13 times between 1901 and 2011. This reflects the concentrated pattern of urbanization. In 2011, the number of urban agglomerations (UAs) /towns was 7935 as against 5161 in 2001. While the number of statutory towns rose from 3799 to 4041 between 2001 and 2011, the number of census towns experienced a

Year	Total Population	Rural Population	Percentage Rural	No of Cities / Towns	Urban Population	Percentage Urban
1901	238.4	212.5	89.2	1916	25.9	10.8
1911	252.1	226.2	89.7	1908	25.9	10.3
1921	251.3	223.2	88.8	2048	28.1	11.2
1931	279.0	245.5	88.0	2220	33.5	12.0
1941	318.7	274.5	86.1	2427	44.2	13.9
1951	361.1	298.6	82.7	3060	62.4	17.3
1961	439.2	360.3	82.0	2700	78.9	18.0
1971	548.2	439.0	80.1	3126	109.1	19.9
1981	683.3	523.9	76.7	4029	159.5	23.3
1991	846.3	628.7	74.3	4689	217.6	25.7
2001	1028.7	742.5	72.2	5161	286.1	27.8
2011	1210.7	833.5	68.8	7935	377.1	31.2

Source: *Census of India for different years [4].*

Table 1.
India: Total, rural and urban population (in million) and level of urbanization (percentage) 1901–2011.

phenomenal jump from 1362 to 3894. About 30 percent of urban population growth in the last decade is accounted for by census towns.

Table 2 presents the distribution of urban population between size classes of towns in India from 1901 to 2011. It reflects a top-heavy urban structure, highlighting the increasing density of large cities. **Table 3** presents the trends in metropolitan population in India and reflects a similar trend.

There are large interstate variations in urbanization patterns in India, having differential implications for urban transport demand and strategy. Among the states, Delhi was the most urbanized in 2011, with 97.5 percent urbanization level, followed by Goa (62.2 percent), Mizoram (52.1 percent) and Tamil Nadu (48.4 percent). **Table 4** presents the percentage of urban population in 1971, 1981, 1991, 2001 and 2011, and decadal annual exponential growth in urban population for 1971–81, 1981–91, 1991–2001 and 2001–11.

India's urban population is projected to more than double between 2011 and 2050 – from 377 million to 814 million. With an estimated rural population of 860 million in 2014, the country would still have 810 million in villages in 2050 [5]. Thus, India would confront the dual challenges of urban and rural development for many decades. The country has to address not only the problems of transportation within cities, it will have to connect villages to cities and towns providing efficient transport services to rural areas.

2.2 Population density in urban areas

Census of India 2011 data reveals that not only many cities, but also urban agglomerations or regions in India have a population density of more than 10,000 – with central city areas being denser than peripheries. **Table 5** provides data on densities of 10 urban districts in India with the highest population density. A simple conclusion from international comparisons relating to population densities of urban regions is that the density patterns of many cities and urban districts in India overwhelmingly support a public transport-led urban development strategy.

	Number of Agglomerations/Towns						Percentage of Urban Population					
Census Year	Class I	Class II	Class III	Class IV	Class V	Class VI	Class I	Class II	Class III	Class IV	Class V	Class VI
1901	24	43	130	391	744	479	26.00	11.29	15.64	20.83	20.14	6.10
1911	23	40	135	364	707	485	27.48	10.51	16.40	19.73	19.31	6.57
1921	29	45	145	370	734	571	29.70	10.39	15.92	18.29	18.67	7.03
1931	35	56	183	434	800	509	31.20	11.65	16.80	18.00	17.14	5.21
1941	49	74	242	498	920	407	38.23	11.42	16.35	15.78	15.08	3.14
1951	76	91	327	608	1124	569	44.63	9.96	15.72	13.63	12.97	3.09
1961	102	129	437	719	711	172	51.42	11.23	16.94	12.77	6.87	0.77
1971	148	173	558	827	623	147	57.24	10.92	16.01	10.94	4.45	0.44
1981	218	270	743	1059	758	253	60.37	11.63	14.33	9.54	3.58	0.50
1991	300	345	947	1167	740	197	65.20	10.95	13.19	7.77	2.60	0.29
2001	393	401	1151	1344	888	191	68.67	9.67	12.23	6.84	2.36	0.23
2011	468	474	1373	1686	1748	424	70.15	8.54	11.11	6.39	3.36	0.45

Note: Class I: 100,000 or more, Class II: 50,000 – 99,999, Class III: 20,000 – 49,999; Class IV: 10,000 – 19,999, Class V: 5000–9999 and Class VI: Below 5000.
Each urban agglomeration, comprising generally a number of cities, towns and outgrowths is considered as one unit.
Source: *Census of India for different years [4].*

Table 2.
India: Number of agglomerations/towns and percentage of urban population by size classes of towns 1901–2011.

Census year	Number	Population (in Million)	Population per city (in Million)	Percentage of urban population
1901	1	1.51	1.51	5.84
1911	2	2.76	1.38	10.65
1921	2	3.13	1.56	11.14
1931	2	3.41	1.70	10.18
1941	2	5.31	2.65	12.23
1951	5	11.75	2.35	18.81
1961	7	18.10	2.58	22.93
1971	9	27.83	3.09	25.51
1981	12	42.12	3.51	26.41
1991	23	70.66	3.07	32.54
2001	35	108.29	3.09	37.85
2011	53	160.70	3.03	42.61

Source: *Census of India for different years [4].*

Table 3.
India: Number of metropolitan cities and their share in urban population 1901–2011.

Sl No	States	Percentage of Urban Population					Annual Exponential Growth Rate			
		1971	1981	1991	2001	2011	1971–1981	1981–1991	1991–2001	2001–2011
1	Andhra Pradesh	19.3	23.3	26.8	27.3	33.4	3.9	3.6	1.4	3.04
2	Arunachal Pradesh	3.7	6.3	12.2	20.4	22.9	8.3	9.3	7.0	3.31
3	Assam	8.8	9.9	11.1	12.7	14.1	3.3	3.3	3.1	2.46
4	Bihar	10.0	12.5	13.2	10.5	11.3	4.3	2.7	2.6	3.03
5	Chhattisgarh	NA	NA	NA	20.1	23.2	NA	NA	3.1	3.49
6	Delhi	89.7	92.8	89.9	93.0	97.5	4.6	3.8	4.1	2.37
7	Goa	26.4	32.5	41.0	49.8	62.2	4.4	4.0	3.3	3.01
8	Gujarat	28.1	31.1	34.4	37.4	42.6	3.4	2.9	2.8	3.07
9	Haryana	17.7	22.0	24.8	29.0	34.9	4.7	3.6	4.1	3.68
10	Himachal Pradesh	7.0	7.7	8.7	9.8	10.0	3.0	3.1	2.8	1.45
11	Jammu &Kashmir	18.6	21.1	22.8	24.9	27.4	3.8	3.4	3.4	3.10
12	Jharkhand	NA	NA	NA	22.3	24.0	NA	NA	2.6	2.80
13	Karnataka	24.3	28.9	30.9	34.0	38.7	4.1	2.6	2.5	2.74
14	Kerala	16.2	18.8	26.4	26.0	47.7	3.2	4.8	0.7	6.56
15	Madhya Pradesh	16.3	20.3	23.2	26.7	27.6	4.5	3.7	2.7	2.28
16	Maharashtra	31.2	35.0	38.7	42.4	45.2	3.4	3.3	3.0	2.12
17	Manipur	13.2	26.4	27.7	23.9	32.5	9.7	3.0	1.2	3.70
18	Meghalaya	14.6	18.0	18.7	19.6	20.1	4.9	3.1	3.2	2.70
19	Mizoram	11.4	25.2	46.2	49.5	52.1	11.8	9.6	3.3	2.59
20	Nagaland	10.0	15.5	17.3	17.7	28.9	8.5	5.6	5.3	5.10

Sl No	States	Percentage of Urban Population					Annual Exponential Growth Rate			
		1971	1981	1991	2001	2011	1971–1981	1981–1991	1991–2001	2001–2011
21	Odisha	8.4	11.8	13.4	15.0	16.7	5.2	3.1	2.6	2.38
22	Punjab	23.7	27.7	29.7	34.0	37.5	3.6	2.6	3.2	2.29
23	Rajasthan	17.6	20.9	22.9	23.4	24.9	4.5	3.3	2.7	2.54
24	Sikkim	9.4	16.2	9.1	11.1	25.2	9.6	−3.2	4.8	9.42
25	Tamil Nadu	30.3	33.0	34.2	43.9	48.4	2.5	1.8	3.6	2.39
26	Tripura	10.4	11.0	15.3	17.0	26.2	3.3	6.2	2.5	5.66
27	Uttar Pradesh	14.0	18.0	19.9	20.8	22.3	4.8	3.3	2.8	2.53
28	Uttaranchal	NA	NA	NA	25.6	30.2	NA	NA	2.8	3.36
29	West Bengal	24.8	26.5	27.4	28.0	31.9	2.8	2.5	1.8	2.60
Union Territories										
1	Andaman & Nicobar Islands.	22.8	26.4	26.8	32.7	37.7	6.4	4.1	4.4	2.10
2	Chandigarh	90.6	93.6	89.7	89.8	97.3	5.9	3.1	3.4	2.38
3	Dadra & Nagar Haveli	0.0	6.7	8.5	22.9	46.7	—	5.3	14.6	11.57
4	Daman & Diu	—	—	46.9	36.3	75.2	—	4.9	1.9	11.59
5	Lakshadweep	0.0	46.3	56.3	44.5	78.1	—	4.5	−0.8	6.24
6	Pondicherry	42.0	52.3	64.1	66.6	68.3	4.7	4.9	2.3	2.73
	All India	20.2	23.7	25.7	27.8	31.2	3.8	3.1	2.7	2.76

Note: a) The figures for the states of Uttar Pradesh, Bihar and Madhya Pradesh for the 1970s and 1980s pertain to the undivided states as existed during that time. The figures for the1990s are, however, for the new states and hence these figures are not temporally comparable.
b) In the absence of the Census data for total and urban population for the year 1981 in case of Assam, the urban and total population growth rates have been assumed to be constant during 1970s and 1980s. The same has been assumed for 1980s and 1990s for Jammu and Kashmir. The percentage of urban population has been arrived for Assam (1981) and Jammu and Kashmir (1991) based on these assumptions.
c) Goa in 1971 and 1981corresponds to Goa, Daman and Diu.
Source: *Census of India for different years [4].*

Table 4.
India: Level of urbanization and growth in urban population across states and union territories 1971–2011.

2.3 Composition of urban population

Apart from the trends and patterns of urbanization and population density, the composition of population and income distribution structure in urban India also favors the use of public transport for living and working. An overwhelming majority in cities belongs to the poor, low and lower-middle income groups. The Global Wealth Report 2015 published by Credit Suisse suggests that more than 90 percent of the adult population in India fall below the bottom of the wealth pyramid (less than $10,000). The middle class population in India, defined as those with annual wealth of about Rs.61,480 or $13,662 is estimated at 23.6 million [6]. About one-fourth of urbanites have been identifies to be below the poverty line. An equivalent number are slum dwellers. More than 65 percent of urban households lives in two rooms or less.

Table 6 presents a picture of urban poverty vis-à-vis rural poverty in India based on the Rangarajan Committee report. According to the Committee, a person spending less than Rs.1407 per month or Rs.47 a day was considered poor in cities in

Rank	Urban District	Area (Sq. Kms)	2001 Census		2011 Census	
			Population (In Lakhs)	Density	Population (In Lakhs)	Density
1	North East Delhi	56	17.68	31,573	22.42	36,155
2	Central Delhi	23	6.46	28,104	5.82	27,730
3	East Delhi	49	14.64	29,869	17.09	27,132
4	Chennai	174	43.44	24,963	46.47	26,553
5	Kolkata	185	45.72	24,718	44.97	24,306
6	Mumbai Suburban	446	86.40	19,373	93.57	20,980
7	Mumbai City	157	33.38	21,261	30.85	19,652
8	West Delhi	131	21.29	16,251	25.43	19,563
9	Hyderabad	217	38.30	17,649	39.43	18,172
10	North Delhi	59	7.82	13,256	8.88	14,557

Source: *Census of India 2001, 2011 [4].*

Table 5.
Most densely populated districts of India 2011.

Year	Poverty ratio (%)			No. of poor (million)		
	Rural	Urban	Total	Rural	Urban	Total
1. 2009–10	39.6	35.1	38.2	325.9	128.7	454.6
2. 2011–12	30.9	26.4	29.5	260.5	102.5	363.0
3. Reduction	8.7	8.7	8.7	65.4	26.2	91.6

Source: *Planning Commission (2014) [7].*

Table 6.
India: Rural and urban poverty estimates 2009–10 and 2011–12.

2011–12. The number of urban poor was estimated at 102.5 million, accounting for 26 percent of the urban population in the same year.

Census 2001 estimated the urban slum population in India at 42.6 million. It reported that 41.6 percent of slum population in the country lived in metropolitan cities. Mumbai had the largest number of slum dwellers, accounting for 54 percent of the population. Census 2011 has placed the number of slum-dwellers in India at 65.5 million. It further reveals that 46 million-plus cities contain 38 percent of the slum households. 9 metropolitan cities have more than 30 percent of households in slums, with Visakhapatnam topping the list at 44.1 percent, followed by Jabalpur Cantonment Board (43.1 percent) and Greater Mumbai (41.3 percent). Among the largest municipal corporations, apart from Greater Mumbai, Kolkata and Chennai have reported more than 25 percent of households living in slums.

The trends of urbanization, patterns of population density and state of slums, poverty and housing in cities suggest that the demographic and income distribution structures of urban India are overwhelmingly suitable for a public transportation-led model of urban development. Transportation planners and traffic engineers advocate the following strategies for urban transportation depending on their peak hour per direction traffic (PHPDT) that significantly depend upon the density of commuters:

PHPDT Recommended strategy

10,000 - 15,000 Bus and Dedicated Busways

15,000 - 30,000 Light Rail Transit
> 30,000 Heavy Rail Mass Transit
Based on the above criteria and other factors, many cities in India qualify for light rail transit and heavy rail transit. The largest metropolitan cities also need high speed rail connecting them to sub-urban centres and regional towns.

2.4 Trends in motorization

The number of registered motor vehicles in India increased from 0.3 million in 1951 to 55 million in 2001 and 210 million in 2015. While the share of two wheelers rose from 8.8 percent in 1951 to 73.5 percent in 2015, the share of busses declined from 11 percent to 1 percent. **Table 7** presents the trends in the number of motor vehicles and the composition of the vehicular population for the period 1951–2015.

The population of motor vehicles reported by million-plus cities in India in 2015 was 66.24 million. Among these, Delhi had the highest number at 88.51 lakhs, followed by Bengaluru (55.60 lakhs), Chennai (49.34 lakhs), Ahmedabad (34.20 lakhs), Greater Mumbai (25.71 lakhs), Surat (24.59 lakhs), Hyderabad (23.69 lakhs), Pune (23.37 lakhs), and Jaipur (22.49 lakhs). The largest number of two-wheelers in 2015 was in Delhi at 56.98 lakhs, followed by Bengaluru (38.41 lakhs), Chennai (35.16 lakhs), Ahmedabad (24.32 lakhs), Surat (19.13 lakhs); Pune (17.65 lakhs); Hyderabad (17.08 lakhs); Jaipur (16.58 lakhs) and Greater Mumbai (14.70 lakhs). Considering the quantum of cars in 2015, Delhi had 27.30 lakhs, followed by Bengaluru (10.89 lakhs), Chennai (8.60 lakhs), Greater Mumbai (7.97 lakhs), Kolkata (5.41 lakhs), Ahmedabad (5.26 lakhs), Hyderabad (4.02 lakhs) and Pune (3.75 lakhs). **Table 8** shows the number and share of two wheelers and cars in the population of motor vehicles for metropolitan cities as of 31st March 2015.

Table 9 presents the growth of motor vehicle population in 22 metropolitan cities in India over the period 2005–15 for which data are available. As the table

Year	Number in Million	Composition (% of Total Vehicle Population)				
		Two Wheelers	Cars, Jeeps and Taxis	Busses	Goods vehicles	Other vehicles
1951	0.3	8.8	52.0	11.0	26.8	1.3
1961	0.7	13.2	46.6	8.6	25.3	6.3
1971	1.9	30.9	36.6	5.0	18.4	9.1
1981	5.4	48.6	21.5	3.0	10.3	16.6
1991	21.4	66.4	13.8	1.5	6.3	11.9
2001	55.0	70.1	12.8	1.2	5.4	10.5
2006	89.6	72.2	12.9	1.1	4.9	8.8
2011	141.8	71.8	13.6	1.1	5.0	8.5
2012	159.5	72.4	13.5	1.0	4.8	8.3
2013	176.0	72.7	13.6	1.0	4.7	8.0
2014	190.7	73.1	13.6	1.0	4.6	7.7
2015	210.0	73.5	13.6	1.0	4.4	7.5

Source: *Government of India, Ministry of Road Transport & Highways, New Delhi: Road Transport Year Book (2013–14 and 2014–15) [8].*

Table 7.
Total number of registered motor vehicles in India (in million) 1951–2015.

Million Plus Cities	Total Number of Registered Motor Vehicles	Two Wheeler		Cars	
		Number	% of Total	Number	% of Total
Agra	9,05,023	7,41,778	81.96	76,107	8.41
Ahmedabad	34,19,828	24,31,839	71.11	5,25,891	15.38
Allahabad	8,97,035	7,30,758	81.46	72,779	8.11
Aurangabad	4,26,246	3,35,725	78.76	19,591	4.60
Bengaluru	55,59,730	38,41,139	69.09	10,88,587	19.58
Bhopal	10,80,477	8,47,334	78.42	1,36,627	12.65
Chandigarh	7,45,520	3,95,565	53.06	2,61,752	35.11
Chennai	49,34,412	35,16,062	71.26	8,60,932	17.45
Coimbatore	19,01,277	15,47,395	81.39	2,32,751	12.24
Delhi	88,50,720	56,98,242	64.38	27,30,071	30.85
Dhanbad	5,63,426	4,27,714	75.91	58,836	10.44
Durg-Bhillai	7,68,922	6,44,138	83.77	49,569	6.45
Ghaziabad	7,51,603	5,33,808	71.02	1,52,256	20.26
Greater Mumbai	25,71,204	14,70,175	57.18	7,97,267	31.01
Gwalior	6,17,681	4,87,259	78.89	52,685	8.53
Hyderabad	23,68,818	17,07,714	72.09	4,02,334	16.98
Indore	17,12,702	13,01,383	75.98	2,08,005	12.14
Jabalpur	6,38,219	4,93,633	77.35	67,445	10.57
Jaipur	22,49,240	16,58,006	73.71	3,05,445	13.58
Jamshedpur	4,72,051	3,51,696	74.50	55,020	11.66
Jodhpur	9,16,172	6,50,097	70.96	71,972	7.86
Kannur	1,88,497	1,12,851	59.87	43,920	23.30
Kanpur	14,61,530	11,72,577	80.23	1,47,072	10.06
Kochi	6,05,689	3,36,316	55.53	1,71,063	28.24
Kolkata	14,01,638	6,00,156	42.82	5,41,432	38.63
Kollam	2,74,006	1,75,528	64.06	58,097	21.20
Kota	6,54,041	5,12,740	78.40	51,749	7.91
Kozhikode	4,12,304	2,89,801	70.29	70,539	17.11
Lucknow	17,09,662	13,61,787	79.65	2,44,121	14.28
Madurai	9,54,893	7,93,510	83.10	68,804	7.21
Malappuram	2,76,765	1,51,351	54.69	59,297	21.43
Meerut	5,25,235	4,24,975	80.91	63,148	12.02
Nagpur	12,75,575	10,67,160	83.66	1,08,951	8.54
Nashik	6,22,206	4,61,628	74.19	62,473	10.04
Patna	10,18,798	7,05,298	69.23	1,35,638	13.31
Pune	23,37,085	17,65,172	75.53	3,75,267	16.06
Raipur	11,11,745	8,45,861	76.08	84,377	7.59
Rajkot	9,79,423	7,87,608	80.42	93,185	9.51
Ranchi	5,47,036	3,56,067	65.09	65,434	11.96

Million Plus Cities	Total Number of Registered Motor Vehicles	Two Wheeler		Cars	
		Number	% of Total	Number	% of Total
Srinagar	2,35,614	1,00,291	42.57	77,043	32.70
Surat	24,59,111	19,12,715	77.78	3,07,540	12.51
Trichy	7,63,396	6,36,961	83.44	58,712	7.69
Thiruvananthapuram	5,71,956	3,49,657	61.13	1,53,674	26.87
Thrissur	3,55,491	2,26,285	63.65	72,994	20.53
Varanasi	7,68,769	6,09,656	79.30	55,727	7.25
Vijayawada	6,10,321	4,52,403	74.13	53,755	8.81
Vadodara	10,41,818	8,03,969	77.17	1,23,509	11.86
Visakhapatnam	7,30,872	5,74,135	78.55	79,592	10.89
Total	6,62,43,782	4,73,97,918	71.55	1,16,53,035	17.59

Source: *Ministry of Road Transport and Highways, Government of India, New Delhi: Road Transport Year Book (2013–2014 and 2014–2015) [8].*

Table 8.
Share of two wheelers and cars in total number of registered motor vehicles in million plus cities of India as on 31st March 2015.

Metropolitan City	No. of Motor Vehicles (in Thousands)		Average Annual Growth (%)
	2005	2015	
Ahmedabad	1632	3420	10.96
Bengaluru	2232	5560	14.91
Bhopal	428	1080	15.23
Chennai	2167	4934	12.77
Coimbatore	682	1901	17.87
Delhi	4186	8851	11.14
Greater Mumbai	1295	2571	9.85
Hyderabad	1433	2369	6.53
Indore	705	1713	14.30
Jaipur	923	2249	14.37
Kanpur	425	1462	24.40
Kochi	166	606	26.51
Kolkata	911	1402	5.39
Lucknow	615	1710	17.80
Madurai	330	955	18.94
Nagpur	770	1276	6.57
Patna	378	1019	16.96
Pune	827	2337	18.26
Surat	692	2459	25.53
Varanasi	366	769	11.01

Metropolitan City	No. of Motor Vehicles (in Thousands)		Average Annual Growth (%)
	2005	2015	
Vadodara	586	1042	7.78
Visakhapatnam	435	731	6.80

Source: *Ministry of Road Transport and Highways, Government of India, New Delhi: Road Transport Year Book (2013–2014 and 2014–2015) [8].*

Table 9.
Growth in number of registered motor vehicles in select metropolitan cities 2005–2015.

shows 16 out of 22 metropolitan cities recorded more than 10 percent annual growth over the period; 3 cities had an annual growth rate exceeding 20 percent.

The car-penetration rate defined as the number of cars per 1000 persons is very small in India compared to that in developed countries and several developing countries. **Table 10** compares data on Gross National Income (GNI) and vehicular penetration rates for select countries with those for India.

The data in the above table suggest that with the rise in GNI, following structural transformation and economic growth, the vehicular penetration rate, with attendant problems of congestion, pollution, noise and carbon emissions in cities, will lead to increased demand for road space and public transport, including rail-based transit.

Ironically, many of India's urban mobility problems can be traced to the lack of an appropriate planning model and public transport development strategy rooted in the economics of cities. In particular, cities have not exploited the links between

Country	GNI per capita (US$) for 2013	Number per 1000 persons		
		Passengers Cars	Total Vehicles	Two-wheelers
Developed Countries				
United States	53,470	360	783	27
United Kingdom	41,680	455	517	19
Japan	46,330	466	598	81
Germany	47,270	544	603	50
Australia	63,390	562	711	32
Developing Countries				
Mexico	9940	203	285	15
Malaysia	10,430	358*	396*	356
South Africa	7190	110**	162**	6
Brazil	11,690	227	290	108
China	6560	76	93	70
South Korea	25,920	300	386	42
India	1570	19	167	123

**Data relates to 2012.*

***Date relates to 2011.*

Source: *Ministry of Road Transport & Highways, Government of India, New Delhi: Road Transport Year Book (2013–2014 and 2014–2015) [8].*

Table 10.
Vehicular penetration rates in select developed and developing countries 2013.

agglomeration externalities and transportation in their spatial planning and development models. Land use planning and transportation planning have been pursued as disjointed exercises in India. Cities had land use planners, but no transport planners. As a result, they have not been able to harness the power of city externalities to guide transport-land use integration and local economic development, address congestion and raise resources to finance public transport. The trends of urbanization, metropolitanization and motorization; patterns of population composition and densities in cities; abysmal state of urban transport with no robust model of financing in sight; emerging energy security and environmental concerns; andthe demands of inclusive economic growth in India call for exploring the principles of New Urbanism, Smart Growth and TOD for restructuring urban planning.

3. New urbanism, smart growth and TOD

New Urbanism and Smart Growth emerged in the last four decades in the United States, Europe and other developed countries in response to their problems of urban sprawl, a consequence of automobile-dependency. They are rooted in a search for alternatives to low-density, single-use and spread-out patterns of urban expansion, increasing traffic congestion and air pollution, and adversely impacting the environment and quality of life.

New Urbanism is a design-oriented with architectural roots. Promoted by architects, it is focused on neighborhood design. Smart Growth is policy-oriented with environmental roots. Spearheaded by planners, it is centered on promoting guided development. Smart Growth is not so much concerned with urban design as it is with growth promotion. It elevates the discourse on urban planning from growth control to issues of how and where growth should be accommodated. It calls for public subsidies for growth, such as infrastructure facilities and land use incentives. Both New Urbanism and Smart Growth advocate TOD.

Transit Oriented Development (TOD) owes its origin to the paradigms of New Urbanism and Smart Growth. It is an urban planning and development approach aimed at creating vibrant, livable and sustainable communities by concentrating growth around one or more transit stations or within a transit corridor. It emphasizes compact, walkable, mixed-use communities with access to high quality transit services within a walking distance. TOD principles are not new; they were introduced by many cities in North America and Australia into their planning models after World War II. However, TOD as a specific policy paradigm has taken root only in the last twenty years.

The concentration of development based on a TOD approach acts against urban sprawl and uneconomic extension of costly infrastructure, catalyzes external economies of agglomeration, mitigates congestion diseconomies, and assists in the mobilization of resources through increases in land and property values and other tax bases. TOD enables lower-stress living without complete dependence on a car for mobility. It is environment-friendly and inclusive. The poor, who do not own automobiles benefit significantly when included under a TOD scheme. As an instrument of inclusive regional and urban planning, TOD promotes the inclusion of the poor in the urban development process. The economic, social and environmental benefits of TOD are briefly presented below:

TOD: Economic Benefits:

The economic benefits of TOD include reduced congestion, agglomeration economies, resource mobilization for financing infrastructure, reduced costs of development, efficiency of investment, etc.

Reduced Congestion: TOD reduces the need to travel and, thus, reduces congestion and stress levels.

Agglomeration Economies: TOD, if designed properly, can augment agglomeration economies by enhancing access to the economic mass and facilitating the collocation of productive economic activities in nodes with potential to engineer growth. These economies lead to benefits of backward and forward linkages, market access, sharing of common infrastructure facilities and resources, specialized labour pooling, human capital accumulation, knowledge spillovers and networking. They lead to economies of sharing, matching and learning; they promote specialization, diversity and competition.

Increased Revenue Yields: Properties around transit hubs are accorded higher values. These higher property values could be converted into revenue for the government through value capture levies.

Efficiency of Investment: TOD directly fosters patronage for growth and helps to optimize existing transit and connectivity infrastructure. It maximizes the efficiency and carrying capacity of the transportation network.

TOD: Social Benefits.

Affordable housing and public transport are key enablers of social inclusion. They increase the accessibility to jobs, health care, education, recreation and socio-cultural interactions.

TOD: Environmental Benefits.

Public transport can help to reduce the proliferation of personal vehicles and thus, reduce the level of emissions. This reduction could be quite significant, especially during the peak hours.

The success of TOD depends on its design. **Box 1** presents some key principles to guide TOD designing.

1. Multimodal Transit Station.
   Transit is the focus of TOD.Transit facilities should not be designed in isolation, rather it should connect the neighborhoods. Further, it should include a mix of modes like two wheelers, car, bicycles, BRT, LRT and NMT.
2. Interconnected Streets.
   Such a pattern not only decreases congestion but also encourages mixed use development along with enhanced travel choices.
3. Mixed Use Development.
   A compact structure involving diverse land use pattern can benefit residents as well as workers to meet their daily requirements including work, shopping and leisure.
4. Walkability.
   In order to encourage walking it is important to design a pedestrian-friendly structure. Such a structure must include sidewalks, shaded pedestrian routes, benches to rest and safe crossing points at transit stations.
5. Compact Development.
   In order to be successful, the structure needs to be compact. The extent of neighborhoods around transit nodes is based on a comfortable walking distance from edge to centre (approximately 400 to 800 meters in radius).
6. Street-facing Buildings.
   Streets can be better defined by placing the buildings near them. Street front retail should be provided to humanize the building wall and activate the sidewalk.
7. Urban Place-making.
   A successful TOD design works on developing public spaces in the neighborhood. It is important for improving social interaction and strengthening community bonds and participation.
8. Neighborhood High Street.
   Retail streets provide the goods and services of daily life, activate the street, reduce auto reliance, and increase ownership and safety of the pedestrian realm.
9. Streetscape Design.
   A beautified street pattern equipped with pedestrian utilities improves the desire to walk and makes it pleasant while shortening the sense of distance.

10. Bicycle-friendly Streets / Parking.
    Bicyles are environment friendly and efficient alternatives to automobiles. Bike lanes, bike routes, and secure parking make the bicycle an easy option.
11. Urban Parks & Plazas with Minimized Ecological Footprint.
    Open spaces enable public interaction and promote healthy communities.
12. A Well-designed Transit Station for a High Quality User Experience.
    The design of the transit station is at the heart of a successful TOD structure. Its design is critical for enhancing customer attraction and ensuring seamless and efficient accessibility to consumers.
13. Reduced Parking Standards.
    Reducing parking standards provides increased site area for alternative public amenities.
14. Safety & Security.
    Ensuring safety and security of transit users especially pedestrians, not only improves the transit experience but also enhances transit ridership.
15. Market Acceptance and Successful Implementation.
    A vibrant and transit supportive space which attracts several jobs and residents is critical for a TOD programme. Flexible strategies along with designs which cater to the needs of the surrounding neighborhood can ensure a successful TOD.

Source: UNDP 2012 [9].

Box 1.
Transit-oriented development: design principles.

While the principles of Smart Growth and TOD originated in developed countries in response to their problems of sprawl, the paradigms make good sense for developing countries like India. However, TOD policies have not been implemented in an appreciable way in India. Only recently Delhi and Haryana have brought out planning guidelines for TOD, calling for the integration of transportation and land use. Bengaluru, Mumbai, Pimpri-Chinchwad, Ahmedabad. Hyderabad, Naya Raipur and Bhubaneswar have embarked on programmes to promote transit-oriented planning and development. TOD presents significant opportunities to India to make the country's urbanization process efficient, inclusive and sustainable. However, the execution of TOD and financing of transit investments are key challenges for Indian cities. Apart from the principles of sustainable development, successful international practices of transport-land use and integrationapproaches to financing public transport investments can guide the design of TOD in India. Section 4 refers to some oft-cited examples of successful international practices of TOD. Section 5 presents the broad approaches to financing of public transport, including transit to guide Indian cities to draw lessons for TOD.

4. Transit oriented development: international practices

TOD is emerging as a preferred paradigm to plan cities, localities and urban extensions and renew old cities and derelict areas within cities in many countries. Some of the successful TOD models practiced internationally that can provide lessons for Indian cities for the integration of transportation and land use are discussed in this section.

4.1 Hong Kong SAR

Hong Kong is internationally known for its successful integration of rail transit investments and urban development. The integrated "rail-property" development model (R + P), plays a vital role in managing and financing railway expansion, advancing high-quality urban designs, creating "one-stop" settings for "live-work--shop-play", guiding regional urban growth, and more. As with all good public-private partnerships, this occurs in a win-win fashion – i.e., the railway corporation reaps financial benefits and society at-large benefits from more sustainable, transit-oriented patterns of development. Maritime Square Residential-Retail Development atop Tsing Yi Station provides a good example of Hong Kong TOD. Maritime Square features hierarchically integrated uses. Shopping mall extends from the ground floor to the 3rd level. Station concourse sits on the 1st floor, with rail lines and platforms above and ancillary/logistical functions (like public transport/bus interchange and parking) at or below. Above the 4th and 5th floor residential parking lies a podium garden and above this, high-rise, luxury residential towers [10].

The Hong Kong Government derives a major proportion of its revenues from land, including premium on new land and modification of existing leases, property taxes, stamp duty, rents, etc. [11]. The Hong Kong MTR has generated many benefits to the community. These include travel time saving, employment gains, environmental health benefits, property value increases and so on. The network obviously generates enormous external benefits as it passes through the densely populated districts, commercial and employment centres and carries large passenger loads.

4.2 Bogota

Bogota, the capital of Colombia, has some of the most progressive public investment initiatives in developing countries, including the first-class TransMilenio BRT; integrated TDM measures; the transit-linked social housing Metrovivienda program; the Alameda Porvenir, the world's longest pedestrian way; and other public projects that incorporate good urban design and innovative financing schemes [12]. Bogota's TransMilenio is one of the world's most successful examples of Bus Rapid Transit (BRT) [13]. It is characterized by dedicated main trunk routes for high speed busses, physically separated from the rest of traffic [14]. The bus stations are well-connected with systematic feeder services. The integrated approach of Bogota addressing affordable housing and affordable transport simultaneously, has improved the access to work, leisure, recreation, shopping.

4.3 Curitiba

Curitiba's bus system is composed of a hierarchical system of services. Minibusses routed through residential neighborhoods feed passengers to conventional busses on circumferential routes around the central city and on inter-district routes. The backbone of the system is composed of the Bus Rapid Transit, operating on the five main arteries leading into the centre of the city like spokes on a wheel hub [15]. Along each of the five arteries there is a trinary road system, comprised of middle express bus lane with vehicle lanes on each side for local auto traffic and parking.

Curitiba's Master Plan integrated transportation with land use planning. It limited central area growth, while encouraging commercial growth along the transport arteries radiating out from the city centre. The city centre was partly closed to vehicular traffic, and pedestrian streets were created. Rush hour in Curitiba has heavy commuter movements in both directions along the public transportation arteries.

4.4 Copenhagen

Danish Town Planning Institute created the "Egnsplan" or the Finger Plan in 1947. It was based on a TOD principle, with mixed land use and high-density areas around the centre [16]. Shopping malls, offices, recreational centres and housing were all planned in pedestrian areas with good bicycle facilities such as cycle lanes and parking and a good connection to public transport. The design includes five fingers or corridors of urban development along the suburban areas which are connected through railway lines and would directly connect the areas to Copenhagen Central Business District (CBD). The neighborhoods around the transit stations were planned to be developed in a TOD fashion with high density housing and amenities. The approach aimed at an ordered and integrated 'green' growth and was developed at the time of extensive and rapid urban development. There were spaces left for the use of farmland and recreational purposes between each finger, known as "green wedges". A ring road was planned at the end of each finger which linked the Copenhagen harbor and inner city to industrial locations. Most of the land was developed by the end of the 1960s and the two southern-most fingers were extended.

Orestadtownship is one of the best examples of successful TOD following the Finger Plan. It combines economic activities, housing and amenities – jobs, housing, retail, leisure and education – all based on TOD. It helped Copenhagen to remain competitive and release pressure on CBD.

Unlike the international cities with global best practices on TOD, Indian cities have grossly neglected transportation planning, public transport investments and transport-land use integration for long. Key issues of financing public transit and development integrated with such transit are typically ignored in public discourses. As a result, a coherent strategy for financing public transport has not emerged in India. Section 5 refers to international practices for financing of transit oriented development in the broader concept of financing public transport to guide Indian cities.

5. Financing transit oriented development

The financing of TOD cannot be artificially divorced from the broader issues of financing urban transport and cities. Both planning and economic considerations are important for designing a financing strategy. The approaches to financing of various types of public transport infrastructure, including TOD internationally include the following methods:

- Equity, including public-private partnerships (PPP), special purpose vehicles, infrastructure debt funds, investment funds, infrastructure financing companies.

- Debt tools, including private debt, commercial bank debt, take-out financing, bond financing – infrastructure bonds, municipal bonds (revenue and general obligation), green bonds, etc.

- Foreign Direct Investment and Foreign Portfolio Investment.

- Grant financing, combining central and state grants with local government resource mobilization and using public funds to leverage market resources and PPP.

- Direct fees, including user fees, utility fees, benefit charges and congestion pricing.

- Using land as a resource - value capture and impact instruments such as land and property taxes, land value tax, land value increment tax, betterment levy, developer exactions, impact fees, special assessment districts, land

Sl. No.	Name	Description	Advantages	Disadvantages
1.	Fare increases	Increase fares or change fare structure to increase revenues	Widely applied. Is a user fee (considered equitable)	Discourage transit use. Is regressive.
2.	Discounted bulk passes	Discounted passes sold to groups based on their ridership	Increases revenue and transit ridership	Increases transit service costs and so may provide little net revenue
3.	Property taxes	Increase local property taxes	Widely applied. Distributes burden widely.	Supports no other objectives. Is considered regressive.
4.	Sales taxes	A special local sales tax	Distributes burden widely.	Supports no other objectives. Is regressive.
5.	Income tax	Special income tax for transit or transportation	Progressive with respect to income. Relatively stable.	May be difficult to implement.
6.	Fuel taxes	An additional fuel tax in the region	Widely Applied. Reduces vehicle traffic and fuel use	Is considered regressive.
7.	Vehicle fees	An additional fee for vehicles registered in the region	Applied in some jurisdictions. Charges motorists for costs.	Does not affect vehicle use.
8.	Utility levy	A levy to all utility accounts in the region	Easy to apply. Distributes burden widely.	Is small, regressive and support no other objectives.
9.	Employee levy	A levy on each employee within a designated area or jurisdiction	Charges for commuters.	Requires administration. Encourage sprawl if in city centers.
10.	Road tolls	Tolls on some roads or bridges	Reduces traffic congestion.	Costly to implement. Can encourage sprawl if only applied in city centers.
11.	Vehicle-Km tax	A distance-based fee on vehicles registered in the region	Reduces vehicle traffic.	Costly to implement.
12.	Parking taxes	Special tax on commercial parking transactions	Is applied in other cities.	Discourages parking pricing and downtown development.
13.	Parking levy	Special property tax on parking spaces throughout the region.	Large potential. Distributes burden widely supports strategic goals.	Costly to implement. Opposed by suburban property owners.
14.	Expanded parking pricing	Increase when and where public parking facilities (e.g. on-street parking) are priced	Moderate to large potential. Distributes burden widely. Reduces parking & traffic problems.	Requires parking meters and enforcement, and imposes transaction costs.
15.	Development or transport impact fees	A fee on new development to help finance infrastructure, including transit improvements.	Charges beneficiaries.	Limited potential.
16.	Land value capture	Special taxes on property that benefit from the transit service	Large potential. Charges beneficiaries.	May be costly to implement. May discourage TOD.

Sl. No.	Name	Description	Advantages	Disadvantages
17.	Station rents	Collect revenues from public private development at stations	Charges beneficiaries.	Limited potential.
18.	Station air rights	Sell the rights to build over transit stations.	Charges beneficiaries.	Limited potential.
19.	Advertising	Additional advertising on vehicles and stations.	Already used.	Limited potential. Sometimes unattractive.

Source: *Todd Litman 2016 [18].*

Table 11.
Potential public transport funding options.

readjustment, town planning scheme, joint development, land monetization including the lease and sale of land and air rights with enhanced Floor Space Index and value-enhancing land use changes in TOD zones, tax increment financing, etc.

- Bullet Bonds and Pooled Finance Fund Scheme.

Land value capture (LVC) instruments take many forms and can be classified into two major types: (i) tax- or fee-based and (ii) non-tax- or non-fee-based, also called "development-based LVC." Tax- or fee-based instruments capture land value increases through, for example, land and property taxes, betterment charges, special assessments, and tax increment financing. In contrast, development-based LVC instruments capture these increments through land-related transactions such as selling or leasing land, development rights and air rights; making land readjustments; and redeveloping urban areas [17]. If adapted well to local contexts, development-based LVC instruments can be an effective finance and planning mechanism for cities in India.

The issues of financing public transit and TOD are intricately connected. However, as Indian cities are struggling to finance the development of mass rapid transit and bus rapid transit systems, not many have focused on TOD funding linked to LVC and non-LVC instruments. Based on international experience, a combination of financing instruments needs to be considered for adoption in India. These have to be suitably customized to fit the context of cities. A summary of various potential options for funding public transport, including transit is presented in the table below (**Table 11**).

International experience suggests that no one size fits all. But it makes clear that public transit and TOD impact on local, regional and national economies and lead to enhanced tax bases of all governments. Thus, if they are financed by borrowed funds with repayment linked to a value creation, capture and recycling strategy, cities in India can hope to get out of their vicious circles and traverse on a path of planned development. Future tax increments can finance current investment programmes which augments tax bases.

6. Towards TOD in India: case studies

Some state governments and urban local bodies in India have resorted to novel initiatives to plan and implement projects aimed at improving urban mobility following the TOD principle and Smart Growth framework. Some case studies are presented below.

6.1 Janmarg: Ahmedabad

The city has decided to develop and implement an integrated public transit system including:

- A Suburban Rail Transit System to connect the city with its industrial suburbs such as Kalol, Naroda, Mehmedabad, etc.
- A Metro Rail System to cater to the high intensity movement between Ahmedabad and Gandhinagar.
- A Bus Rapid Transit System (BRTS) to cater to major mobility needs of the city.
- A regular bus system to support BRTS.
- Decentralized Regional Bus & Rail Terminal.
- Integrate different form of transport, i.e., BRT with other regional and urban transport systems, with bicycles and pedestrian facilities.
- Integration of Land Use -Transport elements like increased FSI along BRTS corridor.

The primary objective of the integrated public transit initiative in Ahmedabad is to make the city more accessible – with physical, social and economic accessibility.

Ahmedabad city has developed a Bus Rapid Transit System under the Jawaharlal Nehru National Urban Renewal Mission (JNNURM) with the name "Janmarg" or "the people's way" and the slogan 'Accessible Ahmedabad'.Janmarg boasts of an innovative plan anddesign which includes pedestrian only sections and one-way bus lane etc.

6.2 "Namma" metro: Bengaluru

Bangalore Metro Rail Corporation Limited (BMRCL), a joint venture of Government of India and Government of Karnataka is a Special Purpose Vehicle entrusted with the responsibility of implementation of the Bengaluru Metro Rail Project. "Namma Metro" is an environment friendly initiative as it aims at reducing carbon emissions in the city. The project has an East–West corridor - 18.10 km long, starting from Baiyappanahalli in the East and terminating at Mysore Road terminal in the West and a 24.20 km North–South corridor commencing at Nagasandra in the North and terminating at Puttenahalli in the South.

In connection with the construction of Bengaluru Mass Rapid Transit System, the Government of Karnataka has introduced a number of innovative measures to create a dedicated resource pool, including special cess to capture land value increments due to transit [19]. The Government has taken up several value capture instruments like development ceases, chess on additional FAR, Transferable Development Right (TDR) etc. to finance Bengaluru mass rapid transit system.

6.3 Delhi MRTS

The Delhi Metro system serves Delhi and its satellite cities of Faridabad, Gurgaon, Noida and Ghaziabad in National Capital Region in India. Delhi Metro is the world's 12th largest metro system in terms of both length and number of stations. The network consists of five color-coded regular lines and the faster Airport Express line, with a total length of 213 kilometers serving 160 stations (including 6 on Airport Express line) [20]. The system has a mix of underground, at-grade, and elevated stations using both broad-gauge and standard-gauge. The metro generated an average daily ridership of 2.661 million passengers.

Delhi Metro has been instrumental in reducing vehicular congestion on the roads. According to a study, Delhi Metro has helped in removing about 3.9 lakh vehicles from the streets of Delhi. The Delhi Metro Rail Corporation has been certified by the United Nations as the first metro rail and rail-based system in the world to get "carbon credits for reducing greenhouse gas emissions" and helping in reducing pollution levels in the city by 630,000 tonnes every year, thus helping in reducing global warming [21].

The Delhi TOD Policy 2013 has provided for significant increases in FSI in transit influence zones to promote intensive development so that TOD can be self-financing adopting a land value capture method and even be surplus-generating. The Delhi Development Authority has proposed to take up TOD to build the East Delhi Hub as a signature destination. This includes the development of 75 acres of land in Karkardooma with FSI raised to a maximum of 4 and maximum density of 2000 persons per hectare on the basis of a TOD model. The project is being taken up on a partnership with the National Building Construction Company (NBCC). Some parcels of land are under development, but Delhi is far from achieving the TOD Policy objectives of inclusive development.

6.4 Delhi TOD policy

The salient features of the Delhi TOD Policy are:

- Development/redevelopment in TOD zone will be incentivized by providing significantly higher FAR of 4.0 on the entire amalgamated plot being developed/redeveloped.

- Additional FAR may be availed only through Transferable Development Rights (TDR), for schemes larger than 1 hectare.

- Entire approved layout plan of a scheme will be included in the influence zone if more than 50 percent of the plan area falls in the influence zone.

- It will be mandatory to use a minimum of 30 percent of overall FAR for residential use, a minimum of 10 percent of FAR for commercial use and a minimum of 10 percent of FAR for community facilities. Utilization of the remaining 50 percent FAR shall be as per the land use category designated in the Zonal Plan.

- There shall be a mix of housing types for a wide range of income brackets within communities with shared public spaces/greens/recreational facilities/amenities, which will minimize gentrification and create more community-oriented developments.

- The mandatory residential component covering 30 percent FAR shall wholly comprise of units of 65 m^2 area or less. Out of the half of the FAR, i.e. 15 percent of the total FAR, has to be used for units of size ranging between 32 and 40 m^2. Over and above this, an additional mandatory FAR of 15 percent, i.e. FAR of 0.6 (out of 4.0) has to be utilized for Economically Weaker Sections (EWS). The size of EWS units will range between 32 and 40 m^2.

- 20 percent of land shall be used for roads/circulation areas. 20 percent area for green open space shall be kept open for general public use at all times. Further, 10 percent of green area may be for exclusive use.

- MRTS agencies are exempted from providing the mandatory 30 percent residential component which is part of the TOD norms applicable to all other developer entities (DEs).

6.5 Hyderabad elevated metro rail

Hyderabad has gone for a metro rail transit based on project report prepared by Delhi Metro Rail Corporation (DMRC) which identified 269 acres of land requirement. Originally the project was conceived as a government-funded project. However, subsequently the city went for a metro based on a PPP mode, adopting a Design, Build, Finance, Operate and Transfer (DBFOT) format. Hyderabad is currently implementing the world's largest elevated metro rail project in PPP mode with L&T as concessionaire.

The revenue model of the concessionaire is: 55 percent passenger fare, 40 percent property development and 5 percent advertisement and parking fees. Thus, the project's revenue is partially fare based. But, a significant portion of the revenue is also non-fare based. The concessionaire hugely relied on the potential of development of property or air space above and around transit stations. It has been provided with some valuable government land at vantage transit stations and is undertaking commercial exploitation of property with engineering innovations. It cannot sell property but can enjoy the rentals during the concession period of 35 years.

6.6 Mumbai metro

Mumbai Metro Line 1 – Versova-Andheri-Ghatkopar Mass Rapid Transit System is the first metro project awarded in the country on a PPP basis. It has provided the much needed connectivity in the financial capital of India linking the East and the West. It has ensured connectivity to Western and Central Railways. Providing modern, fast, clean and caring infrastructure, the Line has carried 100 million commuters in the first year of operation. It has reduced the journey time between Versova and Ghatkopar from 71 minutes to 21 minutes.

While Indian cities have making efforts to promote TOD with new policies, projects and plans emerging, especially in the context of Smart Cities Mission, a study of the Indian initiatives so far suggests robust approaches to financing and execution of TOD have not emerged. The approaches to financing transit also vary considerably as shown in **Table 12** below.

A key lesson from the initiatives of Indian cities towards financing transit and TOD, when compared to international best cases, is that urban policy, spatial planning, city development strategy, city financing framework, transit orientation, zoning, land use and development control regulations and institutional framework to integrate land use and transportation planning, raise resources and execute TOD

Funding Approach	Financing Pattern	Practicing Metro Rail
Government-funded	50–50 Central Government: State Government	Delhi, Bengaluru, Chennai, Kochi, Nagpur
	100 percent State Government	Jaipur, Lucknow (initially)
	100 percent Central Government	Kolkata (North –South) Kolkata (East–West)
Public-Private Partnerships	Private provisioning of operation and maintenance	Delhi Airport Express (initially) – Reliance Infra
	PPP-BOT model (Design, Build, Finance, Operate, Transfer)	Hyderabad Metro (Government of India Viability Gap Funding – 10%, L&T – 20% equity and 70% debt)) Mumbai Metro Line 1 (RInfra –69% MMRDA – 26%, Veolia – 5%)
Private Funding	Complete private funding	Gurgaon Rapid Metro Phase I – Equity 75%, DLF 25% Phase II – Senior Debt from Banks/ Financial Institutions – 70%, Sponsor's Contribution – 30%

Table 12.
Emerging approaches to financing public transit in India.

need to be part of a holistic model of integrated urban development and should not be undertaken disjointedly.

7. Implementing TOD in India: issues and directions

The existing institutional framework in Indian cities is not adequate to tackle the issues associated with planning, financing and implementing TOD. The starting point for successfully implementing TOD in India is to devise an appropriate institutional framework along with clarity in financing mechanisms. Considering the investment and planning efforts demanded by TOD, an effective, extensive and robust institutional framework needs to be put in place. This framework is required at all three levels of governance: centre, state and local. The design, implementation of TOD and enforcement of urban transport pricing and regulatory measures require special attention. Proper co-ordination must be ensured between the several agencies involved at the different levels in order to prevent potential conflicts and delays.

The draft National Transit Oriented Development Policy paper takes into account the above internationally recognized principles and implementation guidelines for TOD. Keeping in view the international best practices and national debate and discussion on TOD as an instrument of sustainable and planned urban development, **Box 2** provides some broad directions for executing TOD in cities and towns of India.

TOD focuses on compact, mixed use development around transit corridors - metro rail, BRTS etc. International best practices have demonstrated that though transit system facilitates transit-oriented development, improving accessibility and creating walkable communities is equally important. Thus, to achieve the goals of TOD, the planning and development principles mentioned earlier in the study need to be adopted. The principles should also be supported by TOD-support policy tools such as right size infrastructure, technology integration, station area planning, land value capture, safety and security, universal accessibility etc. The following key aspects need to be considered for translating TOD principles and policies into practice in India:

1. Approach for TOD Implementation:

1.1 Influence Zone:
Influence zone of any transit corridor or station is the area in its immediate surrounding. It is intended to be developed into a compact, high density structure with mixed land use to cater to the residents' basic needs. It is generally up to a radius of nearly 500–1000 mt of the transit station.

1.2 High Density Compact Development:
TOD calls for the densification of the influence zone. This can be done by providing higher Floor Area Ratio (FAR)/ Floor Space Index (FSI) and higher job and population density in the influence areas.To ensure sustainable and financially viable development, the minimum FAR should be 300–500 percent, and can be higher, depending on the city size.

1.3 Mixed Use Development:
Mixed land use in the TOD zone reduces the need to travel for work, shopping, leisure, education etc. The basic necessities of the residents can be provided within walking distance.

1.4 Mandatory and Inclusive Housing:
The cities should have minimum percentage (30 percent or higher) of allowed FAR for affordable housing in all development/ redevelopment in the influence zones. Housing in the influence zones should have a mix of all economic groups/ sections. The development control regulation should cater housing for EWS as well as LIG and MIG to give an opportunity to the people who depend on public transport for daily commuting to live in walkable neighborhoods.

1.5 Multimodal Integration:
An integrated multimodal network is required for availing various facilities in the influence zone. Seamless physical connectivity, integrated information system and fare integration can provide easy first and last mile connectivity.

1.6 Focus on pedestrians, cyclists and NMT users:
The influence zone should address the needs of pedestrians and NMT users. Sidewalks and amenities like benches, lighting, shops and information signage etc. should be developed.

1.7 Street Oriented Buildings and Vibrant Public Spaces:
Buildings should face the streets so as to define them better. Buildings should be oriented towards facing the pedestrian facilities. Public spaces should be developed to improvesocial interaction and strengthen community bonds and participation.

1.8 Managed Parking:
Use of private vehicles can be discouraged by reducing availability of parking spaces in influence zones and making it expensive. On-street parking should be prohibited within 100 mt of the transit station, except for freight delivery and pick-up or drop-off of the differently abled.

2. Value Capture Financing (VCF) for TOD:
The investment in the transit system as well as increase in FAR and provision for mixed use development would result in increase in value of land within the influence zone. Land Value Capture can be used as a mechanism to finance the required upgradation of infrastructure and amenities within the influence zone and expansion of the public transport system.

3. Statutory Framework:
TOD policy should be notified as part of the Master Plan/ Development Plan of the city whose vision should be resonated by all the stakeholders, especially those involved in infrastructure development and preparation of development plans. The policy document should clearly outline the importance of the high capacity transit network in the city's development.

4. Coordination and Implementation:
Successful implementation of TOD requires the various agencies involved in planning, design and financing to work in coordination with each other. UMTAs need to be operationalized and strengthened.

5. Communication and Outreach:
It is important to create awareness about TOD so as to increase its use. Multiple agencies including both private and public stakeholders must have a collective approach for successful implementation of TOD.

Source: National Transit Oriented Development Policy [22].

Box 2.
Implementing TOD in India.

Table 13 presents a summary of steps to convert the concepts of TOD to micro level implementation and undertake rapid transit station area and transit corridor development in India:

Existing Land Use	
a. Existing Land Use	• Development of existing corridor and station-area. • Making existing station-area pedestrian friendly, including access for persons with disabilities. • Improving parking supply in existing corridor and station-areas.
Transit Supportive Plans and Policies	
a. Growth Management	• Concentration of development around established activity centres and regional transit. • Greater employment opportunities should be provided close to transit stations. • Managing and conserving land.
b. Transit Supportive Corridor Policies	• Development of station-area and increasing transit corridor. • Plans and policies aiming to increase transit-friendliness of transit stations, corridors and areas. • Designs to improve pedestrian amenities, including facilities for the differently abled commuters. • Parking policies.
c. Supportive Zoning Regulations near Transit Stations	• Policies and regulations which encourage development around transit stations. • Zoning ordinances that provide enhanced accessibility topedestrians and encourage transit-oriented characteristics of stations. • Zoning allowances to mitigate traffic and reduce parking. • Ensuring provision of affordable housing units close to transit stations.
d. Tools to implement land use policies	• Outreach to government agencies and the community in support of land use planning. • Regulatory and financial incentives to promote transit supportive development. • Higher FSI should be allowed along transit corridors. • Efforts to engage the development community in station area planning and transit-supportive development.

Table 13.
Summary of steps to execute TOD: station area and corridor development.

8. Conclusions

India is going to experience a multifold rise in the demand for urban transport in the coming years. A strategic approach is required to ensure that the growth momentum is maintained without adversely impacting the quality of environment to urban dwellers. A holistic planning mechanism consolidating urban transport and land use planning is essential for Indian cities, especially metropolitan cities, so that synergies between urban form and functions can be channelized. This will further augment the productivity and efficiency of cities. The present challenges of congestion, pollution, accidents, sprawl etc. can be mitigated by investing in public modes of transport and optimizing multimodal mobility patterns. Urban transport influences the spatial organization of cities. So, urban transport must be approached in a holistic manner integrating pricing, financing regulation and comprehensive land use.

This chapter suggests that TOD is a necessity in India in view of the urbanization, metropolitanization and motorization trends in cities, the numbers and densities therein, income distribution patterns and considerations of sustainability. The country cannot afford auto-centric, sprawling, energy-intensive and an expensive process of urbanization. India needs to move to a public-transportation led, transit-oriented, mixed use, and value capture financing-focused strategy of planned urban development with public transportation investment and transport-land use integration as the key drivers. This also calls for a robust financing strategy. Successful TOD policy requires a robust and integrated framework bestowed with financial independence, responsiveness and competence. TOD policy can be successful by ensuring transparency and accountability towards the users while augmenting the accessibility of Indian cities.

Acknowledgements

We express our sincere thanks to the Housing and Urban Development Corporation (HUDCO), New Delhi for funding support to complete this research under the HUDCO Chair program.

We would like to thank Dr. Prasanna Kumar Mohanty, Chair Professor, Land, Housing, Transport and Urban Economics, University of Hyderabad, Member, Central Board of Directors, Reserve Bank of India and National Housing Bank, India, Ex. Chief Secretary, Undivided Andhra Pradesh, India for his valuable and constructive comments on this paper.

Conflict of interest

The authors declare no conflict of interest.

Author details

Alok Kumar Mishra[1*] and Shibani Mishra[2]

1 School of Economics, University of Hyderabad, Gachibowli, Hyderabad, Telangana, India

2 Odisha Public Service Commission and Doctoral Scholar, School of Economics, University of Hyderabad, Gachibowli, Hyderabad, Telangana, India

*Address all correspondence to: alokmishra@uohyd.ac.in

References

[1] Mishra, A.K. 2019. Cities, transport and agglomeration: Addressing the urban mobility challenges in India. *Growth and Change.* Vol.50. Issue 3. September 2019. Pp. 1115–1133

[2] Cervero, R., and M. Duncan, 2002. Land Value Impacts of Rail Transit Services in Los Angeles County. Report Prepared for the National Association of Realtors, Urban Land Institute.

[3] Suzuki, Hiroaki, Robert Cervero, and KanakoIuchi. 2013. "*Transforming Cities with Transit: Transit and Land-Use Integration for Sustainable Urban Development*". Washington, DC: World Bank Group.

[4] Ministry of Home Affairs, Government of India. *Census of India Report for Various Years.* Accessed at: https://censusindia.gov.in/vital_statistics/SRS_Statistical_Report.html

[5] United Nations. 2015. *World Urbanization Prospects: The 2014 Revision.* United Nations, Department of Economic and Social Affairs, Population Division, New York.

[6] Global Wealth Report 2015. Credit Suisse. October 2015. Accessed at: https://www.credit-suisse.com/about-us/en/reports-research/global-wealth-report.html

[7] Planning Commission, Government of India, 2014. *Report of expert group to review the methodology for measurement of poverty.* New Delhi, Government of India.

[8] Ministry of Road Transport & Highways, Government of India, New Delhi: *Road Transport Year Book* (2013–14 and 2014–15).

[9] UNDP. 2012. *Promoting TODs as an Inclusive Growth Strategy for Indian Cities*. GEF-SUTP Quarterly Newsletter. Vol. 2 No.8. November 2012. Available at: www.undp.org

[10] Leong L. 2016. *The "Rail plus Property" model: Hong Kong's successful self-financing formula*. McKinsey and Company. June 2016.

[11] Cervero R. and Murakami J. 2008. *Rail+Property Development: A Model of Sustainable Transit Finance and Urbanism.* UC Berkeley Center for Future Urban Transport. May 2008.

[12] Jaramillo, Samuel. 2001. *The Betterment Levy and Participation in Land Value Increments: The Colombian Experience.* Cambridge, MA: Lincoln Institute for Land Policy.

[13] Cervero, R. 2013. *Bus Rapid Transit (BRT): An Efficient and Competitive Mode of Public Transport.* Institute of Urban and Regional Development. University of California. August 2013.

[14] Principles of TOD: GEF-SUTP Quarterly Newsletter. 2012. Vol 2 No. 8. GEF-World Bank-UNDP supported sustainable Urban Transport Project.

[15] Goodman J., Laube Melissa and Schwenk Judith 2007. Curitiba's Bus System is Model for Rapid Transit. *The Race Poverty and Environment Journal.* Vol. 12 No.1. Spring 2007.

[16] Richard D. Knowles. 2012. "Transit Oriented Development in Copenhagen, Denmark: from the Finger Plan to Ørestad". *Journal of Transport Geography*.

[17] Peterson, George E. 2009. *Unlocking Land Values to Finance Urban Infrastructure*. The World Bank, Washington, D.C.

[18] Todd Litman. 2016. *Evaluating Public Transit Benefits and Costs: Best*

Practices Guide Book, Victoria Transport Policy Institute (www.vtpi.org).

[19] Iti Vyas, HamendraNath Vyas, Alok Kumar Mishra. 2020 "Land-based financing of cities in India: A study of Bengaluru and Hyderabad and directions for reforms". *Journal of Public Affairs,* 2020.

[20] Delhi Development Authority. 2015. *Master Plan for Delhi – 2021.*

[21] UN-HABITAT.2011. *Innovative Land and Property Taxation,* Global Land Tool Network.

[22] Ministry of Urban Development, Government of India. 2017. *National Transit Oriented Development (TOD) Policy.* Accessed: http://re.indiae nvironmentportal.org.in/files/file/ National%20Transit%20Oriented% 20Development%20%28TOD%29% 20Policy.pdf

Chapter 12

Estimation of the Efficiency Indices for Operating the Vertical Transportation Systems

Yury K. Belyaev and Asaf H. Hajiyev

Abstract

Various lifts' systems with different control rules are considered. It is suggested to use the efficiency indexes: customer's average waiting in lift cabin time and average total time, including the time of delivering the customer to the desired floor. Various control rules are introduced: *Odd-Even*, where one lift serves only customers in *Odd* floors and other lift only does that in *Even* floors *Up-Down* control rule where one lift serves only customers who are going from the first floor to the destination floor $2, 3, \ldots, k$; another lift serves customers from the first floor to the upper floor $k + 1, k + 2, \ldots, n$. The results of simulation, allowing to compare various control rules relatively to the efficiency indexes, are given. It is introduced an optimal number of lifts, which minimizes number of lifts, minimizing a customer's average waiting time. For some systems, the method of finding the optimal number of lifts, is suggested. Necessary figures demonstrating the operation of the lifts' systems and the results of the simulation allow to estimate the efficiency indexes.

Keywords: simulation of various lifts' systems, *odd-even*, *up-down*, situation of control rules, customer's average waiting time and total service time

1. Introduction

The world economy suffered a lot of losses after the coronavirus pandemic and it will take a long time for its rehabilitation. An important role in the development of the world economy will have the transportation and communication systems because it is necessary to renovate the economic communications among countries. For the investigation of the transportation and communication systems, mathematical models of queuing systems with moving servers are widely used. Typical examples of queues with moving servers are the lifts' systems. Lifts and communication systems, the traffic, the airport and the shipping facilities have a lot of similarities. All of them are united by the same principle – servers are moving in these systems. Hence, mathematical models of lifts' systems can be applied for the investigation of other systems with moving servers. As investigation of such systems by analytical approaches faces troubles, the use of the modern computers can allow to simulate their behavior. The simulation of such systems, various systems of programming (*Wolfram Mathematica* and others) allow to get close to reality, the numerical data of the desired parameters and give some advice for applications. The simulation also can also be a hint for the continuation of the analytical research.

Today it is difficult to imagine the development of modern cities, such as New York, Moscow, Shanghai, Istanbul and others, without skyscrapers. The process of designing skyscrapers needs an effective planning and operating of the lifts' systems, which allows to improve various characteristics (customer's waiting and service times) and to reduce energy expenses. An important problem is also to introduce various control rules for the lifts' systems. At a first view, it seems that the lifts' systems have simple structures. In fact, from a scientific point of view, these systems have complicated structures and the construction of mathematical models needs some non-standard approaches. Moreover, although these problems are formulated in the frame of standard queues models with moving servers, for their investigations it is necessary to develop new methods and approaches, using different fields of mathematics. The lifts' systems, the traffic problems, the airport and the shipping facilities have a lot of similarities. The investigation of these systems can allow to estimate the main operating parameters (customers' waiting and service times, energy expenses and others) and to make the necessary recommendations for constructors and engineers. There are many publications in this field e.g. [1–3] and even special scientific journals are published *(International J. Transportation Science and Technology, Research in Transportation Business & Management* and many others*)*. Unfortunately, the complicated lifts' systems, with various control rules, are not yet investigated widely. In [4–7], various mathematical models of lifts' systems with, different control rules, were introduced. The construction of mathematical models of lifts' systems and their research by analytical approaches, face some difficulties, because as it was mentioned above, these models have complicated structures. Hence, one of the effective methods are the simulation and the collection of simulated data, which can be used for estimating the various parameters of such lift systems, by comparing different control rules, finding optimal regimes for their operation. As the customers' arrival process into the lifts' systems has a stochastic structure, hence it leads to constructing and investigating the new stochastic models, approaches and programs for their simulation.

In this paper, the authors consider various lift systems with different parameters and different control rules. This paper can be regarded as a continuation of the authors' investigations presented in [4, 5]. Hence, we follow the notations introduced in these papers.

2. Control policies for the lifts' systems

There are many various control rules for the lifts' systems. We will consider only some of them, for instance, the *Odd-Even* system, where some lifts serve customers at the odd floors and other lifts, at the even floors. Another control rule, we call it following to [4], the *Up-Down* system, where some lifts serve customers going from the first floor to the Down floors $1, 2, \ldots, k$, others serve customers going from the first floor to the Upper floors$1, k+1, k+2, \ldots, n$. This control was introduced in [5]. For some systems by simulation, the numerical values of optimal $k_{opt.}$, which minimize the value of *CWT,* was found. All these control rules can improve the service, i.e. to reduce the customer's waiting and service times and also diminish energy expenses. Some methods of investigation of queues with a finite service capacity can be used for the research of the lifts' systems [8].

An interesting unofficial control policy was created in the seventy years of the XXth century, by the students in the dormitory of the Moscow Lomonosov State University. There are 18 floors in the student dormitory and two lifts' halls with four lifts in each. The first lift hall operates from the *1st to the 12th, 14* th*, 16th* *and 18th* floors. In order that the lifts work more rapidly, it was skipped the odd numbered floors, after the 12th. There is also a second lift hall for serving on the 1st-10th floors. If in the first hall, a lift came to the first floor and the first student yelled the word *"Higher"*, then, the lift would be filled by students who are going up only to the higher floors *(16th and 18th)* and the next lift will be filled by students who are going to the *12th, 14th, 16th* floors and upper. If the first call had been *"LOWER"*, then the lift would have operated between the lower floors (*12th, 14th* and afterward, to the other upper floors). The students called it a *Higher-Lower* system.

In [7], it was introduced the so called "situation control rule" for systems with two lifts. If both lifts are going from up to down, then all arrived customers (at the different floors) will be distributed between lifts. This control rule allows to exclude stopping both lifts almost at the same time, at the same floors. Such systems work effectively for high intensity of customers' flows. For instance, if both lifts are going from up to down, then each lift system defines the floors where the lift must stop and serve the customers. In the case of a customer's arrival at the new floor system, it must be recalculated the number of the floors where the lift must stop. Such a control rule allows using lifts capabilities in a uniform way. Although the "situation control rule" needs some additional software and technical equipment, nevertheless it improves the service (reducing customer's waiting and service times), it saves energy expenses and increase the lifetime of the lifts.

3. The mathematical models of the lift systems

For constructing the mathematical models of the lifts' systems, we use conceptions and parameters introduced in [4, 5]. The followings notations are introduced:

n – is the number of the floors in the building;

k –is the number of the lifts in the building;

$L_kF_nC_{xx}$ – is the systems with k lifts, n floors and control policy xx;

i - is an ordered in time identifying number of a customer during simulation;

$f_a\ (i)$ - is the floor of appearance of the *i-th* customer;

$f_d(i)$ - is the floor of destination of the *i-th* customer.

It is necessary to note that for some different i the $f_a\ (i)$ and $f_d(i)$ can take the same value.

$t_a\ (i)$ - is the instant of appearance of the *i-th* customer;

$t_b\ (i)$ - is the instant of the beginning service of the *i-th* customer in lift cabin;

$t_e(i)$ - is the instant of end service of the *i-th* customer;

$t_{c(j)}$- is the instant when lift on *j-th* cycle is returning to the *1-st* floor;

n – number of the floors in the building;

r – roominess, restriction of maximum possible number of customers, who can be in the lift cabin;

h_f - time necessary for the lift to move up or down, between two neighboring floors;

h_d– time which is spent for opening and closing the floor's door;

Usually, in practice, approximately $h_d = 2h_f$. If we consider the stationary input flow, then, the following parameters are used:

λ_{f1f2} –is the intensity of customers' flow, which appears at the f_1*-th* floor and want to go to f_2*-th* floor;

$\lambda_1 = \sum_{k=2}^{n} \lambda_{1k}$ - is the intensity of customers' flow, which appears at the first floor and are going to upper floors;

$\lambda_2 = \sum_{k=2}^{n} \lambda_{k1}$ – is the intensity of customers' flow, which appears on the upper $\{2, 3, \dots, n_f\}$ floors, who want to go down to the first floor;

CWT(S) – a customer's average Waiting Time in the system *S*, i.e. the mean time from the instant when a customer arrives at the system and waits until the instant when he gets the lift;

CST(S) – a customer's average Service Time in the system *S*, i.e. the mean time from the instant when the customer gets in the lift, until the instant when he gets off the lift;

CTT(S) = CWT(S) + CST(S) – a customer's average Total Time in the system *S*, which is measured as a mean time from the instant when the customer arrives into the system until he gets off the lift (arrival to ordered floor).

For instance, $CTT(L_kF_nC_{xx})$ is a customer's average total time, for a system in a building with *k* lifts, *n* floors and control policy *xx*.

IL–independent lifts' system. It means that all the lifts are operating independently from each other, i.e. if at the preceding instant of a new customer's arrival, several lifts are free (empty), then, all of them will go to this customer's call. Such systems are often used in the buildings with two lifts.

DL – dependent lifts' system (for a customer' call, the nearest lifts going to him);

UD(k)- where one lift serves only customers who are going from the first floor to *2, 3, … , k;* and another one serves customers who are going from the first floor to upper *k + 1, k + 2, … , n;* In such systems, when an *Up* lift is going from j_1*-th* floor $(j_1 > k)$ to down, it can take customers from j_2*-th* floor $k < j_2 < j_1$, if there is an empty space in the cabin. Otherwise, the lift is directly going to the first floor. Similarly, when the *Do(wn)* lift is going from j_3*-th* floor $(j_3 < k)$ to the first floor, it can collect customers from j_4 *-th* floor $j_4 < j_3$, if there is empty space in the cabin.

$T^U (L_2F_nC_{UD(k)})$ - cycle time of the *Up* lift in the system $L_2F_nC_{UD(k)}$;

$T^D(L_2F_nC_{UD(k)})$- cycle time of the *Do* lift in the system $L_2F_nC_{UD(k)}$;

SC – situation control - there is some (robot) software, which depends on new customers' arrivals, gives commands to the lifts where to stop and which floors to pass by. The appearance of a customer at the new floors can change the system of commands;

LRC –Average *Lift Return Cycle* time, i.e. the average time interval between two comings of the lift to at the first floor.

We also introduce the new parameters for the lifts' systems, which describe the *lift energy expenses* and the *single race time*:

LEE_j *(S)* – Average value of the *j-th Lift Energy Expenses* in the system *S*, measured in *Kw* (kilowatt);

Note that Energy Expenses in *Kw* depend not only on the volume and weight of the cabin, but also on its speed, acceleration and deceleration. Empirically, electric Energy Expenses can be shown each day, on the electric counter of each lift.

SRT(t) – Average *Single Rate Time*, i.e. average time when the lift is moving without customers, during time *t*;

SEE(S) – average value of *System Energy Expenses*, i.e. average value of energy expenses of all the lifts in the system *(S)*.

$SEE(S) = LEE_1(S) + LEE_2(S) + \dots + LEE_n(S);$

k_d – coefficient defining the lifts' energy expenses, during a unit time, for opening and closing the doors;

k_f – coefficient defining the lifts' energy expenses, during a unit time, for covering the distance between two neighboring floors.

There are different regimes of operating the lifts' systems.

Loading regimes, where customers from the first floor are going to upper floors. Such regimes are observed in the office buildings, in the morning (08.00–09.30) when customers are going to their offices. Similar regimes are observed in the residence buildings, in the evening (17.30–19.00), when people come back home from their work.

Unloading regimes, in the office buildings, in the evening (17.00–18.00), customers stop working and go back by lifts, from their offices to the first floor.

There also exist **mixed regimes**, when customers from the first floor are going to the upper floors and vice versa. Moreover, there are customers who are going from j_1*-th* floor to the j_2*-th* $(j_1, j_2 = 2,3,..,n)$. In this paper, only loading and unloading regimes will be considered. Some investigations of the mixed regimes can be found in [5].

In the unloading regimes, when lifts are going from the upper j_1*-th* floor to the first floor, the lifts can take customers from j_2*-th* floor $(j_2 < j_1)$, if there is a free space in the cabin. If, at some floor, the number of customers in the cabin became r (roominess), then the lift would go directly to the first floor, without stopping. This policy is observed in all the regimes.

Remind that $L_2F_nC_{n1,n2}$ is the system with *2* lifts, n floors and after completing the customer's service, one lift (with empty cabin) must go to n_1*-th* floor, if there is no lift, otherwise, it should go to n_2*-th* floor. Below, in the **Figures 1** and **2**, axes x means current time;

-  means that the lift is occupied;

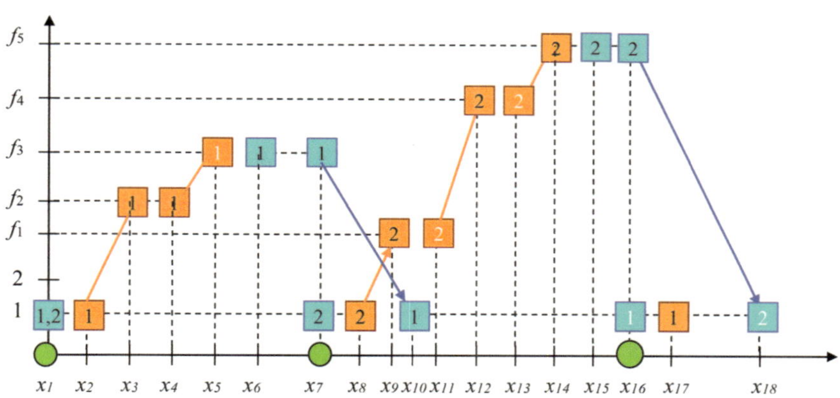

Figure 1.
Example of loading regime for $L_2F_nC_{IL}$.

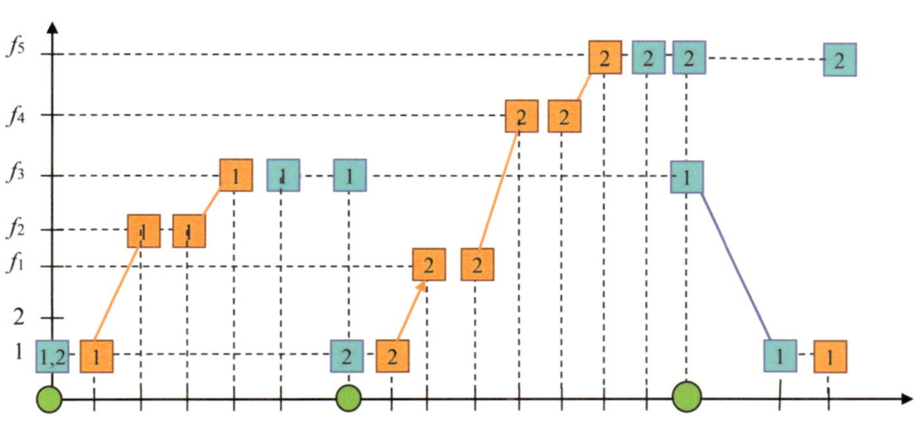

Figure 2.
Example of loading regime for $L_2F_nC_{DL}$.

- means that the lift is empty (free);
- means the instant of the customers' arrival instant.

Definition. The flow of customers is called *rare* for the lift system $L_kF_nC_{xx}$, if at the preceding instant of the customer's arrival, among the k lifts there is at least one (non-occupied) lift, which goes to the customer's call.

4. The systems $L_2F_nC_{IL}$ and $L_2F_nC_{DL}$ in the loading regimes

We will compare both systems $L_2F_nC_{IL}$ and $L_2F_nC_{DL}$ with rare flow of customers, in the loading regime, relatively to a customer's waiting time (CWT). In the **Figures 1** and **2**, axes x means current time and axes y, an ordinal number of the floor, where the lift delivers the customers.

Below, in the **Figure 1**, the lifts' positions at the preceding instants of the customer's arrival are presented (rare flow) (see, **Figure 1**). If the input flow is rare, then, for the system $L_2F_nC_{IL}$ in loading regime, at the preceding instant of a customer's arrival one lift is located at the first floor and another is located at *j-th* floor, where $j = 2,3, \dots, n$. (see, **Figure 1**).

$x_1 = t_a(1) = t_a(2)$, $x_2 = t_b(1) = t_b(2) = x_1 + h_d$, $x_3 = x_2 + (f_2{-}1)h_f$, $x_4 = t_e(1) = x_3 + h_d$, $x_5 = x_4 + (f_3{-}f_2)h_f$, $x_6 = t_e(2) = x_5 + h_d$, $x_7 = t_a(3) = t_a(4) = t_a(5)$,
$x8 = t_b(3) = t_b(4) = t_b(5) = x7 + h_d$, $x_9 = x_8 + (f_1{-}1)h_d$, $x_{10} = x_7 + (f_3{-}1)h_f$,
$x_{11} = t_e(3) = x_9 + h_d$, $x_{12} = x_{11} + (f_4{-}f_1)h_f$, $x_{13} = t_e(4) = x_{12} + h_d$, $x_{14} = x_{13} + (f_5{-}f_4)h_f$,
$x_{15} = t_e(5) = x_{14} + h_d$, $x_{16} = t_a(6)$, $x_{17} = t_b(6) = x_{16} + h_d$, $x_{18} = x_{16} + (f_5{-}1)h_f$.

Consider the system $L_2F_nC_{DL}$ with rare input flow in loading regime. Then, at the preceding instants of a customer's arrival, both lifts occupy the floors $2,3,..,n$. (see, **Figure 2**).

$x_1 = t_a(1) = t_a(2)$, $x_2 = t_b(1) = t_b(2) = x_1 + h_d$, $x_3 = x_2 + (f_2{-}1)h_f$, $x_4 = t_e(1) = x_3 + h_d$,
$x_5 = x_4 + (f_3{-}f_2)h_f$, $x_6 = t_e(2) = x_5 + h_d$, $x_7 = t_a(3) = t_a(4) = t_a(5)$,
$x_8 = t_b(3) = t_b(4) = t_b(5) = x_7 + h_d$, $x_9 = x_8 + (f_1{-}1)h_d$, $x_{10} = t_e(3) = x_9 + h_d$, $x_{11} = x_{10} + (f_4{-}f_1)h_f$,
$x_{12} = t_e(4) = x_{11} + h_d$, $x_{13} = x_{12} + (f_5{-}f_4)h_f$, $x_{14} = t_e(5) = x_{13} + h_d$, $x_{15} = t_a(6)$, $x_{16} = x_{15} + (f_3{-}1)h_f$,
$x_{17} = t_b(6) = x_{16} + h_d$.

Thus, we have $CWT(L_2F_nC_{IL}) = h_d$ and $CWT(L_2F_nC_{DL}) = nh_f/6 + h_d$(1),
$CWT(L_2F_nC_{IL}) < CWT(L_2F_nC_{DL})$. If an intensity of input flow is increasing, then the difference $(CWT(L_2F_nC_{IL})$- $CWT(L_2F_nC_{DL}))$ is decreasing and goes to zero.

After some critical value of intensity λ^* this difference $(CWT(L_2F_nC_{IL})$ - $CWT(L_2F_nC_{DL}))$ is increasing until to some other value of intensity λ^{**}.

Afterward, it is again decreasing and goes to zero, for a high value of intensity. It is clear, that for a high intensity of the input flow, an operating of the systems $L_2F_nC_{IL}$ and $L_2F_nC_{DL}$ is becoming close to each other (see, **Figure 3**). In the **Figures 3–5**, axes x means intensity of the input flow and axes y means the value of the *CTT.* If roominess of the lift is bounded, then for a high intensity of the input flow it is not necessary to introduce any control, because both lifts stop at each floor and the system is operating like deterministic (at each floor there is always at least one customer).

Remark. For small values of intensity of the input flow, the system $L_2F_nC_{IL}$ is preferable than the system $L_2F_nC_{DL}$, i.e. $CTT(L_2F_nC_{IL}) < CTT(L_2F_nC_{DL})$. There exists some interval $(\lambda^*, \lambda^{**})$ of intensity which can be calculated by simulation),

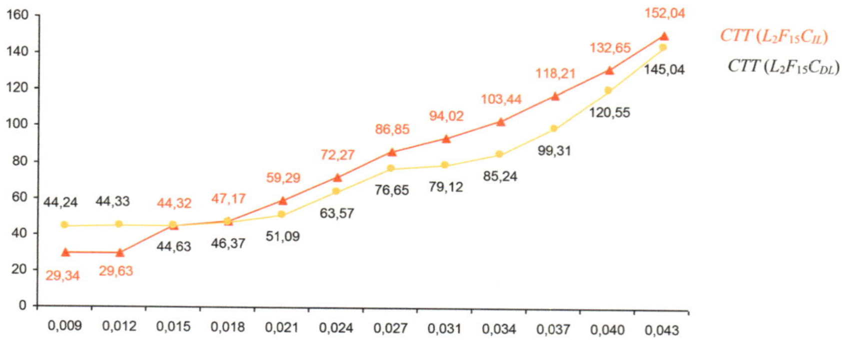

Figure 3.
Graphs of the $CTT(L_2F_{15}C_{IL})$ *and* $CTT(L_2F_{15}C_{DL})$.

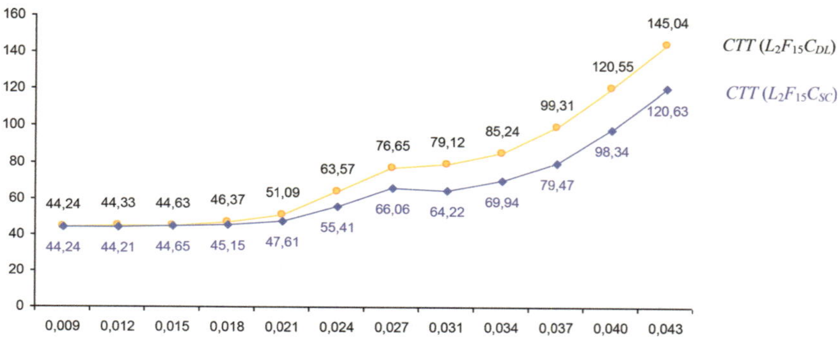

Figure 4.
The graphs of the $CTT(L_2F_{15}C_{DL})$ *and* $CTT(L_2F_{15}C_{SC})$.

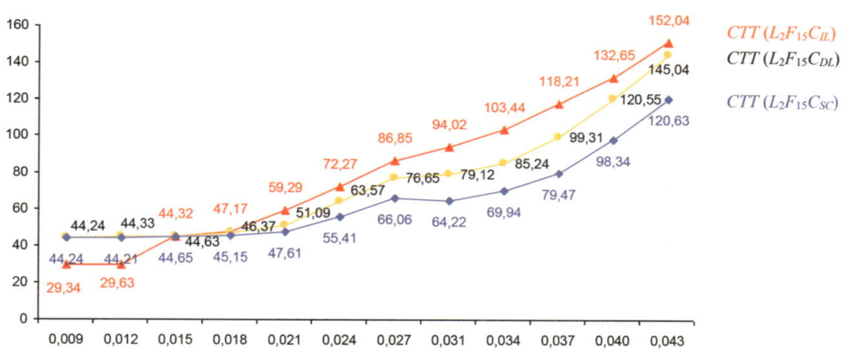

Figure 5.
The graphs of the $CTT(L_2F_{15}C_{IL})$, $CTT(L_2F_{15}C_{DL})$ *and* $CTT(L_2F_{15}C_{SC})$.

where the system $L_2F_nC_{DL}$ is preferable than the system $L_2F_nC_{IL}$, because $CWT(L_2F_nC_{DL}) < CWT(L_2F_nC_{IL})$. Consider the system $L_2F_nC_{UD(k)}$, where one of the two lifts, let us call it *Do-lift,* serves customers who are going from the first floor to *2,3,..,k* floors. Another lift, let us call it *Up-lift,* serves customers from the first floor to the upper floors $k+1, k+2,..,n$. Remind that $T^U(L_2F_nC_{UD(k)})$ is the cycle time for the *Up*-lift, and $T^D(L_2F_nC_{UD(k)})$ is the cycle time for the *Do-lift.* The cycle time of a lift is defined as the time interval between two sequential comings of the lift to the first floor. For this system we also introduce the floor number f_{opt}. (optimal border cut), which can be found from the equation, when the cycle time of the *Up*-lift closes to the cycle time of the *Do*-lift. In other words, f_{opt}. is found from the following ratio.

$f_{opt} = \{k: min/T^U(L_2F_nC_{UD(k)}) - T^D(L_2F_nC_{UD(k)})/\}k$

where /./means the absolute value of (.). Below, as the result of the simulation, various systems are given. In the **Table 1**, for different number of the floors *(n)*, the value of $f_{opt.}$ is given. For simulation, there were used the following lifts' parameters $h_f = 4$, $h_d = 7$(Sec.).

Simulation shows (see, **Table 1**) that typically.

$2n/3 \leq f_{opt.} \leq 3n/4$

Below, in the **Table 2**, the results of simulation for comparison of the systems $L_2F_nC_{IL}$ and $L_2F_nC_{OE}$ relatively to the *CTT*, are given. In the **Figure 3**, the graphical behavior of the *CTT* for both systems $L_2F_nC_{IL}$ and $L_2F_nC_{OE}$ is given. The results of simulation show that relatively to the *CTT*, the system $L_2F_nC_{OE}$ is preferable, than the system $L_2F_nC_{IL}$.

$CTT(L_2F_nC_{OE}) \leq CTT(L_2F_nC_{IL}))$ (see, **Table 2** and **Figure 3**).

Consider the systems $L_2F_{15}C_{IL}$and $L_2F_{15}C_{DL.}$ It is necessary to note denote that by introducing the control rules, can be reduced not only the *CWT* (customer's waiting time) but also the *CST* (customer's service time) and hence, the *CTT* (customer's total time). Below, we will consider the *CTT* for all the systems.

n	$f_{opt.}$	h_d	h_f
12	8	7	4
12	8	7	4
15	11	7	4
15	10	7	4
22	15	7	4

Table 1.
The values of $f_{opt.}$ *For buildings with various floors.*

λ	$CTT(L_2F_{15}C_{IL})$	$CTT(L_2F_{15}C_{DL})$
0,009	29,34	44,24
0,012	29,63	44,33
0,015	44,32	44,63
0,018	47,17	46,37
0,021	59,29	51,09
0,024	72,27	63,57
0,027	86,85	76,65
0,031	94,02	79,12
0,034	103,44	85,24
0,037	118,21	99,31
0,040	132,65	120,55
0,043	152,04	145,04

Table 2.
The values of the CTT $(L_2F_{15}C_{IL})$ *and* CTT $(L_2F_{15}C_{DL})$.

n		CTT($L_2F_nC_{IL}$)	CTT($L_2F_nC_{OE}$)	*Gain (%)*
1	0,075	57,5	57,5	0,0
2	0,3	63,63	61,2	3,8
3	0,45	70,32	64,6	8,1
4	0,6	77,7	69,5	10,6
5	0,75	81,2	71,6	11,8
6	0,9	84,2	74,9	11,0
7	1,05	91,85	78,1	15,0
8	1,2	96,2	80,7	16,1
9	1,5	118,2	91,4	22,7
10	1,8	152,4	118,6	22,2

Table 3.
The values of the CTT($L_2F_{15}C_{IL}$) *and* CTT($L_2F_{15}C_{DL}$).

λ	*CTT($L_2F_{15}C_{DL}$)*	*CTT($L_2F_{15}C_{SC}$)*
0,009	44,24	44,24
0,012	44,33	44,21
0,015	44,63	44,65
0,018	46,37	45,15
0,021	51,09	47,61
0,024	63,57	55,41
0,027	76,65	66,06
0,031	79,12	64,22
0,034	85,24	69,94
0,037	99,31	79,47
0,040	120,55	98,34
0,043	145,04	120,63

Table 4.
The values of the CTT($L_2F_{15}C_{DL}$) *and* CTT($L_2F_{15}C_{SC}$).

The results of simulation (see, **Table 2**) and the graphical behavior (see, **Figure 3**) of the *CTT* (customer's total time) are presented.

The **Table 2** and **Figure 3** show that for a small intensity of the input flow of customers, we have *CTT*($L_2F_{15}C_{IL}$) < *CTT*($L_2F_{15}C_{DL}$)). It follows from the fact that for a small intensity of the input flow at the preceding instant of a customer's arrival in the system $L_2F_{15}C_{IL}$, one lift occupies the first flow and another one, *j-th* floor (j = 2,3,..,n). In the system $L_2F_{15}C_{DL}$, for a small intensity at the preceding instant of a customer's arrival, both lifts occupy the first floor and hence, an average distance from lifts' position to customer call, far than in the system $L_2F_{15}C_{IL}$ (**Table 3**).

In the **Table 3** and **Figure 3**, the values of the *CTT*, depending on the intensity of the input flow for various systems, are shown. For a high intensity of the input flow, a difference between *CTT*($L_2F_nC_{IL}$) and *CTT*($L_2F_nC_{OE}$) is increasing, when the intensity of the input flow goes up (see, **Figure 3**), because in this case all the lifts stop at each floor and the system $L_2F_nC_{IL}$ operates like a system with one lift

λ	$CTT(L_2F_{15}C_{IL})$	$CTT(L_2F_{15}C_{DL})$	$CTT(L_2F_{15}C_{SC})$
0,009	29,34	44,24	44,24
0,012	29,63	44,33	44,21
0,015	44,32	44,63	44,65
0,018	47,17	46,37	45,15
0,021	59,29	51,09	47,61
0,024	72,27	63,57	55,41
0,027	86,85	76,65	66,06
0,031	94,02	79,12	64,22
0,034	103,44	85,24	69,94
0,037	118,21	99,31	79,47
0,040	132,65	120,55	98,34
0,043	152,04	145,04	120,63

Table 5.
The values of the $CTT(L_2F_{15}C_{IL})$, $CTT(L_2F_{15}C_{DL})$ *and* $CTT(L_2F_{15}C_{SC})$.

$(L_1F_nC_{IL})$ but with double roominess (see. **Figure 3**). Consider the systems with two lifts, with a situation control rule and denote it $L_2F_nC_{SC}$, where the *SC* means situation control. At each given time-unit, the software is checking a new customer's arrival into the system and depending on this, each lift gets command, at which floor to stop for the customer's service. Below, in the **Table 1**, the result of simulation of such a system, for a building with 15 floors in an unloading regime, is presented. It is assumed that the roominess of the lifts is quite large and the lifts can take all the customers waiting on the floor. In simulation, we take $h_f = 4$, $h_d = 7$. Below, the results of the simulation (see, **Table 4**) and the graphical behavior (see, **Figure 3**) of the *CTT* (customer's total time), are presented. It is necessary to note that introducing of the control rules, can be reduced not only the *CWT* (customer's waiting time), but also the *CST* (customer's service time) and hence, the *CTT* (customer's total time).

5. Situation control rule

Introducing the *SC* (situation control) allows to reduce the *CTT* for a high intensity of the input flow. Below, the results of the simulation (see, **Table 4**) and the graphical behavior of the $CTT(L_2F_{15}C_{DL})$ and $CTT(L_2F_{15}C_{SC})$ (see, **Figure 4**), are presented:

Data of the **Table 4** show that by increasing of the intensity of the input flow, the gain in the *CTT* is going up. In **Figure 4**, there are given the results of the simulation for the systems $L_2F_{15}C_{DL}$ and $L_2F_{15}C_{SC}$. It is clear that for small and high values of intensity, it is not necessary to introduce the situation control, because both systems are almost the same and moreover, for high values of intensity, they coincide and the efficiency indexes can be calculated. There exists some interval where a difference between efficiency indexes takes maximal value and afterward it goes to zero, because for high values of customers' intensity flows, the lifts must stop at each floor, hence both systems have the same behavior (see **Figure 4**).

Below, the results of simulation (see, **Table 5**) and graphical behavior (see, **Figure 5**) of the *CTT(customer's total time)* for all the three systems, are presented. It is necessary to note that by introducing the control rules, can be reduced not only the *CWT (customer's waiting time)* but also the *CST(customer's service time)* and hence, the *CTT(customer's total time).*

6. Energy expenses

Now we will show that introducing of the control rules, will be to reduced not only the *CWT* and the *CTT,* but also the *LEE* (*lift energy expenses*). Note, as it was mentioned above, for rare input flows it is not necessary to introduce the control rule *DL*, because.

$CTT(L_2F_nC_{IL}) < CTT(L_2F_nC_{DL})$

and moreover, from formula (1), it follows

$LEE(L_2F_nC_{IL}) = k_dh_d$ and $LEE(L_2F_nC_{DL}) = k_f\, nh_f/6 + k_dh_d$

i.e. $LEE(L_2F_nC_{IL}) < LEE(L_2F_nC_{DL})$.

Energy expenses linearly depend on the *CTT* and also on the *SRT* (single rate time). As it follows from **Table 2**, the introduction of the SC (situation control) reduces the value of the *CTT,* by up to 25%. In [4] it was shown that for the *CTT* $(L_2F_nC_{IL})$ in an unloading regime and rare flow of customers, the following ratio is true:

$CWT(L_2F_nC_{IL}) = h_f\,(n\text{-}1)/2 + h_d$, $CST(L_2F_nC_{IL}) = h_f(n\text{-}1)/2 + h_d$ and

$CTT(L_2F_nC_{IL}) = h_f\,(n\text{-}1) + 2h_d$, $LEE(L_2F_nC_{IL}) = k_f(n\text{-}1)h_f + 2k_d\, h_d$,

$SRT = k_f\, h_f\,[3(n\text{-}1)/4] + m\, k_dh_d$. Then, for Poisson flow of customers with intensity λ during the time interval $[0,t)$, we have $SRT(t) = \lambda(k_f\, h_f\, 3(n\text{-}1)/4 + mk_d\, h_d)t$. As λT is an average number of arrivals during the time T, then $\lambda Tk_f\, h_f\,(n\text{-}1)$ is an average energy which lift spends for serving the customers (motion of lift), during time T. As at each arrival instant, there is an average number of customers equal to m, then λTm is the average number of customers who arrived during the time T, into the system. For each customer's arrival, the lift spends the time h_d for opening and closing the door. If we assume that each customer spends the time h_c coming in and getting off a lift, then, the mh_c is the time, which was spent for the m customers (coming in and getting off). Hence, a customer average energy spent for opening and closing the door, for customers coming in the lift and getting off equals $2\lambda_1Tk_dh_d + 2\lambda_1Tmh_c = 2\lambda_1T(k_dh_d + mh_c)$. Thus, we have $LEE(L_1F_nC_{IL}) = \lambda Tk_f\, h_f\,(n\text{-}1) + 2\lambda T(k_dh_d + mh_c)$. Below, for simplicity, we assume $h_c = 0$, which means that during the time h_d, all the customers, who want to come in and get off a lift, can do it.

7. Analysis of the two lifts system in planning office buildings

Suppose it is a plan to construct a 15 floors office building with two similar lifts, which will carry in the morning, the customers to their offices and back, to the 1st floor, at the end of their work. It is necessary to introduce parameters of the lifts, e.g. roominess, velocity of lifts going up and down between floors and times for opening the doors on floors. Here, we consider unloading regime, where all the customers leave offices at the end of work hours. The offices will be placed on the floors *{2, 3, … ,15}*. The number of customers working on these floors will be *{12, 12, 15, 16, 12, 10, 17, 12, 14, 14, 16, 11, 18, 14}* i.e. all together in the building will be $n_c = 193$ customers. They should leave their offices during an interval of *1800 sec* (half an hour) in the evenings. The probability density $p[s] = 2(1800\text{-}s)/1800^2$, $0 < s < 1800$ of customers to leave their offices is given in **Figure 6**:

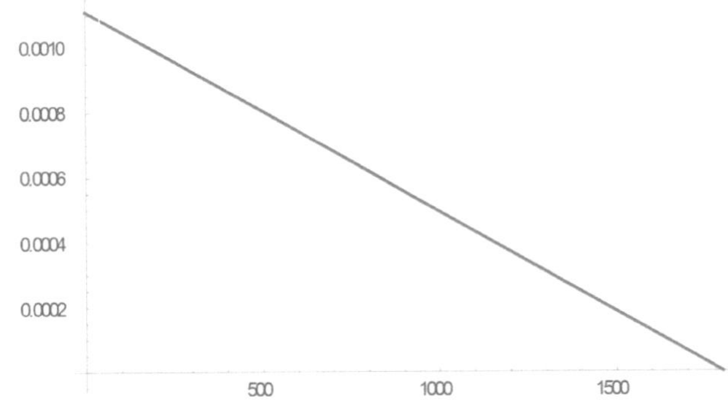

Figure 6.
The decreasing probability density p[s] *during an interval of* 1800 Sec.

The main efficient parameters of the lifts' systems are *the customers' average waiting times (CWT)* and *customers' average total times (CTT)*. Remind that the *CWT* is defined as an average time from the instant when the customer presses the button at the unloading, to get the lift cabin. *The CTT* is defined as the sum of the *CWT* and *CST,* i.e. the average time from the instant when a customer arrives into the system until the instant when it he has left the cabin, at the desired floor. Simulated data can be used to obtain estimates of the *CWT, CST* and *CTT.* The results of the lifts which are operating, can be described by initial histories of customers $\{i, f_a, t_a, f_d\}$, $i = 1, 2, \dots, n_c$, where n_c is the total number of customers in the building, i is the ordinal number of customer, f_d - destination floor. For simulating an unloading regime, it is assumed that $f_d = 1$ and the lift spends $h_f = 2.5$ sec. to cross the distance between two neighboring floors. If the lift stops at some floor, then the time for opening and closing the door is $h_d = 5$ *sec.* Let's assume that only one lift with roominess $r = 10$or$r = 20$, is operating. If the lift is located at the first floor ($f_d = 1$) and its cabin is empty, then it immediately goes up to the highest *15th* floor. It means that we consider an unloading regime, where the lift is going from up to down and collects customers at the lower floors, if roominess allows it. Using our simulated program for unloading regimes, we have obtained six sets with initial customer histories, $\{i, t_a, f_a\}$: three the sets with roominess $r = 10$ and three the sets with roominess $r = 20$.Using programming *Wolfram Mathematica*, we created the program, which transforms initial customer histories $\{i, t_a, f_d\}$ of n_c customers, into full histories $\{i, f_a, t_a, t_b, t_e\}$.Here t_b is the instant when the *i-th* customer goes in the lift cabin and t_e is the instant when the customer leaves the cabin at the *1st* floor. For simulating the 3-dimensional vector $\{i,t_a,f_a\}$, the similar program has been created for loading regime. This program was used by comparing the full histories for a lift with $r = 10$ and $r = 20$.

The estimates of the efficiency of the *CWT* and *CTT* are given in **Table 6**. It follows from **Table 6**, that roominess is the very essential parameter and the *CWT*

Days	*1st*		*2nd*		*3rd*	
r	*CWT*	*CTT*	*CWT*	*CTT*	*CWT*	*CTT*
10	*388.88*	*415.87*	*410.82*	*438.23*	*359.97*	*387.41*
20	*90.04*	*126.91*	*78.28*	*117.12*	*68.05*	*108.96*

Table 6.
The values estimation of the CWT *and* CTT *obtained during the simulation unloading, three times (days), different lifts, with roominess* r = 10, 20.

Days	*1st*		*2nd*		*3rd*	
Lift	*CWT*	*CTT*	*CWT*	*CTT*	*CWT*	*CTT*
L1	*46.86*	*68.90*	*40.28*	*62.53*	*39.59*	*61.00*
L2	*45.73*	*81.55*	*46.58*	*81.65*	*40.78*	*76.05*

Table 7.
The values estimation of the CWT *and* CTT *obtained for lifts* L1 *and* L2, *for data three "days".*

Days	*1st*		*2nd*		*3rd*	
Lift	*CWT*	*CTT*	*CWT*	*CTT*	*CWT*	*CTT*
L1	*59.26*	*87.07*	*48.55*	*76.47*	*51.14*	*79.17*
L2	*57.81*	*85.26*	*60.91*	*87.26*	*50.51*	*78.06*

Table 8.
The estimated values of the CWT *and* CTT *for lifts* L1 *and* L2, *for data three "days".*

and *CTT* are better if $r = 20$. From the engineering point of view, it is more practical to use a lift system with two different lifts *L1* and *L2*, with roominess $r = 10$ each.

Below, in **Table 7**, we illustrate the following control rules: lift *L1* serves lower floors, from floor *2* up to floor *10* and lift *L2* serves all the floors, from floor *11* up to floor *15*.

Note that we obtained in **Table 7**, better parameters, *for three days and two lifts,* than in **Table 6**. The above-considered data, for the *CWT* and *CTT*, correspond to three days. Note that our programs can simulate lifts for many days' operating data.

In **Table 8**, two lifts can stop *L1,* on *{1, 3, 5, 7, 9, 11, 13, 15}*- odd floors, *L2,* on *{1,2, 4, 6, 8, 10, 12, 14}*- even floors, and both lifts have $r = 10$.

We introduce the lifts' systems dispatcher (computer with special control lifts programs), as controller of the traffic of the moving lifts. Then, we can consider essentially many types of control rules for the lifts. For example, we can consider a system with two dependent similar lifts. They can stop, if their cabin contains less than r customers, follow specific rules at the floors with *waiting* customers, and if the system's dispatcher allows it.

8. Conclusion

Several mathematical models of lifts' systems, which have different control rules, are introduced and investigated. By simulation, the data customer's waiting time *CWT* and total time *CTT* were estimated under different control rules and they have been compared relatively to the efficiency indices for the introduced control rules. The result of the calculation shows that relatively to the *CTT* usage of the situation control *SC,* in comparison with *DL* control rule, a gain of around 25% is achieved. If the roominess of the lift cabin is unbounded, then, for a high intensity of the input flow, it is not necessary to introduce any control. Then, it follows that if at each floor there is at least one waiting customer and both lifts stop at each floor, the system is operating like deterministic. The simulation also shows that, in the case of two lifts and a rare customers' flow, it is not necessary to introduce *DL* or *SC* control rules, because the system *IL* (with independent lifts) is preferable than the *DL* and *SC* control rules. It was shown that for a high value of an input flow of customers, the introduced control rule also reduces energy expenses, even by 25%, in some cases, which confirms the advisability of the introduced control rules.

These results allow to make practical recommendations for reducing the various characteristics of the lifts' systems, such as the *CWT, CTT, SRT* (single rate time) and *LEE* (lift energy expenses). We completed the paper by examples with the calculation of a customer's waiting time *CWT* and a customer's average total time *CTT* for customers after work, for non-stationary cases, when there is an intensity of the customers' flow. It can be used for planning of the construction of the new office buildings with two similar lifts. The program can be extended for the case of several (more than two) lifts. We would like to underline that for the simulation of non-stationary cases, it is necessary to prepare a special program, which has a more complicated structure. In **Tables 6–8**, the results of the simulation for non-stationary cases, are given. We used the programming system *Wolfram Mathematica,* to create the programs for the simulation data and for a possible operation of two lifts. The results show that using simulation can help to estimate the appropriate values of roominess and find the optimal control rules, which can optimize the choice of the lifts' parameters (customer's waiting time, energy expenses and others). It can help for planning high floors buildings and future lifts' systems.

Author details

Yury K. Belyaev[1] and Asaf H. Hajiyev[2*]

1 Department of Mathematical Statistics, Umeå University, Umeå, Sweden

2 Azerbaijan National Academy of Sciences, Institute of Control Systems, Baku, Azerbaijan

*Address all correspondence to: asaf.hajiyev@gmail.com

References

[1] S.Markon, H.Kita, H.Kise, T.Bartz-Beielstein, "Control of Traffic Systems in Buildings", Springer, 2006.

[2] G.C.Barney, "Elevator Traffic Handbook", Taylor & Francis, 2003.

[3] Allen, A.O., Probability, statistics and queueing theory with computer science applications. New York, Academic Press, 1978.

[4] BelyaevYu.K., HajiyevA.H. Mathematical models of lift systems with various control rules. Gnedenko Forum. RT&A, No. 2(57), Volume 15, June, 2020.

[5] BelyaevYu.K., HajiyevA.H. Mathematical models of the lift systems and their simulation. Proceed. XIII ICMSEM, Canada. Springer, Ser. Electrical Engineering, 2019. p.507–519.

[6] Al-Sharif L., Abu-Alqumsan A., Ghanem W. Modelling of Elevator Traffic Systems Using Queuing Theory. Symposium on Lift and Escalator Technologies. Northampton, United Kingdom. Volume: 4

[7] БеляевЮ.К., ГаджиевА.Г., ГромакЮ.И., ДугинаТ.Н. Сравнительныйанализпростейших системвертикальноготра нспорта. Изв. АНСССР. Т ехническаякибернетика, №3, 1977, с. 97-103. (in Russian). (English transl.: Soviet Journal Computer and Systems Sciences).

[8] Aalto S. Some methods of investigation of queues with finite service capacity can be used for research of lifts' systems. Mathematical Methods of Operations Research volume 51, p.263–285. 2000

Chapter 13

Passive Safety of Children Carriages on Busses

Juan Dols, Enrique Alcalá and Luis Martínez

Abstract

The safe mobility of young children traveling with carriages in public transportation vehicles is a problem that has not yet been satisfactorily resolved. The lack of national and international standards in this area in the past led to the development of a research developed jointly by the Universitat Politécnica Valencia and the Universidad Politécnica Madrid (Spain). This book chapter shows the results of a research program developed to evaluate the dynamic behavior of occupied children carriages (ChC) during typical driving maneuvering—sudden braking, acceleration and cornering—and in case of low-g accidents reproducing frontal impacts resembling real traffic events (deceleration 2 g, ΔV 20 km/h). In the dynamic trials, three ChC-restraint prototypes and a typical wheelchair (WhCh) back-restraint system combined with two representatives of up-to-date ChC models in misuse and correct use configurations were tested. The results demonstrated the need for preventing children injuries as a consequence of low-g accidents.A Code of Good Practice was proposed jointly with the use of a new ChC-restraint system considering R 107–06 series of amendments. The new design improves the latest revision of regulation R107 regarding the use of back-restraint systems for the transport of WhCh and ChC passengers traveling on busses.

Keywords: transportation safety, children carriages, children injury, restraint system, public transport

1. Introduction

The proper use of public transport systems is currently a vital need that must be pursued by public administrations to maintain quality standards that facilitate the adequate mobility of the majority of the population. These minimum standards should include both accessibility, comfort, and safety when traveling, so that their use can be extended to groups with disabilities or reduced mobility. Among these groups, we must not only include people with physical disabilities but also those who suffer from cognitive or sensory disabilities—auditory or visual. In the same way, accessibility and safety problems were important for passengers traveling in road transport vehicles without leaving their wheelchairs; there are actually passenger groups that experience similar difficulties at the time of using bus transport systems. This is the case of passengers traveling accompanied by children's carriages.

There are not many statistical studies that analyze the accident rate related to the transport of children's carriages (ChC) on busses, but their behavior, from the

IntechOpen

transport safety perspective, is very close to that suffered by passengers traveling without leaving their wheelchair (WhCh). In this sense, the scientific community has introduced the term *incident* in accident studies, more in line with the interaction of the passenger—with/without reduced mobility—with the transport vehicle environment, considering the source of the accident, not only the vehicle's own impact but also other causes such as the boarding-disembarking action, the use/misuse of restraint systems, or the vehicle's critical or emergency maneuvers.

Most studies related with the concurrence of *incidents* in bus transport have shown that these occurred without impacts and in most cases in urban areas [1, 2]. Some studies have concluded that the injuries produced during these incidents were due to sudden changes in speed (60% in braking and 25% in acceleration) [3] or when the vehicle is stopped, during the boarding/disembarking phases [4], affecting the elderly and the disabled to a greater extent. Most of the damage suffered occurs as a combination of a WhCh overturning and the occupant's fall, followed by incidents without falls and situations in which the occupant falls from the WhCh [5]. In [6] the behavior of WhCh passengers who suffered incidents (with or without vehicle impact), through in-depth analysis of real accident reconstructions, were analyzed. Results showed that the greater severity damages suffered by WhCh users were caused due to the inefficient or nonexistent use of the restraint systems by the WhCh and his occupant, without disregarding the influence of the vehicle's driver when performing emergency maneuvers during normal traffic. The driver's training whose collaboration will be fundamental, both to perform adequate assistance and to ensure boarding-disembarking safe operations, is also important [7].

Taking into account the results obtained in all the aforementioned previous research, it is clear that the transport service operating agencies are aware of the risk that these passengers suffer under normal vehicle conditions. Some data recorded by the EMT Madrid operator during the last 18 months (since April 2018 to date) indicated the occurrence of about 10 incidents with ChC, that is, one every 2 months, in which the baby or child was dropped from the carriage. Most of the incidents were due to sudden braking or closed curve driving, and others due to ChC badly positioned [8]. The EMT Madrid has a fleet composed by approximately 2300 low-floor vehicles.

It should be noted that mobility problems generated by the transport of ChC not only affect the user groups themselves but also the transport service operators, which must face situations of social rejection due to the refusal to accept admission to the transport services of certain types of busses, either due to the absence of specific regulations or due to the ignorance of the possible solutions to the problem generated with this type of users.

The absence of specific standards that regulate the accessibility of ChC to the vehicle constitutes one of the most common difficulties. **Figure 1** shows some examples of typical cases of the troubles that users traveling with ChC in road public transportation vehicles have faced up to. To overcome some difficulties related to access to vehicle interior, some carriage must do so through the front door with the folded carriage (**Figure 1A**). If the carriage cannot be folded, access must be made through the central door to overcome the difference height between the bus stop and the floor of the vehicle (**Figure 1B**). Finally, the maneuverability of the ChC itself inside the transport vehicle constitutes another one of the barriers to mobility that will have to be faced (**Figure 1C**). Another common problem that users' groups and transport operators have had to face consists of the absence of safety systems to hold the ChC during transport (**Figure 2A**), as well as compatibility between market types of ChC and WhCh models, when sharing the space reserved (**Figure 2B**).

(A) (B) (C)

Figure 1.
Examples of the challenges faced by ChC in the access to public transportation vehicles: (A) access with the ChC folded through front door, (B) access with the ChC unfolded thorough central door, (C) difficulties in the maneuverability *inside the vehicle for accessing to the space reserved to ChC (source: EMT Madrid).*

(A) (B)

Figure 2.
Examples of the safety conditions in the mobility of ChC in passenger public transportation vehicles: (A) absence of safety systems to hold the ChC unfolded during travel, (B) compatibility of ChC with the WHCHs' space occupied (source: EMT Madrid).

There have not been many studies related to the problem of transporting ChC in urban busses. One of the few and most prominent has been the ASUCAR[1] research project. In the first phase of this project were analyzed the accessibility and safety conditions which ChC had to face up to in this type of mobility. The study was based on a survey of companies operating public and private transportation services in Spain, mainly in urban areas, with the collaboration of the Association of Urban and Road Transports (ATUC) [9]. Of the more than 70 surveys launched in 2008, the responses of 44 companies were analyzed. The results obtained showed that companies used to regulate the access of ChC to urban transport vehicles by applying internal, local, or regional regulations in 70% of cases; the rest, 30%, did

---

[1] The research project, *Determination of accessibility and safety requirements in the use of children's carriages in public transport vehicles (ASUCAR)*, was funded by the Spanish Ministry of Science and Innovation during the period 2008–2010, within the Program PETRI (REF No. PET2008_0328_01 and PET2008_0328_02), coordinated by the Polytechnic University of Valencia (UPV-IDF), and participated by the Polytechnic University of Madrid (UPM-INSIA) and the Municipal Transport Companies of Madrid (EMT Madrid) and Valencia (EMT Valencia).

not apply any rule or left the final decision at the discretion of the driver. Eighteen percent of the companies did not allow access to tandem-type ChC (twin). Only 45% allowed access to unfolded ChC, which should be located in the area reserved for WhCh (34%), of which only 23% established the preference of WhCh over ChC. Only 25% of companies' limited access to one ChC and 34% facilitated the transport of a maximum of two ChC. Only 25% of them forced the ChC to apply their brakes during transport.

At present, there are companies in Spain that operate urban public transport services that have implemented their own accessibility policy and have regulated access to the different vehicle configurations of their fleet to all types of passengers with reduced mobility, including WhCh, electric scooters, ChC, passengers with walkers or suitcases, bicycles, etc. Such is the case, for example, of the Municipal Transport Company of Madrid (EMT Madrid), which developed in 2015 is an internal regulation that facilitates the mobility of different passengers with reduced mobility with different configurations of transport vehicles [10]. In the case of ChC, and depending on the type of vehicle and the existence or not of other PMR users, access of up to four single or double ChC (tandem) is allowed.

1.1 Legal framework for children carriages' travel mobility

Currently, one of the few international regulations that states the mobility of ChC in road transport vehicles is the UNECE Regulation R107 [11], which explicitly establishes the technical requirements that the spaces reserved for the displacement of ChC in M2 and M3 vehicles must comply. In the UNECE Regulation R107, it is established that transport vehicles legalized according to these regulations must allocate at least one space reserved for unfolded ChC, with minimum dimensions of 1300 mm in length and 750 mm in width. The area reserved for the ChC may be the same as that occupied by a WhCh (if a single reserved space is available), or adjacent to it, in which the carriage must travel in a plane parallel to the longitudinal axis of the vehicle, but without specifying, explicitly, if the orientation of the ChC must be forward or rearward facing. UNECE R107 [3] does not explicitly require the use of any safety system to prevent the carriage and its occupant from being damaged due to sudden maneuvers of the transport vehicle (turning, braking, or accelerating), or in worst-case scenarios, when an impact as a result of an accident occurs. The approach established in this transport configuration, therefore, is based on the use of passive safety systems for the transport of ChC based on back-restraint walls, as due for WhCh.

The accessibility conditions of the ChC must be the same as those necessary for a WhCh, that is, there must be at least one door that allows access to this area, and the carriage must be able to maneuver and move through the interior without problems and without any steps, holes, or uprights that make it difficult to access the reserved place. To prevent the ChC from moving inside the passenger compartment, conditions similar to those required for a WhCh are established when traveling with an orientation rearward facing, that is, the reserved area must be contiguous to one side or the wall of the vehicle, while on the other side, a retractable bar—or equivalent rigid device—that limits the lateral displacement of the ChC must be allocated. At the front end of the area reserved for the ChC, a backrest perpendicular to the longitudinal axis of the vehicle shall be placed, which must meet the same structural and resistant requirements required in the case of WhCh transport.

In addition, the companion person must be able to be attached during the trip to a bar or handle, anchored to the side or wall of the vehicle. The reserved area must have a pictogram indicating that the area is reserved, as well as pictograms on the vehicle access doors through which the ChC has to enter and exit. The reserved area will be completed with a warning device (button or similar), which allows the passenger to request the driver of the transport vehicle to stop at the next stop.

2. Accessibility analysis of children carriages in urban busses

The compatibility between the different market typologies of ChC with the different urban busses designs requires the analysis of some habitability parameters, such as ChC dimensions (maximum and minimum gauges of the folded and unfolded ChC) and constructive characteristics of urban busses (dimensions in the passenger compartment, number and location of accessible doors).

Based on results obtained in the development of the ASUCAR project, these constructive parameters were defined with some clarity in previous research [12]. In the aforementioned work, two transport configurations of the ChC inside the vehicle were determined: those configurations in which the ChC is folded and those in which it is used unfolded. When the ChC is used folded inside the vehicle, the child will travel in the arms of a companion or in a special seat for the children's transportation, if there is one installed on the bus (**Figure 3A**); in that case, the ChC is transported as a luggage. To do this, a special area inside the cabin must be enabled so that folded ChC can be deposited as if they were luggage; thus preventing unwanted maneuvering of the vehicle could cause discomfort or damage to other passengers. There is, however, the possibility of transporting the folded ChC in an inappropriate place inside the vehicle (**Figure 3B**), where, in case of sudden movements due to unusual maneuvers of the vehicle, the carriage instability can cause inconvenience to other passengers.

In the case of ChC being transported unfolded in the vehicle, it is usual for the child to travel inside, and in that situation the ChC is used as a motor vehicle seat. **Figure 4A** shows an example of good transport configuration, in which the carriage is located occupying the space reserved for the WhCh, oriented in a rearward-facing direction with the backrest attached to the back restraint. However, it is also common to find incorrect situations in which the carriage occupies the space

(A) (B)

Figure 3.
Configuration of transport for the ChC folded in public transportation busses: (A) with the ChC folded in a reserved space, (B) with the ChC folded in a space not reserved for transport (source: EMT Madrid).

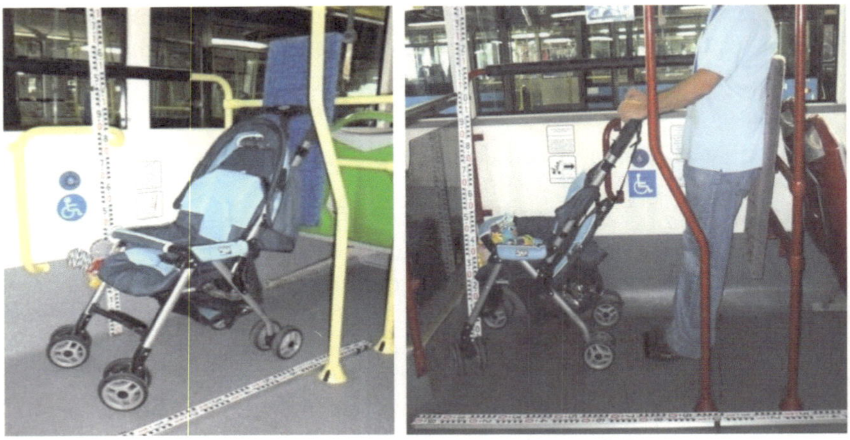

(A) (B)

Figure 4.
Configuration of transport for the ChCs unfolded in public transportation busses: (A) example with the ChC unfolded in the space reserved for the WHCH, (B) example with the ChC unfolded with the space reserved for the WHCH occupied by a companion (source: EMT Madrid).

reserved for the WhCh facing rearward, with a companion touching the back restraint instead of the ChC (**Figure 4B**). This problem occurs as a result of ignorance or misinformation from the users and operators, which do not have internal standards or codes of good practice that regulate the conditions in which this type of mobility must occur.

Finally, in the study conducted by [9, 13, 14], the maximum and minimum dimensional gauges corresponding to more than 150 types of ChC were determined, both in the folded and unfolded configurations. The values obtained allowed to define the most extreme gauges from the dimensions measured on the up-to-date market ChC. These gauges were classified according to the type of carriage to be used, as stroller, twin chair, or tandem. For defining these gauges, two facts in the use of these products were also being considered:

- Firstly, carrycot type is only used during the first months of the child's life or until he reaches 10 kg of weight or 76 cm in height.

- Secondly, the use of car seats is limited to their permanent installation in private vehicles, although in some public transport busses they could be installed as an alternative (**Figure 3A**) when the ChC is folded.

During the development of the ASUCAR project, up to 77 models of continuous low-floor busses interiors were analyzed, representing the majority of large capacity urban accessible vehicles to transport ChC [13]. The objective of this analysis was based on determining which configurations of accessible vehicles to WhCh minimum and maximum dimensions of the front and rear bus aisles, the number of access doors and the length of space reserved for WhCh of the vehicle, and the number of referenced WhCh which fit in it were established—1200 mm length, 700 mm width, and 1350 mm height [11, 15].

To determine if the ChC is dimensionally compatible with the different low-floor busses designs, a comparative analysis was carried out between the maximum gauges obtained for the deployed carts. **Figure 5** shows the maximum and minimum gauges obtained for unfolded (A) and folded (B) carriages, respectively. This study was intended to verify if an occupied ChC could circulate deployed from

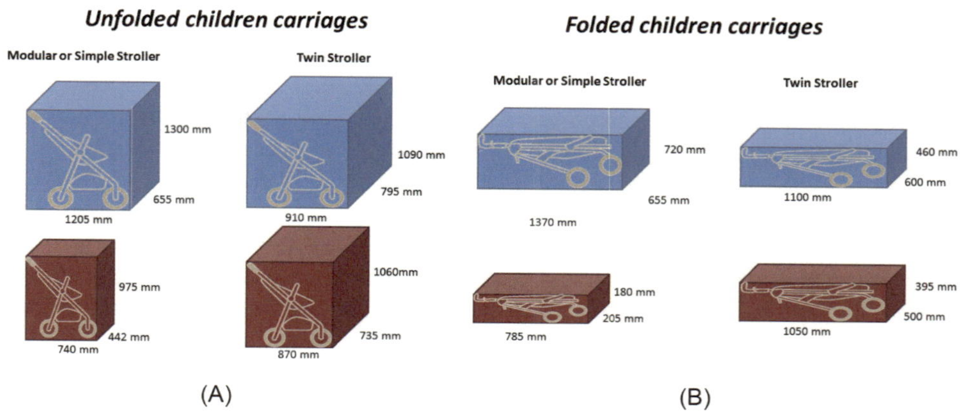

Figure 5.
Maximum and minimum gauges of the (A) unfolded and (B) folded ChCs [12–14].

the front door to the area reserved for WhCh, or if it should do so from the central door. Also, the maximum number of ChC that could be located in the area reserved for the ChC was determined. The results of the comparative study are shown in [13] and can be summarized as follows:

- In the case of the transportation vehicle which has only a space reserved for WhCh:
    - If the space reserved for WhCh is not occupied, a deployed ChC can be placed in the longitudinal direction, or two ChC in the transverse direction.
    - If the space reserved for WhCh is occupied, there is only one possible configuration in which a WhCh can travel simultaneously with a ChC in the longitudinal direction.
- When the transport vehicle has two spaces reserved for WhCh:
    - If the reserved space for WhCh was unoccupied, one or two ChCs deployed in the longitudinal direction, or up to four ChCs in the transverse direction, can be placed.
    - If the reserved space is occupied by only one WhCh, a WhCh and a ChC can be traveled simultaneously in the longitudinal direction, or the WhCh and two ChCs in the transverse direction.
    - If the reserved space is occupied by two WhChs, no ChC deployed inside the vehicle can be moved.

3. Children carriages' kinematics during driving vehicle maneuvers

During the development of the ASUCAR project, a series of experimental tests were carried out whose objective was to reproduce the dynamic behavior of a ChC inside a public transport vehicle, under typical traffic maneuvers, such as sudden accelerations, critical braking, and extreme turns at maximum speed. The tests tried to reproduce urban and suburban traffic conditions and were implemented both in closed and open circuits [14]. The closed-track tests were carried out at the facilities of the Institute for Automobile Safety Research (INSIA), from the Polytechnic University of Madrid, and at the facilities of the municipal transport operators of Valencia and Madrid. The open circuit tests took place in the streets of the cities of Valencia and Madrid (Spain). These open circuit tests were carried out with congested traffic, representing routes where frequent stops, traffic lights, roundabouts, tunnels, and slopes could be found on the bus tracks.

Transport vehicles of category M3 were selected. The vehicle model was a SCANIA N230 E4 with a CARSA City body. The maximum authorized weight was 19,000 kg, with a length of 11.99, 2.5 wide and 2.845 m high. During the trials three representative ChCs were used among the most up-to-date market models, based on their accessibility and safety features; thus, the most bulky, the heaviest, and the most unstable carriage models were defined as Model A, B, and C, respectively. **Table 1** shows the technical characteristics of the ChC used in the experimental trials. The ChCs were oriented during the tests in a direction parallel to the longitudinal axis of the vehicle, both in forward- and rearward-facing directions, and also perpendicular to the longitudinal axis of the vehicle. **Figure 6** shows a couple of examples of the configurations used in the field trials [9, 14].

In the trials three types of dummies were used to reproduce different ages of children: a newborn model to rehearse the ChC with horizontal carrycot, a 9-month mannequin to rehearse strollers and car seats, and a dummy of 3 years to test strollers. The newborn and 9-month mannequins represented, inertially, an equivalent mass of 3.4 and 9 kg, respectively. These two dummies did not use any type of instrumentation to measure accelerations as established by UNECE Regulation R44 [16]. In contrast, the 3-year-old dummy consisted of a TNO P3 model dummy, weighing 15 kg, which contained triaxial accelerometers on the head and chest, as required by the R44. The instrumentation used for data acquisition during the tests consisted of different sensors and signal conditioning systems. Specifically, a triaxial AHRS400CC gyroscope with a range of *2 g* was used, located in the center of gravity of the bus. With the gyroscope, the angles, angular velocities, and accelerations of the three axes could be measured. In each of the centers of gravity of the ChC, a Kistler K-Beam 8390A10 triaxial accelerometer with a range of *10g* was installed. For the data acquisition, an HBM MGCplus AB22 system connected to a portable PC for the configuration, control, and storage of the recorded information was used. All data was acquired at a frequency of 12 Hz, and several 12 V batteries were used to ensure system power.

All trials were recorded by several digital video cameras, anchored to the structure of the vehicle, for further analysis of the images. In some of them, several tests were reproduced simultaneously with different ChC and configurations, to save costs and time. The developed test battery tried to reproduce different behaviors of the ChC when traveling inside the vehicle. In that sense, three possible situations were identified for each of the ChC tested:

- The ChC with the wheels spinning freely.
- The ChC with the brakes applied on the wheels.
- The ChC held with the hands of an adult.

The first closed-track test consisted of a slalom test, consisting of passing the bus between five cones separated with a distance of 15 m from each other. The speed of the bus was around 25 km/h. The second closed-track test was to make the vehicle follow a circular path, similar to the passage through a roundabout. The turning

DOI: http://dx.doi.org/10.5772/intechopen.90613

Children carriages' technical data	Model A	Model B	Model C
Frame type	 Three-wheel stroller	 Rectangle frame	 Telescopic port
Wheel type	 Multi-radial	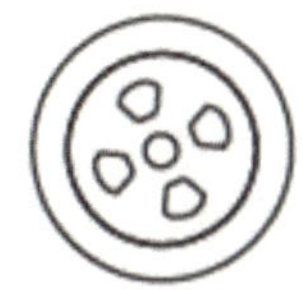 Solid	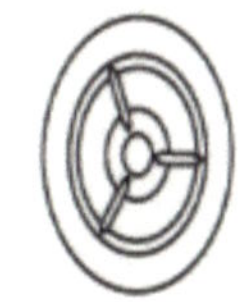 Radial
Track width	Different track	Same track	Different track
Handle type	 Complete	 Simple	 Complete
Modes	 Flatbed/car seat/stroller	 Stroller	 Flatbed/car seat/stroller
Folding type	 Book	 Umbrella	 Book

Children carriages' technical data		Model A	Model B	Model C
Length (mm)		1050	890	1130
Width (mm)		655	442	560
Height (mm)		1105	1035	1140
Weight (kg)*		15.28	6.48	13.76
Front track (mm)		292	345	385
Rear track (mm)		594	345	510
Wheelbase (mm)		650	545	625
Folded	Length (mm)	1105	1040	1370
	Width (mm)	655	310	560
	Height (mm)	540	270	720

**Note: maximum weight between possible configurations—flatbed, car seat, or stroller.*

Table 1.
Technical characteristics of ChC used in dynamic tests (source: [14]).

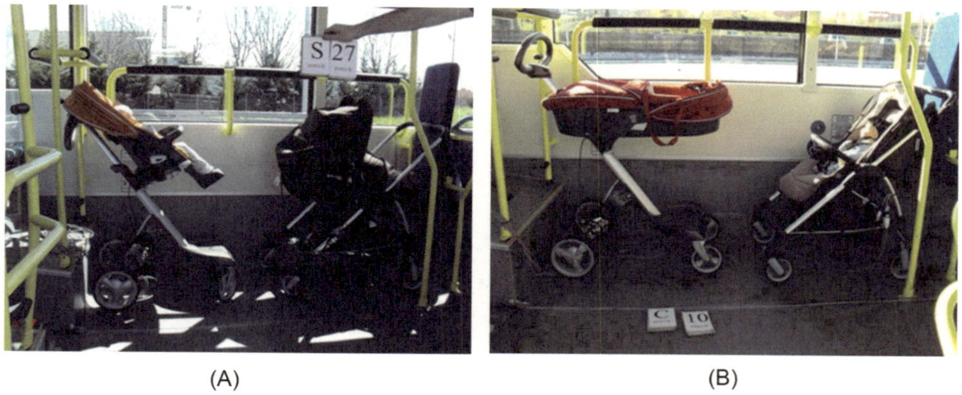

Figure 6.
Provision of ChC deployed during experimental open circuit trials: (A) unfolded rearwards facing Model A carriage and forwards facing Model C carriage; (B) unfolded rearwards facing Model B carriage and forwards facing Model C carriage (source: [14])

radius drawn by the vehicle ranged between 15 and 20 m, and the circulation speed reached between 20 and 25 km/h. This test was consistent with other similar trials such as those developed according to SAE J266 [17] or SAE J2181 [18] standards, both aimed at measuring the dynamic stability of trucks and busses. The third closed-track tests consisted of a braking test, representative of an emergency braking. The vehicle had to be driven in a straight line until reaching a speed of 50 km/h and then proceed to braking in the shortest possible time and the shortest distance, without the driver losing control of the vehicle.

The combination of the different configurations tested, both in open and closed tracks, allowed us to obtain a battery of 61 tests (17 combinations for each of the slalom, circular, and braking path tests and 10 combinations for open circuit tests), using different models of ChC, with different testing dummies, carriage orientations, and brake application status. For all tests, a sign criterion was used for the signals obtained, both for the vehicle and for the ChC, based on the reference system of ISO 4130 [19]. **Figure 7** shows three examples of the behavior of ChC during experimental testing in closed circuit. As can be seen, the dynamics of the vehicle movement during the tests performance can cause carriage, when it is not held by any restraint system, and the wheels are braked, to tip over, and its occupant could hit the ground and the interior parts of the transport vehicle. The overturn, in addition, could occur also when the ChC moves in a rearward-facing orientation, which is the transport configuration recommended by current legislation, UNECE Regulation R107 [11], for this type of products.

From the analysis of the results obtained, and the video recordings made, some conclusions could be drawn about the dynamic behavior of the ChC when subjected to typical maneuvers of urban transport vehicles. From each of the trials, it was concluded that:

A. Slalom test:

- *ChC restrained*: the maximum acceleration corresponds to the Model C carriage (most unstable). The average acceleration is *0.41 g* in X and Y. On Z axis, considering acceleration of gravity, the highest value reaches *1.26 g*.

- *ChC unrestrained*: the acceleration values are higher and correspond to Model A (more voluminous); in many tests, the ChC falls or hits an object, reaching values of *1.72 g* on the X axis and *1.98 g* on the Z axis.

B. Circular test:

- *ChC restrained*: highest accelerations correspond to Model A (more voluminous). On Z axis, considering acceleration of gravity, the highest value reaches *1.49 g*.

- *ChC unrestrained*: wheel brakes applied, the acceleration values are much higher, since in many tests the ChC falls or hits an object, corresponding to Model A (more voluminous), and reaching values of up to *5 g* in the Y direction.

C. Braking test:

- *ChC restrained*: the highest acceleration values correspond to Model A (more voluminous) and reach *1 g* on the X axis. Y values are much smaller, since we are in a braking test where the most important acceleration is longitudinal. In the vertical direction, Z axis, the maximum acceleration reaches *1.9 g*.

- *ChC unrestrained*: with ChC moving freely, the acceleration values are very high since the carriage is violently struck against the interiors of the vehicle, accelerations of up to *6.5 g* are being obtained in the X direction; the most unstable model is Model B (heavier). Even with the Model C (more unstable) of ChC, longitudinal accelerations are achieved: *3.7 g* in X axis and *3.03 g* in Z axis.

D. Open road test:

- *ChC restrained*: in the open circuit the maximum accelerations correspond to the longitudinal direction X, where values of up to *0.43 g* for the Model C (more unstable) are reached; in the vertical direction Z, considering the acceleration of gravity, it reaches *1.56 g*.

- *ChC unrestrained*: the maximum accelerations for the longitudinal axis X are for the most unstable ChC, the Model C, where *1.82 g* is reached; for Model C, *1 g* is reached on the Y axis, and *1.85 g* in the vertical direction Z. The maximum vertical acceleration on the Z axis when the ChC is moving freely is reached with Model A (more voluminous), where it reaches *2.05 g*.

In general terms, it could be concluded that when the ChC is unrestrained and moving freely without the wheel brakes applied, it moves inside the bus, hitting the different parts of the passenger compartment (**Figure 7C**). In these tests it was found that if the ChC is not restrained by an adult or any other system, in the event of normal or emergency movement maneuvers of the vehicle, it tends to slip and hit the bus interiors, regardless of its orientation, even if the wheel brakes are applied (**Figure 7A** and **7B**). When wheel brakes are not applied, the ChC moves but has less tendency to overturn, and the resulting acceleration is somewhat lower.

Many of the ChCs in the current market have a tendency toward lateral overturning that can be easily achieved for vehicle maneuvers with accelerations of less than 5 m/s^2 (*0.5 g*) [12]. When the ChC hits the vehicle's interior without a direct impact of the child, the acceleration of the dummy's head and chest can be up to 10–20% of the ChC deceleration. In case of lateral overturn, in some models of

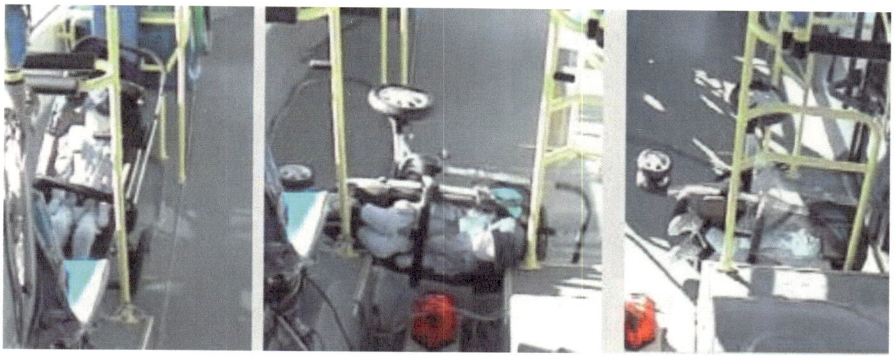

(A)

(B)

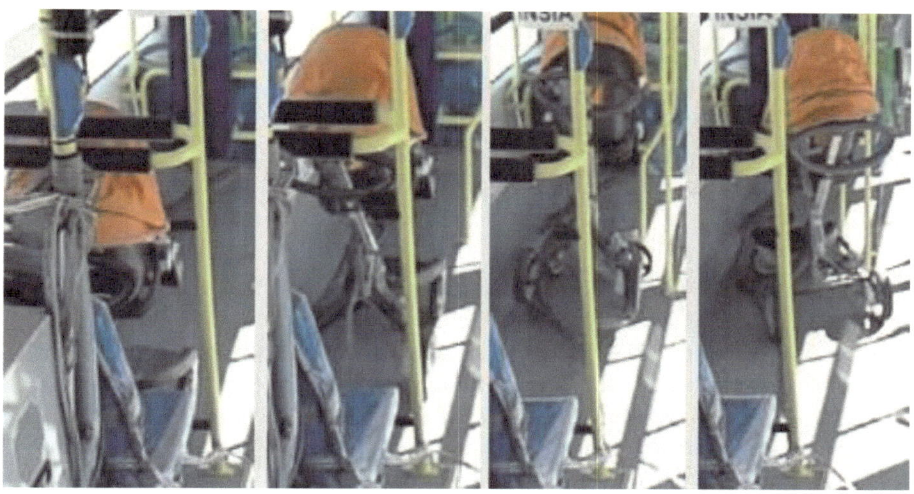

(C)

Figure 7.
Examples of the dynamic behavior of the ChC unfolded during the closed-track tests developed in the ASUCAR project (source: [20]). (A) Circular test: wheel brakes applied. (B) Slalom test: wheel brakes applied. (C) Braking test: wheels spinning freely.

ChC, the child could suffer an impact onto the head when hitting against other parts of the body. In these cases, the deceleration could generate a high risk of damage to the child head.

It was possible to verify that with ChC unrestrained, the safer position for traveling is to place it longitudinally in a rearward-facing direction, resting on the backrest installed in the area reserved for a wheelchair user, as recommended by the UNECE Regulation R107 [11]. However, this position could also be dangerous when the ChC is not restrained in any way to the backrest (as it is a passive safety

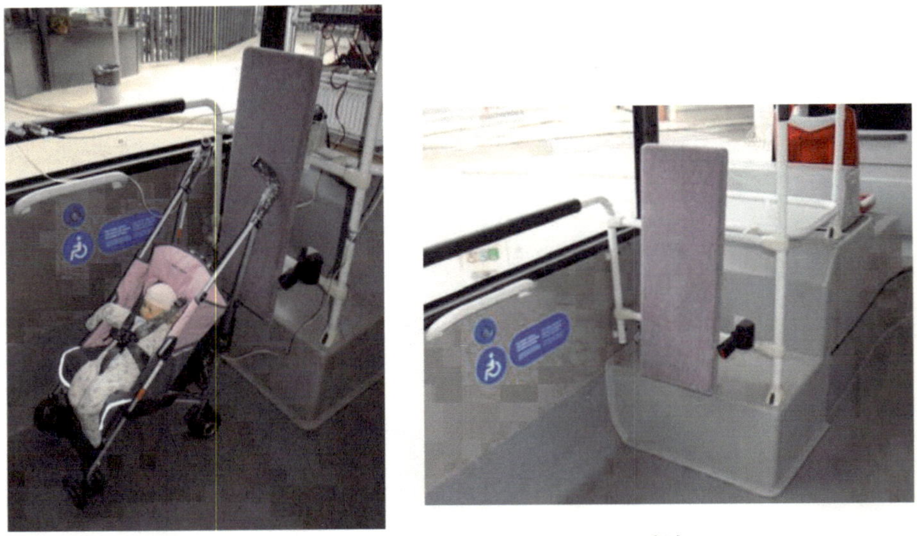

(A) (B)

Figure 8.
Most adequate position of unfolded ChC for traveling in M3 vehicles: (A) with space occupied by ChC; (B) with space unoccupied by ChC (source: [14]).

system), so it can rotate and be launched toward the corridor with a relatively high deceleration ($2.5\,g$) and jump to the front of the vehicle ($6.5\,g$) during an emergency braking. In these cases, the ChC should be held by a restraint system such as the one shown in **Figure 8**, with a safety lap belt. The only drawback of this traveling position is that in case of an emergency braking or a frontal impact, the passengers of the bus, or other ChCs located in front of it, could be thrown against the carriage, causing physical damage to the child or baby. Finally, it should be noted that, for all tests, the transversal position of the carriage to the direction of travel is inadvisable.

3.1 Configurations of children's carriages traveling inside the bus

Taking into account the dynamic analysis of the transport configurations using different ChC models and considering the interior designs of passenger compartment of large capacity passenger vehicles (M3), the areas where the ChC would be traveling and the most appropriate restraint system for each one of them were identified and characterized. In that sense, three possible zones can be identified inside the passenger compartment of the transport vehicle, as shown in **Figure 9**. These areas have the following characteristics:

- **Zone 1**: The area near the backrest of the space reserved for the wheelchair user. If the space reserved for the wheelchair is unoccupied, the ChC must be placed against the backrest, in a rearward-facing direction, with the wheel brakes applied. The use of a seat belt is recommended to secure the carriage.

- **Zone 2**: The area near the central entrance door of the bus, and located next to the first row of rear seats, where the ChC and an accompanying adult can travel next to the crossbar located there. In the event that the space reserved for wheelchair is occupied, the ChC could be placed in this area, facing forward or rearward, with the wheel brakes applied and using a flexible restraint system, such as a seat belt.

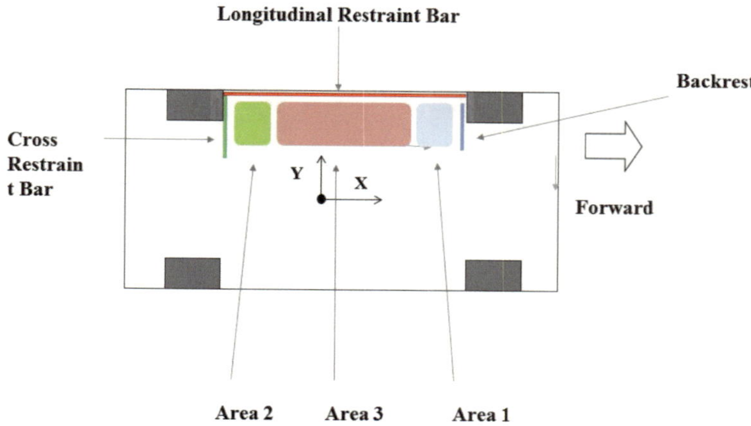

Figure 9.
Characterization of the zones for the location of the ChC unfolded during transport in M3 vehicles (source: [14]).

- **Zone 3**: The central area of the bus between the two zones 1 and 2, in which the ChC and the accompanying adult could travel along the bus sidewall. In that case, it is necessary to use a specific restraint system to hold the ChC in their place following the instructions provided by the transport operator. The child must have the harness attached. The ChC should apply the wheel brakes and should be traveling facing rearward.

As a result of the experimental tests, it was concluded that, among the three zone alternatives, the safest for traveling with ChC is Zone 1, provided that it is not occupied by a wheelchair user, and where the ChC is directly supported on the backrest facing rearward, with the wheel brakes applied and secured by means of a safety belt, that some busses already have incorporated. **Figure 8** shows an example of this configuration.

3.2 Strength analysis for children's carriage restraint systems

Finally, considering the value of the maximum accelerations obtained in the experimental field trials, it was possible to establish the order of magnitude of the forces that the restraint system should withstand to retain the ChC and its occupant. For this calculation, the accelerations generated when the ChC suffered a fall against the floor or the interior parts of the vehicle were not taken into account. The results were obtained considering the total mass of the ChC and its occupant, the maximum deceleration suffered by them during the braking test (which is the most unfavorable). Thus, the maximum force that the restraint system would have to withstand occurs in those cases in which the heaviest ChC is used, reaching the value of 782.46 N. If a safety factor of 2 is applied to the ChC-user set [9], it can be stated that the maximum load that the restraint system would have to bear to withstand the maximum deceleration generated during a braking force would reach the value of approximately 1565 N (1.56 kN).

Analyzing these experimental results, and taking into account that for Zone 3 (located between the backrest of the area reserved for wheelchairs passengers and the first row of rear seats), it was verified that it does not exist currently in the market safety systems designed to facilitate the retention of ChC when traveling in high capacity transport vehicles (M3). So, a new restraint system designed to hold the ChC unfolded during transportation in road vehicles was developed.

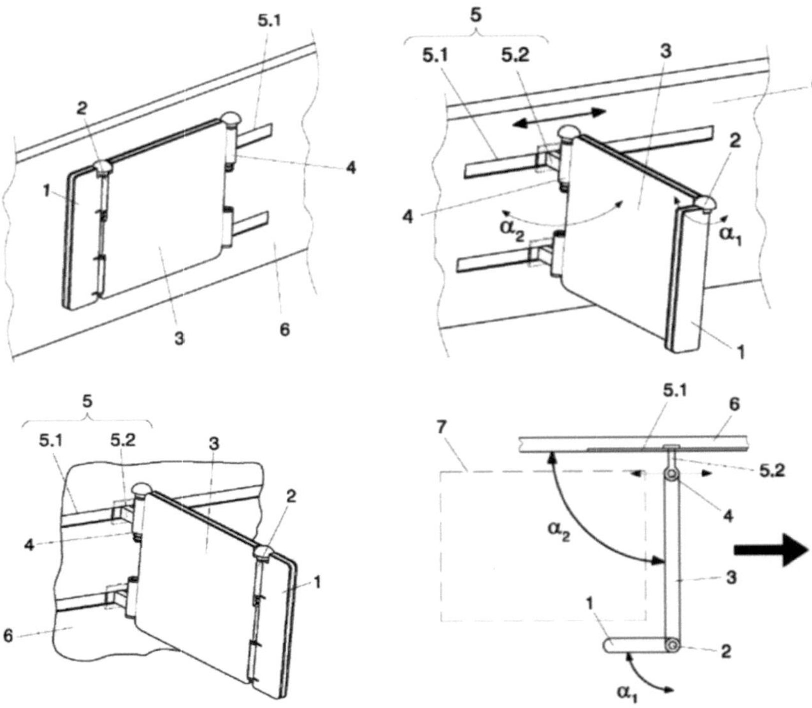

Figure 10.
Scheme of patent ES2403161 defining the operation of a folded-unfolded passive restraint system for the transport of ChC in M3 vehicles (source: [21]).

The new safety system was registered by the Polytechnic University of Valencia and the Polytechnic University of Madrid, through the Spanish Patent ES2403161 [21]. The originality of this invention is that, unlike other restraint systems applied to assure the mobility of wheelchair users in public transportation vehicles, the passive restraint system developed prevents involuntary displacements of ChC, not only longitudinally but also laterally. **Figure 10** shows a simplified scheme of the assembly, which presents the device when it is in the rest position (folded) and when it is used with the ChC in transport position (unfolded). The passive restraint system can be used simultaneously with wheelchair users according to current regulations (UNECE Regulation R107). The safety system has been manufactured in lightweight materials and is able to withstand forces in impacts of up to $2\,g$.

4. Safety requirements of children carriages traveling on busses

To date there have not been many experimental works aimed at obtaining the dynamic behavior of ChC when traveling in road transport vehicles subjected to low speed impact. One of the few references of this type of research was carried out in the development of the ASUCAR project. This project was the first scientific study that has been carried out in Spain in the field of transportation safety of ChC on busses, with no precedents for research projects similar in the rest of Europe [22]. The ASUCAR project continued in a second phase called ASUCAR-2[2], whose main

[2] The ASUCAR-2 research project, *Validation of the usability of a retention system for the safety of children's carriages in public transport vehicles*, was funded by the Polytechnic University of Valencia in the INNOVA 2012 Program (Contract No. 20120579).

objective was the strength and usability validation of the safety system developed to facilitate the restraint of ChC in transport vehicles [23, 24]. The design and manufacture of the new safety system was experimentally validated by performing a battery of representative tests of low speed impact (≈ 20 km/h), with an equivalent deceleration level of *2 g*.

As the Annex 8 of UNECE/UN R107 does not explicitly define any dynamic test to verify the behavior of the ChC under situations of impact, this research was considered an experimental work that could be used for defining the technical requirements for testing the structural behavior of this type of safety systems. In that sense, during the ASUCAR-2 project, a battery of six impact tests were developed on a sled-test platform, whose characteristics were defined based on the results obtained in the first phase of the project. During these tests different prototypes of safety systems were used for the ChC. First, a backrest panel was selected, representative of those used in urban busses, which complied with the technical requirements of UN/ECE Regulation R107. The backrest was provided by the Municipal Transport Company of Madrid. In total, three different prototypes of the safety system developed were used, inspired on the design shown in **Figure 10** [21]. The characteristics of each prototype tested were based on the following aspects:

- *Prototype #1*: folding screen 500 mm high and metal structure covered with wooden panels

- *Prototype #2*: folding screen 400 mm high and metal structure covered with wooden panels

- *Prototype #3*: folding screen 400 mm high and metal structure covered with foam-padded wooden panels

The ChC tested were selected among the most representative models of the market (**Figure 11**), according to the results obtained in previous phases of the ASUCAR project [9, 12], and were characterized by:

- **SX model**: trolley with an adjustable height carrycot. It has four independent wheels, with front steering wheels, and rear wheels with brakes; front and rear tracks are different.

- **QB model**: trolley with adjustable seat angle and used in its highest position. It has four wheels, in which the front ones are twin, and the rear equipped with brakes; front and rear tracks are different.

The sled tests were designed to reproduce an acceleration up to a defined speed (V ≈ 20 km/h) and programmed to stop in a controlled manner with a deceleration of *2 g*, reproducing the deceleration pulse shown in **Figure 11**. Testing took place between July and September 2013 at the facilities of the University Institute of Automobile Research (INSIA), belonging to the Polytechnic University of Madrid (Spain). The ChC restraint system and the WhCh back-restraint were installed on a representative module of the space reserved for a WhCh user, dimensionally and geometrically, similar to that of a standard urban transport M3 vehicle (**Figure 11**). All the tests were carried out with the carriage facing rearward. An accelerometer was installed on the platform module to measure the longitudinal deceleration, according to SAE J211 [25]. All trials were recorded with two high-speed cameras (1000 fps) and a third conventional camera at 30 fps. In

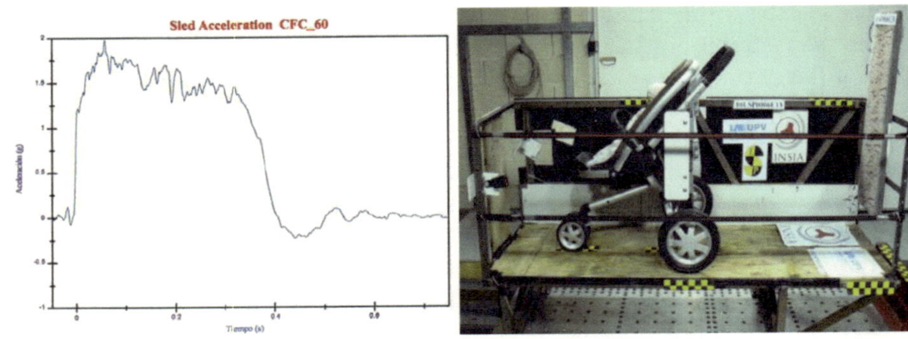

(A) (B)

Figure 11.
(A) Deceleration pulse applied during sled tests. (B) Configuration for sled-test platform (source: [24]).

the trials two types of dummies were used to represent the occupants of the carriage:

- *A TNO P3 dummy* representative of a 3-year-old child equipped with two triaxial accelerometers: one for the head (filter CFC_1000) and one for the chest (filter CFC_180).

- *A Q1 dummy* representative of a 1-year-old child. This dummy was equipped with sensors to measure triaxial acceleration in the head (CFC_1000), chest (CFC_180), pelvis (CFC_1000), and a load cell at the top of the neck to measure forces (CFC_1000) and torques (CFC_600) as well as chest deformation (CFC_600).

These dummies were approved to comply with UNECE Regulations R129 Rev.00 (2014) and R44 Rev.04 [16] for the approval of child safety systems. At the time the study was conducted, there was no international regulation defining the criteria for damage to be applied to Q1 dummies. Therefore, the damage criterion used for this dummy was defined based on the information in UNECE Regulation R94 Rev.01 [26], applying the parameters defined in the work of Mertz, Irwin, and Prasad [27].

The different test configurations were designed to validate the structural behavior of the safety systems developed and analyze the usability of different configurations of the space reserved for wheelchair users and ChC, under low impact conditions. The terms *correct use* and *misuse* as per defining the ideal mobility conditions of ChC during transport were established. All testing was representing worst-case situations regarding ChC model and travel orientation, direction of impact, brakes applied to ChC, and distance from ChC to backrest panel. The configurations of the dynamic impact tests performed are shown in **Table 2**.

The damage criteria for children to define the validity of the results were established by scaling the factors described in [27]. These scale factors had previously been used as reference values in UNECE Regulation R94 Rev.01 [26] for adults of average size as well as in the Federal Motor Vehicle Safety Standards (FMVSS) regulations. The "limit values" established in these regulations, and defined in the analysis of the results of this research, correspond to the probability of generating damage to the different parts of the body. An analysis of the results obtained in the dynamic impact tests allowed us to reach a series of conclusions about the *correct use* and *misuse* configurations in the road transport of ChC. These

Test No.	Test description						Distance from ChC to backrest (mm)	Config.
	Impact direction	ChC orient	Restraint system prototype	Dummy	ChC model	ChC brakes applied		
E01-urban bus module	Frontal	RF	Prototype #1	TNO P3	SX	NO	89 mm	*Misuse*
E02-wheelchair backrest	Frontal	RF	Wheelchair backrest system	Q1	SX	NO	150 mm	*Misuse*
E03-prototype #2	Frontal	RF	Prototype #2	Q1	SX	NO	150 mm	*Misuse*
E04-prototipe #3	Frontal	RF	Prototype #3	Q1	QB	SI	0 mm	*Correct use*
E05-prototype #3	Frontal	RF	Prototype #3	Q1	QB	NO	150 mm	*Misuse*
E06-wheelchair backrest	Frontal	RF	Wheelchair backrest system	Q1	QB	SI	0 mm	*Correct use*

RF: rearward facing.

Table 2.
Configuration for sled testing of ChC (source: [24]).

conclusions were drawn from the battery of tests carried out in the ASUCAR and ASUCAR-2 projects and can be summarized as follows:

a. The safest transport configurations for the mobility of ChC in large capacity road passenger vehicle (M3) are those in which the carriage travels facing rearward, with its back in contact with the backrest panel or bulkhead and with the wheel brakes applied. This configuration is considered as *correct use* to minimize the effects of accelerations on the occupant during the occurrence of a low severity impact.

b. In the case of ChC traveling in an area different from the space reserved for the wheelchair user due to this being occupied, the new restraint system developed (ASUCAR) is capable of supporting the dynamic loads generated during a low speed impact ($2\,g$, ΔV = 20 km/h), representative of the most unfavorable design conditions (more unstable (SX), wider (QB)), when transporting a child up to 15 kg in a large capacity road passenger vehicle (M3).

c. In configuration of *misuse*, with no wheel brakes applied and a gap of at least 150 mm between the ChC and backrest panel, the dynamic parameter values could be increased by two to five times in comparison with *correct use* and may exceed critical values or damage tolerance for ChC occupants, mainly in neck vertical movements.

d. The strength behavior of the restraint system for ChC based on folding rear panels (ASUCAR) has been proven as effective against low severity impacts as the system based on backrest panels established by UNECE/UN Regulation R107, for the transport of users in wheelchairs in large capacity road passenger vehicles (M3). So, it can be considered a useful restraint device compatible with the rigid backrest panel established in actual regulation [11].

5. Conclusions

The recommendations developed in this section have been inspired and grouped based on the *Code of Good Practice* developed as a result of the ASUCAR and ASUCAR-2 projects [13, 20, 22], so that it can be used by the different agents involved in public transportation services: operators and public transport companies, manufacturers, user and consumer associations, as well as public administrations. The *Code of Good Practice* has been structured in three categories:

5.1 Good practices for transportation services operators

Recommendations when ChC travel folded:

A. If there is a space reserved for folded ChC, it must have minimum dimensions necessary to house the majority of existing ChC in the market.

B. If there is a space reserved for folded ChC, it is recommended that for the transport of children, seated seats and/or special seats (groups 0 and 1) be installed facing rearward.

Recommendations when ChC travel unfolded:

A. If there is an unoccupied space reserved for wheelchair users, the ChC must travel in that space facing rearward in contact with the backrest, with the brake wheels applied. The child must have the harness attached.

B. If the space reserved for wheelchair users is occupied, the ChC must travel in the area located next to the panel of the first row of rear seats, facing forwards and using a flexible restraint system (seat belt).

C. If the space reserved for wheelchair users is occupied, and the area closest to the first row of rear seats has insufficient space, there must be a specific restraint system for ChC in the vehicle, which would be used following the operator's instructions. The child must have the harness attached.

5.2 Good practices for ChC manufacturers

A. The ChC manufacturer must inform its customers about the technical requirements of ChC to be used in public transportation vehicles, considering the stability and the protection of the child in the event of a fall or low speed impact.

B. The ChC manufacturer shall design the carriage structurally resistant to support the use of specific restraint systems for transport on busses. The strength of the restraint system must be at least 1600 N.

5.3 Good practices for ChC users

A. Whenever ChC is going to be used for traveling in public transportation vehicles, it must incorporate a harness to hold the child, wheel brakes, and structures as stable as possible.

B. When the vehicle is accessed, it should be preferably travelling in the space reserved for the wheelchair, rearward facing with the wheel brakes and the carriage in contact with the backrest. The companion person must travel in the aisle next to the carriage.

C. If the space reserved for the wheelchair is occupied, it should be travelling in the area near the first row of rear seats, facing forward. In this case, it is recommended to use a flexible restraint system (belt) to hold the ChC structure.

D. When the vehicle is accessed, if the space reserved for the wheelchair is occupied, and there is a special restraint system for ChC, it is recommended to use it with the ChC oriented facing rearward, with wheel brakes applied and the carriage in contact with the folding panel.

E. When the vehicle is accessed, if you cannot travel with the cart unfolded, and there is a reserved space for folded ChC inside the vehicle (provided that the ChC can be folded), it is recommended to locate the carriage in the reserved area and use the special seat (group 0 and 1) for the child, if available.

Acknowledgements

This research was developed under the auspices of the Spanish Ministry of Science and Innovation —who funded the project ASUCAR (PET2008_0328) in the 2008–2011 National Research Programme—and the ASUCAR-2 research project *Validation of the usability of a retention system for the safety of children's carriages in public transport vehicles*, funded by the Polytechnic University of Valencia in the INNOVA 2012 Program (Contract No. 20120579). The development of the injury criteria used in this study was supported by the Spanish Ministry of Science and Innovation who funded the project *Children and Elderly Safety in Bus Accidents* (CESBA/SANCA) (TRA2011-26313) in the 2008–2011 National Research Programme.

Author details

Juan Dols[1]*, Enrique Alcalá[2] and Luis Martínez[2]

1 Institute for Design and Manufacturing (IDF), Universitat Politècnica de València, Valencia, Spain

2 University Institute for Automobile Research (INSIA), Universidad Politécnica de Madrid, Madrid, Spain

*Address all correspondence to: jdols@mcm.upv.es

References

[1] Kirk A, Grant R, Bird R. Bus and Coach Passenger Casualties in Non-collision Incidents. UK: Vehicle Safety Research Centre & ICE Ergonomics, Loughborough University; 2001

[2] ECBOS. 2003. Enhanced Coach and Bus Occupant Safety. Final Report. 5th FM EU Project No. 1999-RD.11130

[3] Berntman M, Wrestrand A, Holmberg B. Bus travel safety–A travel chain perspective. In: 12th International Conference on Mobility and Transport for Elderly and Disabled Persons (TRANSED 2010). Australia: Institute of Transport Studies, Monash University. p. 2010

[4] Fitzgerald SG, Songer T, Rotko AK, Karg P. Motor vehicle transportation use and related adverse events among persons who use wheelchairs. Assistive Technology. 2007;**19**(4):180-187

[5] Frost KL, Bertocci G. Retrospective review of adverse incidents involving passengers seated in wheeled mobility devices while traveling in large accessible transit vehicles. Medical Engineering & Physics. 2010;**32**:230-236

[6] Schneider LW, Klinich KD, Moore JL, MacWilliams JB. Using in-depth investigations to identify safety issues for wheelchair-seated occupants of motor vehicles. Medical Engineering & Physics. 2010;**32**:237-247

[7] COST 349. Accessibility of Coaches and Long-Distance Buses for People with Reduced Mobility. Final Scientific Report. Brussels, Belgium: Directorate General for Transport. European Commission; 2006

[8] EMT Madrid. 2019. Incidents Related with the Transport of Children Carriages in the EMT Fleet during 2018–19 Period. Phone Consult. Confidential. Not published

[9] Alcalá E, Martín A, Vallés B, Martínez L, Dols J, Pons V, et al. Kinematics of children prams in emergency maneuvers of urban buses. In: Proceedings of the 13th EAEC European Automotive Engineers Congress, Valencia, Spain. 2011. ISBN: 978-84-615-1794-7

[10] EMT Madrid. Política de accesibilidad a los autobuses de la EMT. Empresa Municipal de Transportes de Madrid, SA. España. 2015. Available from: https://www.emtmadrid.es/RSC/Accesibilidad [Accessed: September 02, 18]

[11] Regulation No. 107. Addenda 106. Rev. 6 to 1958 Agreement. Date into force 10.06.14. Uniform Provisions Concerning the Approval of Category M2 or M3 Vehicles with Regard to Their General Construction. United Nations; 2009

[12] Alcalá E, Aparicio F, Martín A, Vallés B, Dols J, Terrón J. Caracterización de los carritos de transporte de niños para su uso en autobuses y autocares. In: IX Congreso de Ingeniería del Transporte, Madrid. 2010

[13] Alcalá E, Dols J, Martín A, Vallés B, Aparicio F, Pons V. Good Practices and Universal Accessibility to Urban Buses of Children Prams. Vol. 3. Madrid, Spain: European Magazine for Traffic, Transport and Road Safety. Springer-Etrasa, Securitas Vialis; 2011. pp. 7-27. DOI: 10.1007/s12615-012-9047. ISSN: 1886-9697

[14] Dols J, Pons V, Alcalá E, Vallés B, Martín A. Analysis of Dynamic Behaviour and Safety of Baby Carriages in Public Transportation Buses. Transportation Research Part A: Policy and Practice. Vol. 49. Elsevier Ltd.; 2013. pp. 1-9. ISSN: 0965-8564. DOI: 10.1016/j.tra.2013.01.004

[15] DPTAC. Accessibility Specification for Small Buses Designed to Carry 9 to 22 Passengers (Inclusive). Disabled Persons Transport Advisory Committee. Department for Transport: United Kingdom; 2001

[16] Regulation No 44 rev. 2. Uniform Provisions Concerning the Approval of Restraining Devices for Child Occupants of Power-Driven Vehicles ('Child Restraint Systems'). Economic Commission for Europe of the United Nations (UN/ECE); 2011. Available from: http://www.unece.org/trans/main/wp29/wp29wgs/wp29gen/wp29fdocstts.html

[17] SAE J266. Steady-State Directional Control Test Procedures for Passenger Cars and Light Trucks. 1996

[18] SAE J2181. Steady-State Circular Test Procedures for Trucks and Buses. 1993

[19] ISO 4130. Road Vehicles. Three-dimensional Reference system and Fiducial Marks. Definitions. 1978

[20] Valles B, Dols J. Código de buenas prácticas para la utilización de carritos de niños en vehículos de transporte público. XIX Jornada de Autobuses y Autocares, Madrid; 2012

[21] Dols J, Pons V, Alcala E, Martínez L, Martín A, Valles B. Sistema de Seguridad Pasivo para carritos de niños en vehículos de transporte público. University Polytechnic of Valencia and University Polytechnic of Madrid; 2011. Patent No. P-201131557. Available from: https://lp.espacenet.com/publicationDetails/biblio?DB=lp.espacenet.com&II=0&ND=3&adjacent=true&locale=es_LP&FT=D&date=20130514&CC=ES&NR=2403161A2&KC=A2

[22] Dols J, Aparicio F, Alcalá E, Pons V, Martínez L, Martín AL, et al. Code of Good Practice for Transporting Children Carriages on Buses. Valencia, Spain: Universitat Politècnica de València; 2012. ISBN:978-84-8363-810-1. Available from: http://hdl.handle.net/10251/16991

[23] Martínez L, García A, Espantaleón M, Alcalá E, Dols J. Analysis of children strollers and prams safety in urban buses. In: 24th Enhanced Safety of Vehicles Conference (ESV). Gothenburg, Sweden: NHTSA; 2015. pp. 599-608. ISBN 978-1-5108-1013-6

[24] Dols J, Pons V, Alcalá E, Vallés B, Martín A. Transportation safety requirements in the use of children carriages on buses. Journal of Traffic and Transportation Engineering. David Publishing Company. 2017;5(2):63-76. DOI: 10.17265/2328-2142/2017.02.002. ISSN: 2328-2142

[25] SAE. Instrumentation for Impact Test-Part 1-Electronic Instrumentation. J211–1. Society of Automotive Engineers; 2007. Available from: https://law.resource.org/pub/us/cfr/ibr/005/sae.j211-1.1995.pdf

[26] Regulation No. 94 rev. 1. Uniform Provisions Concerning the Approval of Vehicles with Regard to the Protection of the Occupants in the Event of a Frontal Collision. United Nations; 2007. Available from: http://www.unece.org/trans/main/wp29/wp29wgs/wp29gen/wp29fdocstts.html):http://www.unece.org/trans/main/wp29/wp29wgs/wp29gen/wp29fdocstts.html

[27] Mertz HJ, Irwin AL, Prasad P. Biomechanical and scaling bases for frontal and side impact injury assessment referent values. The Stapp Association. Stapp Car Crash Journal. 2003:155-188. Available from: http://ncbi.nlm.nih.gov

Chapter 14

T-S Fuzzy Observers to Design Actuator Fault-Tolerant Control for Automotive Vehicle Lateral Dynamics

Naoufal El Youssfi and Rachid El Bachtiri

Abstract

This article presents a fault-tolerant control (FTC) procedure for the automotive vehicle lateral dynamics (AVLD) described by the Takagi-Sugeno (T-S) fuzzy models. This approach focuses on actuator faults, which requires knowledge of the system parameters and the faults that are occurring. For this reason, T-S fuzzy observers are suitable for simultaneously estimating system states and actuator faults. The proposed control makes it possible to maintain vehicle stability even in the presence of faults. The design of fuzzy observers and fuzzy controllers is mainly based on the one-step method of Lyapunov, which is provided in the form of linear matrix inequalities (LMIs). The simulation results clearly illustrate the effectiveness of the applied controller strategy to maintain vehicle stability.

Keywords: fault-tolerant control (FTC), automotive vehicle lateral dynamics (AVLD), linear matrix inequality (LMI), fault estimation (FE), continuous Takagi-Sugeno (T-S) fuzzy models

1. Introduction

The vast majority of road accidents are due to faults or incorrect driving reflexes. This is why the automotive industry and researchers in the field of road safety have dedicated themselves to developing and producing vehicles that are more reliable, more relaxed, and safer, which are discussed in Refs. [1–3]. In recent decades, many new solutions have been suggested by the introduction and development of new passive safety systems such as airbags and driver-assisted active safety systems such as adaptive cruise control (ACC), antilock braking system (ABS), dynamic stability control (DSC), and electronic stability program (ESP) [4–6]. Furthermore, the dependence of the control of these systems on actuator components is becoming increasingly complicated. Generally, these systems can be exposed to certain catastrophic faults such as unknown actuator faults. It is well known that conventional control strategies are unable to adapt when system failures happen.

To address this challenge, some previous research has based on estimation techniques for vehicle dynamics, road bank angle, and faults in order to develop control laws capable of ensuring that the system maintains strict stability even when various faults happen. This task is commonly referred to as fault-tolerant control

(FTC), which is widely studied in modern control systems [7–9]. In fact, the safety of persons and the preservation of system performance are crucial requirements that have to be considered in the control design. The problem of fault tolerance has long been addressed from many angles. The FTC synthesis approaches are categorized as passive or active FTC. For the passive FTC approach, we consider possible fault situations and take them into account in the control design step; this approach is similar to the robust control design. It is pointed out in many books that this strategy is generally restrictive. While active FTC improves post-fault control performance and addresses serious faults that break the control loop, it is generally advantageous to switch to a new controller that is either online or designed offline to control the faulty plant. The FTC process is based on two theoretical steps: fault estimation (FE) and setting the controller so that the control law is reconfigured to meet the performance requirements due to the faults.

The FTC procedure was first adjusted for linear systems. Most engineering systems include vehicle dynamics that have nonlinear behaviors. The T-S fuzzy representation is widely known to be a successful solution to approximate a large class of nonlinear dynamic systems. T-S fuzzy models are nonlinear systems represented by a set of local linear models. By mixing the representations of linear systems, the global fuzzy model of the system is obtained, which makes it much easier for the observer and controller to synthesize. A major advantage is that it provides an efficient design strategy for representing a nonlinear system. As a result, many researchers have become interested in the FTC approach for T-S fuzzy systems (see [10–12]).

In this regard, great efforts have been made to improve the stability of the vehicle lateral dynamics to enhance the safety and comfort of the passengers in critical driving conditions. Since the vehicle is a very complex system, the challenge is to achieve more precise control designs and to increase the effectiveness of a vehicle dynamic control system, which necessitates an accurate knowledge of vehicle parameters, in particular, sideslip, roll, and yaw angles [12, 13].

Our main objective in this chapter is to apply a fault-tolerant control method in the vehicle lateral dynamics system. The model used contains four states: lateral slip angle, yaw rate, roll rate, and roll angle. The problem of the design of fault-tolerant active control of nonlinear systems is studied by using the T-S representation which combines the simplicity and accuracy of nonlinear behaviors. The idea is to consider a set of linear subsystems. An interpolation of all these sub-models using nonlinear functions satisfying the convex sum property gives the overall behavior of the described system over a wide operating range. The stability of the T-S models of the vehicle lateral dynamics is mainly studied using the Lyapunov function, and sufficient asymptotic stability conditions are given in the form of linear matrix inequalities (LMIs).

The chapter is structured as follows: the second section deals with the model of vehicle lateral dynamics considering roll motion, the third section presents this model through the T-S fuzzy model, the fourth section presents the fault-tolerant control strategy based on the T-S fuzzy observer, and the fifth section is devoted to simulations and analysis of the results. Finally, the conclusion will be included in the last section.

Notation: A real symmetrical negative definite matrix (resp. positive definite matrix) is represented by $X < 0$ (resp. $X > 0$). X^{-1} indicates the inverse of the matrix X. The marking "*" signifies the transposed element in the symmetrical position of a matrix.

2. Description of the lateral dynamics of the automotive vehicle

The automotive dynamics model is very challenging to use in controlling and monitoring applications due to its complexity and number of degrees of freedom.

For this purpose, a nominal model for the synthesis of observers and controllers is needed. The vehicle motion is characterized by a combination of translational and rotational movements (see **Figure 1**) [14], generally six principal movements. The model used in this document outlines the lateral dynamics of the automotive vehicle, taking into account the rolling motion (**Figure 2**) [15]. This model is achieved by considering the widely known "bicycle model" with a rolled degree of freedom. Here, the lateral velocity $v_y(t)$, the yaw angle $\psi(t)$, and the roll angle $\phi(t)$ of the vehicle are the differential variables. The suspension is modeled as a spring and torsion damping system operating around the roll axis, as illustrated in **Figure 2**. The pitch dynamics of the vehicle is ignored or even neglected.

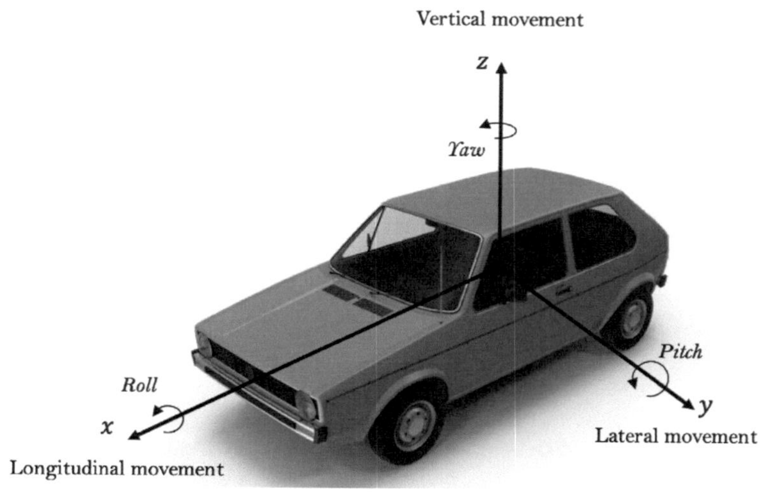

Figure 1.
The movements of the automotive vehicle bodywork.

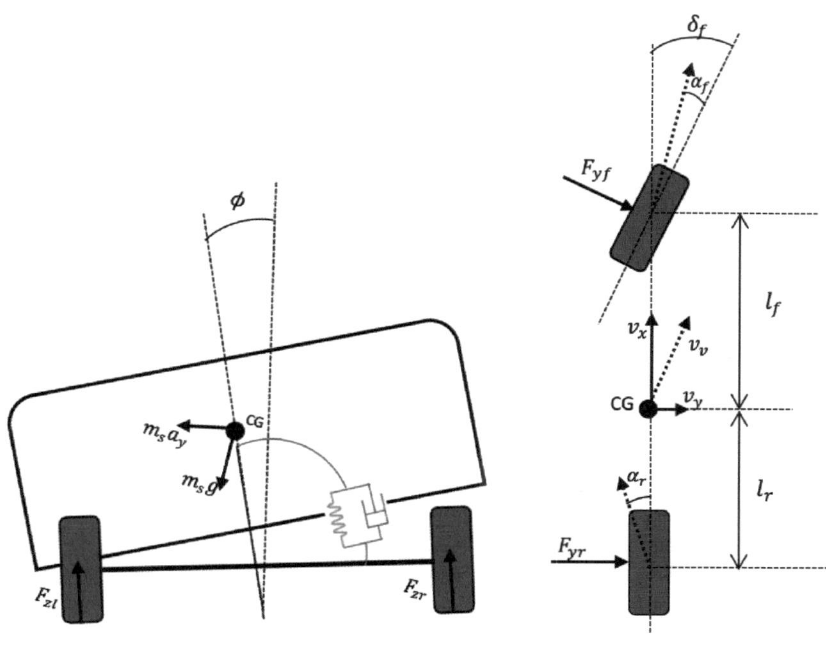

Figure 2.
Bicycle model with roll behavior of automotive vehicle dynamics (3-DOF).

The nonlinear model of the automotive vehicle lateral dynamics considering the roll angle to be small is given in [12] by the following simplified differential equations:

$$\begin{cases} m\dot{v}_y(t) = -m\dot{\psi}(t)v_y(t) + 2\left(F_{yf} + F_{yr}\right) \\ I_z\dot{\psi}(t) = 2\left(l_rF_{yr} - l_fF_{yf}\right) \\ I_x\ddot{\phi}(t) + K_\phi\phi(t) + C_\phi\dot{\phi}(t) = mh_{roll}\left(v_y(t) + v_x(t)\dot{\psi}(t)\right) \end{cases} \tag{1}$$

All system parameters are defined in **Table 1**.

Lateral forces can be given according to Pacejka's magic formula referenced in [16], depending on the sliding angles of the tires. To simplify the vehicle model, we assume that the forces F_{yf} and F_{yr} are proportional to the slip angles of the front and rear tires:

$$\begin{cases} F_{yf} = C_f\alpha_f(t) \\ F_{yr} = C_r\alpha_r(t) \end{cases} \tag{2}$$

where C_f and C_r are the front and rear wheel cornering stiffness coefficients, respectively, which depend on the road adhesion coefficient σ and the parameters of the vehicle. The linear model functions very successfully in the case of small slip angles; however, in the case of increasing slipping, a nonlinear model should be envisaged.

3. T-S fuzzy representation of the automotive vehicle lateral dynamics

The challenge in modeling vehicle dynamics accurately is that contact forces are complex to measure and to model. By using the T-S models' method proposed in [15, 17], it is a highly useful mathematical representation of nonlinear systems, as they can represent any nonlinear system, regardless of its complexity, by a simple structure based on linear models interpolated by nonlinear positive functions. They have a simple structure with some interesting properties. This makes them easily exploitable from a mathematical viewpoint and makes it possible to extend certain results from the linear domain to nonlinear systems.

Parameters	Description	Unit
m	Vehicle total mass	[kg]
v_x/v_y	Vehicle speed/vehicle lateral velocity	[m/s]
I_x/I_z	Roll/yaw inertia moment at the center of gravity	[kg m^2]
l_f/l_r	Distance from the center of gravity to front/rear axles	[m]
h_{roll}	Center of gravity height from the roll axis	[m]
C_ϕ/K_ϕ	Combined roll damping/stiffness coefficients	[N.ms/rad] [N.m/rad]
F_{yf}/F_{yr}	Front/rear lateral forces	[N]
α_f/α_r	Front/rear tire slip angles	[rad]

Table 1.
Parameters of the automotive vehicle lateral dynamics system.

Tire characteristics are usually assumed that the front and rear lateral forces (2) are modeled by the following rules:

$$\text{if } |\alpha_f| \text{ is } M_1 \text{ then} \begin{pmatrix} F_{yf} = C_{f1}\alpha_f(t) \\ F_{yr} = C_{r1}\alpha_r(t) \end{pmatrix} \tag{3}$$

$$\text{if } |\alpha_f| \text{ is } M_2 \text{ then} \begin{pmatrix} F_{yf} = C_{f2}\alpha_f(t) \\ F_{yr} = C_{r2}\alpha_r(t) \end{pmatrix} \tag{4}$$

The front and rear tire slip angles are given in this instance as cited in [12], by

$$\begin{cases} \alpha_f(t) \approx \delta_f(t) - \dfrac{a_f\dot{\psi}(t)}{v_x(t)} - \beta(t) \\ \alpha_r(t) \approx \dfrac{a_r\dot{\psi}(t)}{v_x(t)} - \beta(t) \end{cases} \tag{5}$$

The proposed rules are only made for $\alpha_f(t)$; this assumption reduces the number of adhesion functions and considers the rear steering angle to be ignored; $\alpha_f(t)$ and $\alpha_r(t)$ are considered to be in the same fuzzy ensemble. The combined front and rear forces are generated by

$$\begin{cases} F_{yf} = \sum_{i=1}^{2} h_i(\xi(t))C_{fi}\alpha_f(t) \\ F_{yr} = \sum_{i=1}^{2} h_i(\xi(t))C_{ri}\alpha_r(t) \end{cases} \tag{6}$$

With $h_i(i = 1.2)$ as membership functions, relating to the front tire slip angles $\alpha_f(t)$, which are considered to be available for measurement, they satisfy the following assumptions:

$$\begin{cases} \sum_{i=1}^{2} h_i(\xi(t)) = 1 \\ 0 \le h_i(\xi(t)) \le 1, \quad i = 1, 2 \end{cases} \tag{7}$$

The membership function $h_i(\xi(t))$ expressions are the following:

$$h_i(\xi(t)) = \frac{\omega_i(\xi(t))}{\sum_{i=1}^{2}\omega_i(\xi(t))} \text{with} : \xi(t) = |\alpha_f(t)| \tag{8}$$

and

$$\omega_i(\xi(t)) = \frac{1}{\left(1 + \left|\frac{\xi(t) - c_i}{a_i}\right|\right)^{2b_i}} \tag{9}$$

To determine the membership function parameters and the stiffness coefficient parameters, an identification method based on the Levenberg–Marquardt algorithm in combination with the least squares method is used as in [18], the values listed in **Table 2**.

Nominal stiffness coefficients	C_{f1}	C_{f2}	C_{r1}	C_{r2}
Values	55,234	15,544	49,200	13,543
Membership function coefficients				
$a_1 = 0.0785$	$b_1 = 1.7009$		$c_1 = 0.0284$	
$a_2 = 0.1126$	$b_2 = 12.0064$		$c_2 = 0.1647$	

Table 2.
Nominal stiffness and membership function coefficients.

The nonlinear lateral dynamics of the automotive vehicle, described by Eq. (1), can be written as follows:

$$\begin{cases} \dot{x}(t) = \sum_{i=1}^{2} h_i(\xi)(A_i x(t) + B_i u(t)) \\ y(t) = \sum_{i=1}^{2} h_i(\xi) C_i x(t) \end{cases} \tag{10}$$

where $x(t) = \begin{bmatrix} v_y & \dot{\psi} & \phi & \dot{\phi} \end{bmatrix}^T$ is the system state vector, $y(t)$ is the measured output vector, and $u(t)$ is the input vector. In this study, we consider that the input signal is the front wheel steering angle given by the driver $u(t) = \delta_{fd}$. $\{A_i, B_i, C_i\}$ are sub-model matrices, which are given as follows:

$$A_i = \begin{bmatrix} a_{11} & a_{12} & 0 & 0 \\ a_{21} & a_{22} & 0 & 0 \\ 0 & 0 & 0 & 1 \\ a_{41} & a_{42} & a_{43} & a_{44} \end{bmatrix}; B_i = \begin{bmatrix} b_{11} \\ b_{21} \\ 0 \\ b_{41} \end{bmatrix}; C_i = \begin{bmatrix} c_{11} & c_{12} & 0 & 0 \\ 0 & 1 & 0 & 0 \end{bmatrix}$$

with

$$a_{11} = -\frac{2(C_{fi} + C_{ri})}{mv_x},\ a_{12} = \frac{2(a_f C_{fi} - a_r C_{ri})}{mv_x} - v_x,\ a_{21} = -\frac{2(a_f C_{fi} + a_r C_{ri})}{I_z v_x}$$

$$a_{22} = \frac{2\left(a_f^2 C_{fi} - a_r^2 C_{ri}\right)}{I_z v_x},\ a_{41} = -\frac{2h_{roll}(C_{fi} + C_{ri})}{mv_x I_x},\ a_{42} = \frac{2m_s h_{roll}(C_{fi} + C_{ri})}{mv_x I_x}$$

$$a_{43} = \frac{h_{roll} m_s g - K_\phi}{I_x},\ a_{44} = \frac{C_\phi}{I_x},\ c_{11} = -2\frac{C_{fi} + C_{ri}}{mv_x},\ c_{12} = -2\frac{C_{fi} a_f - C_{ri} a_r}{mv_x}$$

$$b_{11} = \frac{2C_{fi}}{m},\ b_{21} = \frac{2a_f C_{fi}}{I_z},\ b_{41} = \frac{2m_s h_{roll} C_{fi}}{mI_x}$$

Lemma 1. [19] *Let the matrices N_{ij} and the condition be*

$$\sum_{i=1}^{r} \sum_{j=1}^{r} h_i(\xi(t)) h_j(\xi(t)) N_{ij} = \sum_{i=1}^{r} h_i^2(\xi(t)) N_{ii}$$
$$+ \sum_{i=1}^{r} \sum_{i<j}^{r} h_i(\xi(t)) h_j(\xi(t)) \left(N_{ij} + N_{ji}\right) < 0\Xi$$

true if there exist matrices Ξ_{ii} and Ξ_{ij} such that the following conditions are fulfilled:

$$N_{ii} < \Xi_{ii}, \quad i = 1, \dots, r,$$
$$N_{ij} + N_{ji} \le \Xi_{ij} + \Xi_{ij}^T, \quad i,j = 1, 2, \dots, r, \quad i < j$$
$$\begin{bmatrix} \Xi_{11} & \Xi_{12} & \cdots & \Xi_{1r} \\ * & \Xi_{22} & \cdots & \Xi_{2r} \\ \vdots & \vdots & \ddots & \vdots \\ * & * & \cdots & \Xi_{rr} \end{bmatrix} < 0$$

4. Fault-tolerant control strategy based on T-S fuzzy observers

The purpose of this section is to design a faulty T-S fuzzy system for the vehicle model. Then, observers are trained to estimate system states and actuator faults. Next, a fault-tolerant control law is developed from the information provided by the observers.

4.1 Faulty vehicle lateral dynamics system description

From the T-S zero fault model (10), the system in the presence of actuator faults $f(t)$ is described as follows:

$$\begin{cases} \dot{x}_f(t) = \sum_{i=1}^{2} h_i(\xi(t))\left(A_i x_f(t) + B_i(u_{FTC}(t) + f(t))\right) \\ y_f(t) = \sum_{i=1}^{2} h_i(\xi(t)) C_i x_f(t) \end{cases} \tag{11}$$

where $x_f(t)$ is the faulty system state vector, $y_f(t)$ is the measured output vector of the faulty system, and $f(t)$ are the actuator faults. $u_{FTC}(t)$ is the control law to be conceived thereafter; it is considered to be equivalent to a fault-tolerant control law added to the front wheel steering angle given by the driver.

4.2 T-S fuzzy observers design

In this subsection, the T-S fuzzy observers are constructed to simultaneously estimate states and actuator faults. Consider the following T-S observers:

$$\begin{cases} \dot{\hat{x}}_f(t) = \sum_{i=1}^{2} h_i(\xi(t))\left(A_i \hat{x}_f(t) + B_i\left(u_{FTC}(t) + \hat{f}(t)\right) + L_i\left(y_f(t) - \hat{y}_f(t)\right)\right) \\ \hat{y}_f(t) = \sum_{i=1}^{2} h_i(\xi(t)) C_i \hat{x}_f(t) \\ \dot{\hat{f}}(t) = \sum_{i=1}^{2} h_i(\xi(t)) G_i\left(y_f(t) - \hat{y}_f(t)\right) \end{cases} \tag{12}$$

where $\hat{x}_f(t)$ and $\hat{f}(t)$ are the estimates of the state vector and the faults, respectively. $\hat{y}_f(t)$ is the estimate of the output vector. L_i and G_i are the gain matrices with the appropriate dimensions to be resolved.

The output error between faulty T-S fuzzy systems (11) and T-S fuzzy observers (12) is given by

$$e_y(t) = y_f(t) - \hat{y}_f(t) \tag{13}$$

$$= \sum_{i=1}^{2} h_i(\xi(t)) C_i \left(x_f(t) - \hat{x}_f(t)\right) \tag{14}$$

$$= \sum_{i=1}^{2} h_i(\xi(t)) C_i e_x(t) \tag{15}$$

The dynamics of the estimation error $e_x(t) = x_f(t) - \hat{x}_f(t)$ can be written as follows:

$$\dot{e}_x(t) = \sum_{i=1}^{2}\sum_{j=1}^{2} h_i(\xi(t)) h_j(\xi(t)) \left((A_i - L_i C_j) e_x(t) + B_i e_f(t)\right) \tag{16}$$

with $e_f(t) = f(t) - \hat{f}(t)$.

The tracking error $e_t(t)$ is expressed by

$$e_t(t) = x(t) - x_f(t) \tag{17}$$

and its dynamics is given by

$$\dot{e}_t(t) = \sum_{i=1}^{2} h_i(\xi(t))(A_i x(t) + B_i u(t)) - \sum_{i=1}^{2} h_i(\xi(t)) \left(A_i x_f(t) + B_i(u_{FTC}(t) + f(t))\right) \tag{18}$$

$$= \sum_{i=1}^{2} h_i(\xi(t))(A_i e_t(t) + B_i(u(t) - u_{FTC}(t)) - B_i f(t)) \tag{19}$$

4.3 Design fault-tolerant control based on T-S fuzzy observers

A fault-tolerant control is proposed in this section to ignore the impact of faults and to maintain the stability of the faulty vehicle dynamics system. The suggested control procedure is outlined in **Figure 3** [20].

Consider the following control law:

$$u_{FTC}(t) = -\sum_{i=1}^{2} h_i(\xi(t)) F_i \left(x(t) - \hat{x}_f(t)\right) + u(t) - \hat{f}(t) \tag{20}$$

where F_i are the controller gains.

Substituting Eq. (20) in Eq. (19) gives:

$$\dot{e}_t(t) = \sum_{i=1}^{2}\sum_{j=1}^{2} h_i(\xi(t)) h_j(\xi(t)) \left(A_i e_t(t) + B_i F_j \left(x(t) - \hat{x}_f(t)\right) - B_i e_f(t)\right) \tag{21}$$

$$= \sum_{i=1}^{2}\sum_{j=1}^{2} h_i(\xi(t)) h_j(\xi(t)) \left((A_i + B_i F_j) e_t(t) + B_i F_j e_x(t) - B_i e_f(t)\right) \tag{22}$$

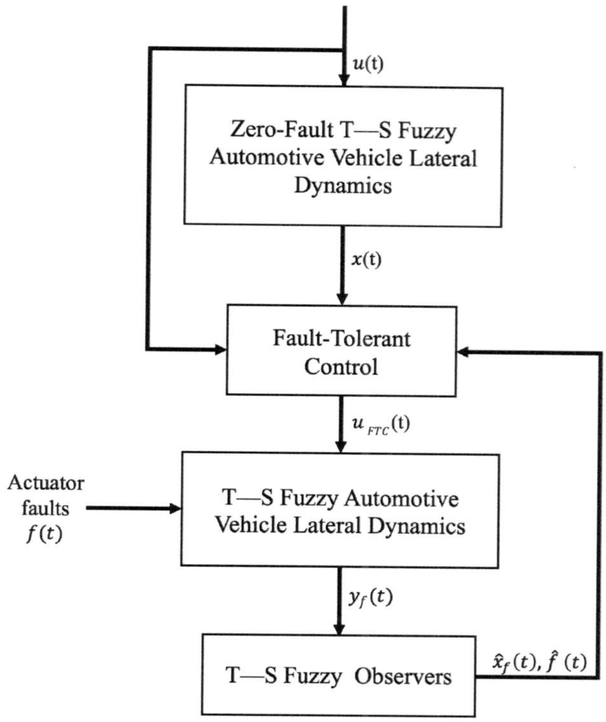

Figure 3.
Observer-based FTC scheme.

Assume that $\dot{f}(t) = 0$, so the dynamics of the fault estimation error $e_f(t)$ is given as follows:

$$\dot{e}_f(t) = \dot{f}(t) - \sum_{i=1}^{2} h_i(\xi(t)) G_i \left(y_f(t) - \hat{y}_f(t) \right) \tag{23}$$

$$= -\sum_{i=1}^{2}\sum_{j=1}^{2} h_i(\xi(t)) h_j(\xi(t)) G_i C_j e_x(t) \tag{24}$$

Consider the following augmented system:

$$\dot{\bar{e}}(t) = \sum_{i=1}^{2}\sum_{j=1}^{2} h_i(\xi(t)) h_j(\xi(t)) \overline{A}_{ij} \bar{e}(t) \tag{25}$$

where $\bar{e} = \begin{bmatrix} e_t(t) \\ e_x(t) \\ e_f(t) \end{bmatrix}$ and

$$\overline{A}_{ij} = \begin{bmatrix} A_i + B_i F_j & B_i F_j & -B_i \\ 0 & A_i - L_i C_j & B_i \\ 0 & -G_i C_j & 0 \end{bmatrix}$$

Theorem 1.1. The state tracking error $e_t(t)$, the system state estimation error $e_x(t)$, and the fault estimation errors $e_f(t)$ converge asymptotically toward zero if there exist positive scalar $\rho > 0$, symmetrical matrices $P > 0$, $Q_2 > 0$, $Y > 0$, and Υ_{ii} $(i = 1, 2)$, as well as other matrices with appropriate dimensions U_i, V_i and Υ_{ij} $(i, j = 1, 2\&i < j)$, such that the following conditions are satisfied:

$$\Delta_{ii} < \Upsilon_{ii}, \quad i = 1, 2. \tag{26}$$

$$\Delta_{ij} + \Delta_{ji} \leq \Upsilon_{ij} + \Upsilon_{ij}^T, \quad i,j = 1, 2, \quad i<j \tag{27}$$

$$\begin{bmatrix} \Upsilon_{11} & \Upsilon_{12} \\ * & \Upsilon_{22} \end{bmatrix} < 0 \tag{28}$$

where

$$\Delta_{ij} = \begin{bmatrix} A_iP + PA_i^T + B_iV_j + V_j^TB_i^T & B_i\overline{X}_j & 0 \\ * & -2\rho Y & \rho I \\ * & * & Q_2\overline{A}_i + \overline{A}_i^TQ_2 - U_i\overline{C}_j - \overline{C}_j^TU_i^T \end{bmatrix} \tag{29}$$

with $\overline{X}_j = \begin{bmatrix} V_j & -Z \end{bmatrix}$ and $Y = \begin{bmatrix} P & 0 \\ 0 & Z \end{bmatrix}$.

The gains of T-S fuzzy observers L_i and G_i and T-S fuzzy controllers F_i are calculated from

$$F_i = V_iP^{-1}, \ \overline{E}_i = \begin{bmatrix} L_i \\ G_i \end{bmatrix} = Q_2^{-1}U_i \tag{30}$$

Proof: the gains L_i, G_i, and F_i are calculated by analyzing the system stability outlined in differential Eq. (25) by using the Lyapunov method with a quadratic function.

Let us select the following quadratic Lyapunov function:

$$V(\overline{e}(t)) = \overline{e}^T(t)Q\overline{e}(t) \tag{31}$$

where Q is divided as follows:

$$Q = \begin{bmatrix} Q_1 & 0 \\ 0 & Q_2 \end{bmatrix}$$

The time derivative of $V(t) = V(\overline{e}(t))$ can be shown to be

$$\dot{V}(t) = \overline{e}^T(t)Q\dot{\overline{e}}(t) + \dot{\overline{e}}^T(t)Q\overline{e}(t) \tag{32}$$

$$= \sum_{i=1}^{2}\sum_{j=1}^{2} h_i(\xi(t))h_j(\xi(t))\overline{e}^T(t)Q\overline{A}_{ij}\overline{e}(t) + \left(\overline{A}_{ij}\overline{e}(t)\right)^T(t)Q\overline{e}(t) \tag{33}$$

$$= \sum_{i=1}^{2}\sum_{j=1}^{2} h_i(\xi(t))h_j(\xi(t))\overline{e}^T(t)\left(Q\overline{A}_{ij} + \overline{A}_{ij}^TQ\right)\overline{e}(t) \tag{34}$$

$\overline{A}_{ij}$ can be defined as follows:

$$\overline{A}_{ij} = \begin{bmatrix} A_i + B_i F_j & B_i \overline{F}_j \\ 0 & \overline{A}_i - \overline{E}_i \overline{C}_j \end{bmatrix} \tag{35}$$

where

$$\overline{F}_j = \begin{bmatrix} F_j & -I \end{bmatrix},\ \overline{A}_i = \begin{bmatrix} A_i & B_i \\ 0 & 0 \end{bmatrix},\ \overline{C}_i = \begin{bmatrix} C_i & 0 \end{bmatrix},\ \text{and}\ \overline{E}_i = \begin{bmatrix} L_i \\ G_i \end{bmatrix}$$

Inequality (34) is negative if the following conditions are satisfied:

$$\dot{V}(t) = \sum_{i=1}^{2}\sum_{j=1}^{2} h_i(\xi(t))h_j(\xi(t))\Lambda_{ij} < 0 \tag{36}$$

with

$$\Lambda_{ij} = \begin{bmatrix} Q_1 A_i + A_i^T Q_1 + Q_1 B_i F_j + F_j^T B_i^T Q_1 & Q_1 B_i \overline{F}_j \\ B_i \overline{F}_j^T Q_1 & Q_2 \overline{A}_i + \overline{A}_i^T Q_2 - Q_2 \overline{E}_i \overline{C}_j - \overline{C}_j^T \overline{E}_i^T Q_2 \end{bmatrix} \tag{37}$$

Using the lemma of congruence, we have

$$\Lambda_{ij} < 0 \Leftrightarrow X \Lambda_{ij} X^T < 0 \tag{38}$$

with

$$X = \begin{bmatrix} Q_1^{-1} & 0 \\ 0 & Y \end{bmatrix},\ Y = \begin{bmatrix} Q_1^{-1} & 0 \\ 0 & Z \end{bmatrix},\ Z = Z^T > 0$$

Then, this inequality is obtained

$$\begin{bmatrix} A_i Q_1^{-1} + Q_1^{-1} A_i^T + B_i F_j Q_1^{-1} + Q_1^{-1} F_j^T B_i^T & B_i \overline{F}_j Y \\ Y B_i \overline{F}_j^T & Y S_{ij} Y \end{bmatrix} < 0 \tag{39}$$

The negativity of (39) enforces that

$$S_{ij} < 0$$

with $S_{ij} = Q_2 \overline{A}_i + \overline{A}_i^T Q_2 - Q_2 \overline{E}_i \overline{C}_j - \overline{C}_j^T \overline{E}_i^T Q_2$.
which can be analyzed using the following property:

$$\left(Y + \rho S_{ij}^{-1}\right)^T S_{ij} \left(Y + \rho S_{ij}^{-1}\right) \le 0 \Leftrightarrow Y S_{ij} Y \le -\rho\left(Y + Y^T\right) - \rho^2 S_{ij}^{-1} \tag{40}$$

Accordingly, (39) can then be delineated as follows:

$$\begin{bmatrix} A_iQ_1^{-1} + Q_1^{-1}A_i^T + B_iF_jQ_1^{-1} + Q_1^{-1}F_j^TB_i^T & B_i\overline{F}_jY & 0 \\ * & -2\rho Y & \rho I \\ * & * & Q_2\overline{A}_i + \overline{A}_i^TQ_2 - Q_2\overline{E}_i\overline{C}_j - \overline{C}_j^T\overline{E}_i^TQ_2 \end{bmatrix} < 0 \tag{41}$$

Using lemma 1, and with some manipulations, we can obtain easily (26)–(28). This completes the proof.

Remark 1. *The calculation of the observer and controller gains is done independently in* [11, 21], *which is restrictive. Therefore, in this study, the resolution of the LMIs is carried out in one step.*

5. Vehicle simulation results

In this section, the computer testing displays a critical driving situation. Initially, the system states and actuator faults will be estimated. Next, to demonstrate the effectiveness of the proposed FTC, the states of the faulty system will be simulated considering the fault-tolerant control law.

Simulations are performed with the forward steering angle profile given in **Figure 4**. This shows a sequence of right and left turns. The vehicle data is shown in **Table 3** [22].

Each (A_i, C_i) is observable. The resolution of the LMIs of the theorem above, using the LMI toolbox [23] and selecting $\rho = 4$, results in the following matrices:

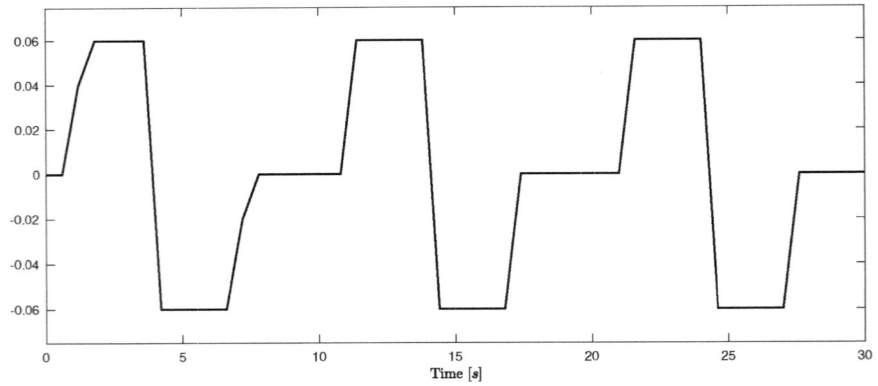

Figure 4.
Steering wheel angle $\delta_f(t)$ by [rad].

Parameters	Value	Parameters	Value
g	9.806 m/s^2	l_f	1.18 m
m	1832 kg	l_r	1.77 m
v_x	23 m/s	h_{roll}	0.90 m
I_x	614 kg m^2	C_ϕ	6000 Nms/rad
I_z	2988 kg m^2	K_ϕ	140,000 Nms/rad

Table 3.
Values of the vehicle parameters used in the simulations.

$$P = 10^3 \begin{bmatrix} 0.3145 & 0.0416 & 0.0057 & 0.0696 \\ 0.0416 & 0.0197 & 0.0026 & -0.0219 \\ 0.0057 & 0.0026 & 0.0844 & -0.3410 \\ 0.0696 & -0.0219 & -0.3410 & 3.5255 \end{bmatrix}, \quad F_1 = \begin{bmatrix} 0.2112 \\ -0.8970 \\ -0.0188 \\ -0.0094 \end{bmatrix}^T$$

$$Q_2 = 10^3 \begin{bmatrix} 0.0844 & -0.1222 & 0.0146 & -0.0014 & 0.0250 \\ -0.1222 & 1.5054 & -0.0286 & -0.0433 & -0.0480 \\ 0.0146 & -0.0286 & 3.2562 & 0.0068 & -0.0039 \\ -0.0014 & -0.0433 & 0.0068 & 0.0464 & 0.0001 \\ 0.0250 & -0.0480 & -0.0039 & 0.0001 & 1.3162 \end{bmatrix},$$

$$F_2 = \begin{bmatrix} 0.2642 \\ -1.1348 \\ -0.0227 \\ -0.0117 \end{bmatrix}^T$$

$$L_1 = \begin{bmatrix} -2.0916 & 14.8942 \\ -4.8065 & -1.8332 \\ -0.1687 & 1.0810 \\ 19.9900 & -6.3497 \end{bmatrix}, \quad L_2 = \begin{bmatrix} 22.4158 & -43.4496 \\ 0.8851 & -1.4886 \\ 0.5984 & 0.3418 \\ 18.2517 & -11.5319 \end{bmatrix}$$

$$G_1 = [40.1735 \quad 39.7987], \quad G_2 = [5.8133 \quad 6.7343]$$

To see the effectiveness of the proposed scheme, we compare the states of the zero-fault T-S model with the states of the faulty system both with and without the FTC strategy.

Figure 5 illustrates the time evolution of the fault $f(t)$ and its estimate $\hat{f}(t)$, which shows that the fault estimate followed closely the fault, whereas **Figures 6–9** show the response of the automotive lateral dynamics reference system states simultaneously with its states in the presence of faults, in two scenarios: with and without the application of the FTC approach.

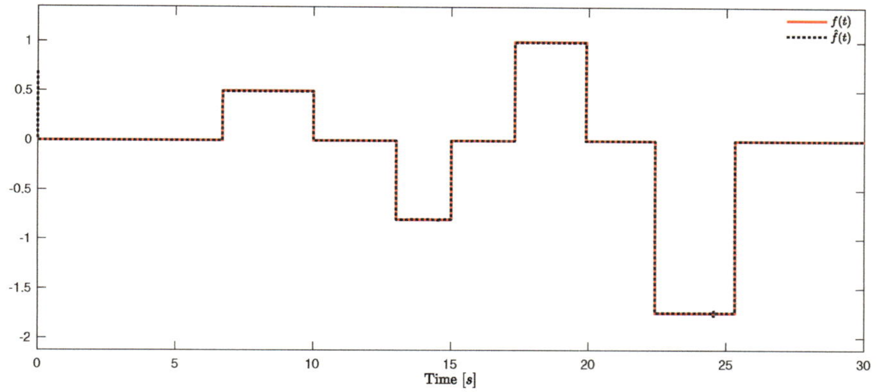

Figure 5.
Fault $f(t)$ and its estimate $\hat{f}(t)$.

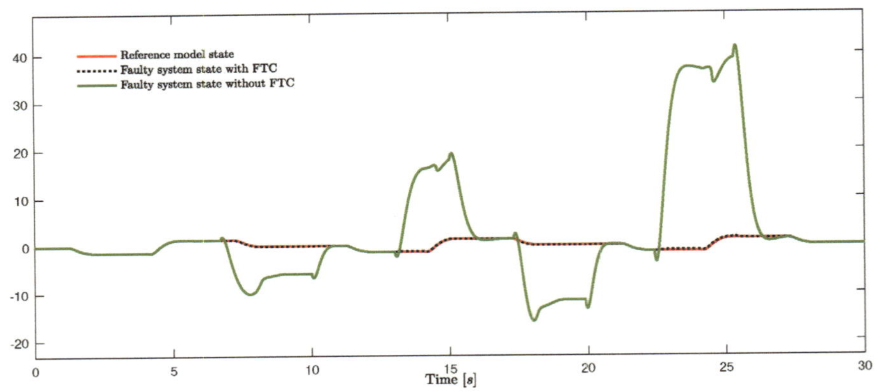

Figure 6.
Comparison of the lateral velocity state $v_y(t)$ response in the presence of actuator faults with the reference.

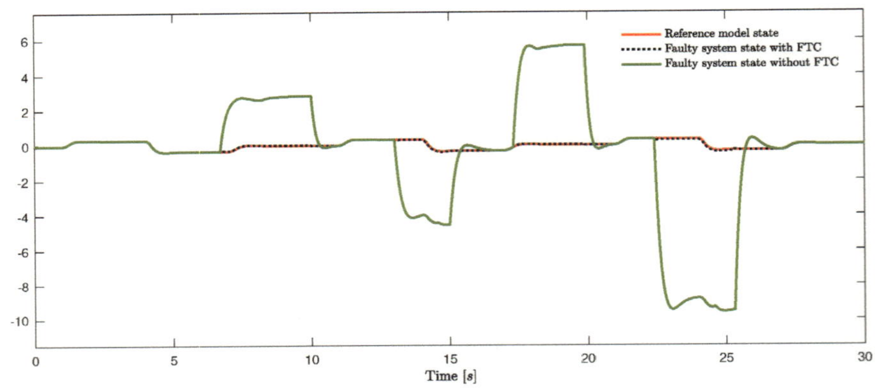

Figure 7.
Comparison of the yaw rate state $\dot{\psi}(t)$ response in the presence of actuator faults with the reference.

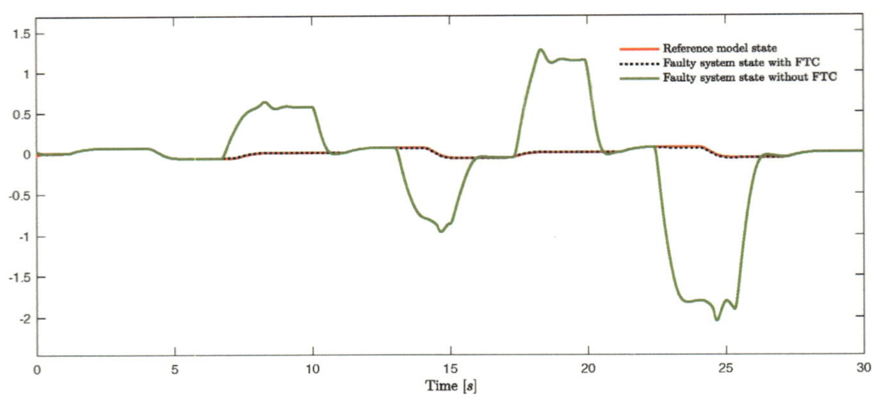

Figure 8.
Comparison of the roll angle state $\phi(t)$ response in the presence of actuator faults with the reference.

In **Figures 6–9**, simulations of faulty system states in the absence of FTC clearly show that the performances of the vehicle has been lost and that its states reached unsustainable levels immediately after the actuator became faulty, but applying fault-tolerant control law, the vehicle stays stable throughout the simulation

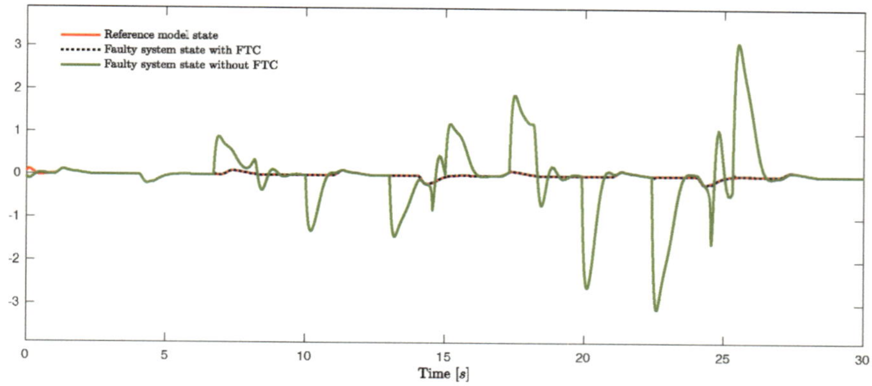

Figure 9.
Comparison of the roll rate state $\dot{\phi}(t)$ response in the presence of actuator faults with the reference.

without losing performances despite the presence of faults, demonstrating the success of the proposed FTC approach.

6. Conclusion

In this chapter, a fault-tolerant nonlinear control law scheme is proposed to reform the states of the lateral dynamics system of the automotive vehicle when it becomes faulty at the actuator side. The nonlinear model of the lateral dynamics of the automotive vehicle is first represented by a T-S fuzzy models, and then the fault-tolerant control design based on the T-S fuzzy observers is presented. The mentioned strategy is based on the use of the T-S reference models and the information provided by the T-S fuzzy observers. This control law is developed to reduce the deviation of the faulty system from the reference; it uses the steering angle, the estimation error, and the tracking error. The stability of the whole system is investigated in one step with the Lyapunov theory and by solving the LMIs constraints. The simulations of the automotive vehicle strongly demonstrate that the engineered FTC is very successful and can be adapted to the specific driving situations. These results will be applied to the CarSim software in future work.

Conflict of interest

The authors declare no conflict of interest.

Author details

Naoufal El Youssfi* and Rachid El Bachtiri
Industrial Technologies and Services Laboratory, Higher School of Technology, CEDoc-Engineering Sciences and Techniques, Faculty of Science and Technology, Sidi Mohamed Ben Abdellah University, Fez, Morocco

*Address all correspondence to: ely.naoufal@gmail.com

References

[1] Delgrossi L, Zhang T. Vehicle safety communications: protocols, security, and privacy. Vol. 103. 2012. Available from: https://trid.trb.org/view/1246249 [Accessed: 14 January 2020]

[2] Mastinu G, Plchl M, editors. Road and off-Road Vehicle System Dynamics Handbook. CRC Press; 2014. Available from: https://www.taylorfrancis.com/books/9780429129810 [Accessed: 22 January 2020]

[3] Du H, Zhang N, Naghdy F. Velocity-dependent robust control for improving vehicle lateral dynamics. Transportation Research Part C: Emerging Technologies. 2011;**19**(3):454-468. Available from: https://www.sciencedirect.com/science/article/pii/S0968090X10000884 [Accessed: 22 January 2020]

[4] Ye X, Hu J, Qi J, Fang Y, Tian P, Shao Q. An integrated multidisciplinary optimization search strategy for automotive passive safety optimization. In: Proceedings of the 2nd International Conference on Intelligent Systems, Metaheuristics & Swarm Intelligence. 2018. pp. 89-97. DOI: 10.1145/3206185.3206209 [Accessed: 22 January 2020]

[5] Anderson ZM, Giovanardi M, Tucker C, Ekchian JA. U.S. Patent No. 10,377,371. Washington, DC: U.S. Patent and Trademark Office. 2019. Available from: https://patents.google.com/patent/US10377371B2/en [Accessed: 25 January 2020]

[6] Genta G, Morello L. The Automotive Chassis: Volume 2: System Design. Springer Nature; 2019. DOI: 10.1007%2F978-3-030-35709-2 [Accessed: 25 January 2020]

[7] Yang H, Xia Y, Geng Q. Fault-tolerant control with effectiveness loss. In: Analysis and Synthesis of Delta Operator Systems with Actuator Saturation. Singapore: Springer; 2019. pp. 331-349. Available from: https://link.springer.com/chapter/10.1007/978-981-13-3660-7_18 [Accessed: 22 January 2020]

[8] Marx B, Koenig D, Georges D. Robust fault-tolerant control for descriptor systems. IEEE Transactions on Automatic Control. 2004;**49**(10):1869-1876. Available from: https://ieeexplore.ieee.org/abstract/document/1341595 [Accessed: 22 January 2020]

[9] Zhang Y, Jiang J. Bibliographical review on reconfigurable fault-tolerant control systems. IFAC Proceedings Volumes. 2003;**36**(5):257-268. Available from: https://www.sciencedirect.com/science/article/pii/S1474667017365035 [Accessed: 31 January 2020]

[10] Shen Q, Jiang B, Cocquempot V. Fault-tolerant control for TS fuzzy systems with application to near-space hypersonic vehicle with actuator faults. IEEE Transactions on Fuzzy Systems. 2011;**20**(4):652-665. Available from: https://ieeexplore.ieee.org/abstract/document/6111465 [Accessed: 19 January 2020]

[11] Kharrat D, Gassara H, El Hajjaji A, Chaabane M. Adaptive observer and fault tolerant control for Takagi-Sugeno descriptor nonlinear systems with sensor and actuator faults. International Journal of Control, Automation and Systems. 2018;**16**(3):972-982. DOI: 10.1007/s12555-017-0546-8 [Accessed: 27 January 2020]

[12] El Youssfi N, Oudghiri M, Bachtiri RE. Control design and sensors fault tolerant for vehicle dynamics (a selected paper from SSD'17). International Journal of Digital Signals and Smart Systems. 2018;**2**(1):50-67. DOI: 10.1504/IJDSSS.2018.090874 [Accessed: 27 January 2020]

[13] El Youssfi N, Oudghiri M. Fault estimation and tolerant control for vehicle lateral dynamics. In: 2018 7th International Conference on Systems and Control (ICSC). IEEE. 2018. pp. 213-218. Available from: https://ieeexplore.ieee.org/abstract/document/8587619 [Accessed: 30 January 2020]

[14] Barton DC, Fieldhouse JD. Vehicle mechanics. In: Automotive Chassis Engineering. Cham: Springer; 2018. DOI: 10.1007/978-3-319-72437-9_1 [Accessed: 27 January 2020]

[15] El Youssfi N, Oudghiri M, Aitouche A, El Bachtiri R. Fuzzy sliding-mode observer for lateral dynamics of vehicles with consideration of roll motion. In: 2018 26th Mediterranean Conference on Control and Automation (MED); IEEE. 2018. pp. 861-866. Available from: https://ieeexplore.ieee.org/abstract/document/8442946 [Accessed: 27 January 2020]

[16] Pacejka HB, Bakker E. The magic formula Tyre model. Vehicle System Dynamics. 1992;**21**(S1):1-18. DOI: 10.1080/00423119208969994 [Accessed: 28 January 2020]

[17] El Youssfi N, El Bachtiri R, Chaibi R, et al. Static output-feedback H_∞ control for T–S fuzzy vehicle lateral dynamics. SN Applied Science. 2020;**2**:101. Available from: https://link.springer.com/article/10.1007/s42452-019-1897-y [Accessed: 02 February 2020]

[18] Geng C, Mostefai L, Denaï M, Hori Y. Direct yaw-moment control of an in-wheel-motored electric vehicle based on body slip angle fuzzy observer. IEEE Transactions on Industrial Electronics. 2009;**56**(5):1411-1419. Available from: https://ieeexplore.ieee.org/abstract/document/4776489 [Accessed: 28 January 2020]

[19] Chang XH. Takagi-Sugeno Fuzzy Systems Non-fragile H-Infinity Filtering. Vol. 282. Berlin: Springer; 2012. Available from: https://link.springer.com/book/10.1007%2F978-3-642-28632-2 [Accessed: 29 January 2020]

[20] Ichalal D, Marx B, Ragot J, Maquin D. Observer based actuator fault tolerant control for nonlinear Takagi-Sugeno systems: An LMI approach. In: 18th Mediterranean Conference on Control and Automation, MED'10; IEEE. 2010. pp. 1278-1283. Available from: https://ieeexplore.ieee.org/abstract/document/5547874 [Accessed: 01 February 2020]

[21] Han J, Zhang H, Wang Y, Liu X. Robust state/fault estimation and fault tolerant control for TS fuzzy systems with sensor and actuator faults. Journal of the Franklin Institute. 2016;**353**(2):615-641. Available from: https://www.sciencedirect.com/science/article/pii/S0016003215004494 [Accessed: 29January 2020]

[22] El Youssfi N, Oudghiri M, El Bachtiri R. Vehicle lateral dynamics estimation using unknown input observer. Procedia Computer Science. 2019;**148**:502-511. Available from: https://www.sciencedirect.com/science/article/pii/S1877050919300651 [Accessed: 30 January 2020]

[23] Gahinet P, Nemirovskii A, Laub AJ, Chilali M. The LMI control toolbox. In: Proceedings of 1994 33rd IEEE Conference on Decision and Control, Vol. 3. IEEE. 1994. pp. 2038-2041. Available from: https://ieeexplore.ieee.org/abstract/document/411440 [Accessed: 30 January 2020]